Advances in Neuroethics

Advances in neuroscience research are bringing to the forefront major benefits and ethical challenges for medicine and society. The ethical concerns related to patients with mental health and neurological conditions, as well as emerging social and philosophical problems created by advances in neuroscience, neurology and neurotechnology are addressed by a specialized and interdisciplinary field called neuroethics.

As neuroscience rapidly evolves, there is a need to define how society ought to move forward with respect to an ever growing range of issues. The ethical, legal and social ramifications of neuroscience, neurotechnology and neurology for research, patient care, and public health are diverse and far-reaching — and are only beginning to be understood.

In this context, the book series "Advances in Neuroethics" addresses how advances in brain sciences can be attended to for the benefit of patients and society at large.

More information about this series at https://link.springer.com/bookseries/14360

Hans-Georg Dederer • David Hamburger
Editors

Brain Organoids in Research and Therapy

Fundamental Ethical and Legal Aspects

Springer

Editors
Hans-Georg Dederer
Faculty of Law
University of Passau
Passau, Germany

David Hamburger
Faculty of Law
University of Passau
Passau, Germany

ISSN 2522-5677 ISSN 2522-5685 (electronic)
Advances in Neuroethics
ISBN 978-3-030-97640-8 ISBN 978-3-030-97641-5 (eBook)
https://doi.org/10.1007/978-3-030-97641-5

This Springer imprint is published by the registered company Springer Nature Switzerland AG
The registered company address is: Gewerbestrasse 11, 6330 Cham, Switzerland

Preface

This volume elucidates the pivotal ethical and legal issues arising from the use of brain organoids for research, therapeutic and enhancement purposes. The ethical reflections extend to the status of brain organoids, informed consent, human-to-animal chimeras and neuro-enhancement. They are mirrored by corresponding legal analyses. The ethical and legal assessments are preceded by an introduction to the scientific and medical background of the brain organoid technology. A final chapter is devoted to the issue of whether international harmonization of normative standards for brain organoid research and therapy is feasible and advisable.

The volume forms a collection of articles which have been presented on the occasion of an international online symposium titled 'Brain Organoids in Research and Therapy: Emerging Ethical and Legal Issues' which took place on 25–26 February 2021 and was organized by the University of Passau. The symposium was part of the interdisciplinary research project 'Interaction of Human Brain Cells', its acronym being 'ForInter', which has received funds from the Bavarian State Ministry of Science and the Arts (BayStmWK) for the period from April 1, 2019, to March 31, 2023.

The hosting of our conference and the publication of this volume were only possible thanks to the cooperation and participation of a large group of individual contributors. Therefore, our gratitude is due to, first of all, the authors, i.e., in alphabetical order, Silvia Deuring, Veljko Dubljević, Nils Hoppe, Insoo Hyun, Andrea Lavazza, Maria Lorenz, Eric Schneider, Yanni Schneider, Tade M. Spranger, Jeremy Sugarman, Jochen Taupitz, Johannes Teller, Soeren Turan, Silja Voeneky, Jeanette Wihan, Jürgen Winkler, Beate Winner and Naime Zagha. We would also like to thank the members of our research team at the University of Passau, i.e. Anna Kunz, Carla Löwenstein, Jana Pecikiewicz and Hannes Wolff, who greatly facilitated our internal deliberations and the organizational realization of the project. Furthermore, we would like to express our special gratitude towards our sponsor, i.e. the BayStmWK, as well as the University of Passau and the University of Erlangen-Nürnberg for ensuring the seamless and success-oriented management of our project. Finally, we would like to thank our publishing house, i.e. Springer Nature, especially the series editors Veljko Dubljević, Fabrice Jotterand, Ralf J. Jox and Eric Racine, as well as Sylvana Freyberg, the publishing editor, Reshmi Rema, the project coordinator, and Vishal Anand, the project manager, for their always outstanding and reliable support.

Passau, Germany Hans-Georg Dederer
Passau, Germany David Hamburger
May 2022

Contents

Introductory Remarks 1

Hans-Georg Dederer and David Hamburger

1.1 Introduction

The function of the human brain is still, at least to a great extent, quite a mystery. Until recently, only postmortem tissue was available for a structural examination of the human brain.[1] Consequently, the examination results could only reflect the state at the end of life. However, in order to better understand the development and functionality of the human brain, dynamic and functional investigations of different human brain cells are indispensable.

This is where brain organoids, i.e. artificially grown in vitro miniature brain models, come into play because they provide the opportunity for more flexible and versatile research approaches and scenarios. At the same time, the use of brain organoids in research and therapy raises the question how these new entities are to be treated from an ethical and legal point of view. An answer to this question is by no means of a purely academic nature but will have an immense practical impact.

Right now the application of brain organoid technology is not specifically regulated. However, in the not all too distant future, legislators or regulators around the world may have to face the task to regulate this newly emerged and continuously and rapidly developing technology. In this important endeavor, lawmakers will seek, or be in need of, advice and guidance from both ethical and legal experts. For any legislation on, or regulation of, brain organoids and their use for research and therapeutic purposes needs not only to be scientifically well-informed but also thoroughly ethically reflected as well as in conformity with domestic (e.g.,

[1] Koo et al. (2019), Chiaradia and Lancaster (2020).

H.-G. Dederer (✉) · D. Hamburger
Faculty of Law, University of Passau, Passau, Germany
e-mail: Hans-Georg.Dederer@uni-passau.de; David.Hamburger@uni-passau.de

H.-G. Dederer, D. Hamburger (eds.), *Brain Organoids in Research and Therapy*, Advances in Neuroethics, https://doi.org/10.1007/978-3-030-97641-5_1

constitutional) law and in line with any applicable legally binding international instruments or international soft law standards.

It is the very purpose of this contributed volume to map the ethical and legal field of those normative issues which seem to be most pressing at the moment and to spearhead the international ethical and legal debate on human brain organoids. It is, in particular, the law which still faces an almost blank area as regards human brain organoids, whereas ethics has already started to explore and chart this landscape some years ago[2] being, thus, ahead of the law.

1.2 Background to This Volume

1.2.1 The ForInter Consortium

This multi-author volume forms part of a legal research project carried out within the framework of the research consortium "Interaction of Human Brain Cells", the acronym of which is "ForInter" and which is funded by the Bavarian State Ministry of Science and the Arts.[3] The ForInter consortium consists of eight subprojects carried out at the universities of Erlangen-Nuremberg, Munich (TUM), Passau, and Regensburg. It also cooperates with the ETH Zurich. ForInter research covers such diverse, albeit closely interrelated, fields as neurology, neurobiology, neuropathology, bioinformatics, and biomedical law and ethics. It is not restricted to basic research but includes translational research as well.

Broadly speaking, the ForInter research consortium investigates the interaction between the different cell types of the human brain using multidimensional cell culture systems. It is these 3D structures which are popularly called "mini brains". The ethical and, especially, the legal aspects of that research are analyzed and evaluated by a team of legal researchers at the University of Passau.

1.2.2 Brain Organoids in Research and Therapy

Brain organoid research has become a highly dynamic and rapidly developing field yielding ever increasing scientific insights into the fascinating capacities of human brain cells as well as into their stunning interplay. Accordingly, brain organoid research advances our understanding of the complex development and intricate functions of the human brain in an unprecedented way.

Besides, or by, providing in-depth insights into the amazing development of the human brain, human brain organoid research also allows for the detection of

[2] See, e.g., Farahany et al. (2018), Lavazza and Massimini (2018), Bredenoord et al. (2017), Munsie et al. (2017); Boers et al. (2016).

[3] For full information on the ForInter consortium, see https://www.bayfor.org/en/bavarian-expertise/bavarian-research-associations/associations/association/forinter.html (last accessed on 2 November 2021).

the origins, pathways, and effects of neural disorders. For the purposes of such research, brain cells artificially derived from induced pluripotent stem cells (iPSCs) are assembled in vitro to form 3D structures which mimic certain brain regions representing, e.g., the cortex or the hippocampus. In addition, 3D models of the whole brain may permit to investigate the complicated interaction and interdependence of the different brain regions.[4]

Against this background, it is understandable that brain organoid research raises tremendous hopes as regards the cure of severe and hitherto incurable neurological, especially neurodegenerative diseases. Most prominent examples are, inter alia, autism, Chorea Huntington, multiple sclerosis, Parkinson's disease, schizophrenia, or different forms of dementia. For the time being, brain organoid-based therapies of neural disorders still belong to the unforeseeable future. Nevertheless, given the ever increasing pace of neurosciences, the realization of such therapies is reasonably conceivable.

1.2.3 The Relevance and Interrelatedness of Ethics and Law

Research on human brain organoids, and possible future therapies using such organoids, raise, however, serious ethical and legal questions since brain organoids are, or at least might be, inextricably related to fundamental human capabilities or attributes. It is these fundamental human capabilities, or attributes, which we consider specifically human and which we ascribe exclusively to our own species (*homo sapiens*) and which form, therefore, the basis for likewise fundamental normative concepts such as personhood and personality. Among these specifically human capabilities, or attributes, are: moral autonomy, rationality, reasoning, sentiment, empathy, imagination, memorization, awareness, or consciousness. These capabilities, or attributes, which are, or which we consider to be predominantly specifically human, are primarily linked directly to the human brain. Therefore, brain organoid research as well as possible future therapies based on brain organoids raise issues of deepest ethical and legal concern. In particular, the advent of brain organoid technology tests concepts, ideas, and assumptions underlying the current normative order of state and society.

It is the aim of this volume to explore and, at least tentatively, clarify these normative issues arising from brain organoid research and their potential therapeutic use. Accordingly, the unsettled moral or legal status of brain organoids[5] as well as the contentious issues of informed consent,[6] human-animal-chimeras,[7] neuro-enhancement,[8] and international standardization[9] will be examined from an ethical

[4] On the current state of brain organoid research see, e.g., Tanaka and Park (2021).

[5] Chapters 4 (by Andrea Lavazza) and 5 (by Jochen Taupitz).

[6] Chapters 6 (by Jeremy Sugarman) and 7 (by Silvia Deuring).

[7] Chapters 8 (by Insoo Hyun) and 9 (by Nils Hoppe).

[8] Chapters 10 (by Eric Schneider and Veljko Dubljevic) and 11 (by Tade Spranger).

[9] Chapter 12 (by Silja Voeneky).

and legal point of view. Due to the close interrelation of ethics and law, it is, to a certain extent, possible to compare and contrast the findings from both normative disciplines.

In fact, there is a special "intimate" relationship between these two disciplines. This relationship is threefold:

First: ethics reflects the law. This means that ethics guides, or may guide, the lawmaker to enact new laws or to amend existing laws. Ethical reflections may pertain, e.g., directly to already existing law. In this case, ethicists may detect serious moral flaws of the current law or identify morally unacceptable lacunae within the applicable legal or regulatory regime. In contrast, ethical reflections may also take place without directly assessing the moral strength of the law as it exists at present. Rather, ethicists may deliberate and review the moral implications of certain human activities independent of the existing legal order. In that case, legal scholars may take up ethical concerns, debates, and arguments and try to mirror them in light of existing legal or regulatory frameworks and legal doctrine. Any such efforts may also result in recommendations to legislators and regulators to enact new rules or to amend existing rules.

Second: ethics complements the law. This means that ethics guides, or may guide, action taken within the margins opened up by the law. For example, on the constitutional level, the law may provide for freedom of research. This constitutional guarantee[10] does not tell, even less direct, the researcher how to make use of his or her freedom. However, ethics may guide the researcher in the exercise of his or her scientific freedom. Likewise, within the outer limits of the constitution demarcated, e.g., by constitutional basic rights of the individual or rule of law principles such as the principle of proportionality, members of parliament may base their voting for, or against, a bill on ethical considerations.

Third: ethics may be incorporated by the law. This means that the law explicitly refers to ethics. For example, the law may provide that an administrative authority must take into account the advice of an ethics committee before handing down a binding decision on the authorization of a particular activity. Such a procedural arrangement usually corresponds to a substantive rule which stipulates, e.g., that a particular activity may be carried out only if it is ethically acceptable.[11] Moreover, certain activities may be subject to both prior approval by an ethics committee and prior authorization by a public authority. This may apply, e.g., to clinical trials.[12]

[10] For example, in Germany, freedom of research is guaranteed by Article 5(3)(1) of the Basic Law. In the EU, it is guaranteed by Article 13(1) of the Charter of Fundamental Rights of the European Union.

[11] An important example in this regard is the German Stem Cell Act which provides that a research project involving the use of embryonic stem cells may be authorized only if it is ethically acceptable (Section 6(4)(2) of the Stem Cell Act).

[12] See, e.g., Dederer and Frenken (2021, p. 92).

1.2.4 Structure of the Volume

For the aforementioned reasons of interrelatedness of ethics and law, the dialogue specifically between these two disciplines is of peculiar interest. It is one of the objectives of this book to foster such dialogue in a mutually supportive manner. Therefore, we tried to identify the most topical, intriguing and hotly debated normative issues surrounding brain organoids and their use for research and potential therapeutic purposes and to bring together both an ethical scholar and a legal scholar in order to discuss one and the same issue from their respective perspectives.

Looking at the legal contributors to this volume, one may raise the question why did we choose to ask German lawyers in the first place as regards the presentations on the legal issues. Indeed, we have asked German scholars to comment on the legal perspectives and foreign scholars to comment on the ethical perspectives.

This is due to the fact that the legally binding rules related to biomedicine and biomedical research are primarily derived from domestic legal systems. In order to gain a reasonably consistent and coherent picture of what is the binding law in the field of brain organoids, it seemed advisable to us to refer to one specific national legal system only. In contrast, the ethical questions can be addressed, at least to a very great extent, in isolation from a particular national legal system because ethical arguments can be developed independent, and even in disregard, of the law.

Since the legal questions are dealt with by German legal experts, a foreign ethical perspective seemed particularly valuable to us. Introducing ethical expertise from abroad may allow us to question understandings, interpretations and assumptions underlying our own law, i.e., German law, and to critically reflect our national legal framework from an ethical and at the same time foreign perspective.

Whether the concept of contrasting German legal views with ethical views from abroad will bear fruits is an open-end question. In this regard, each reader will draw his or her own conclusions. In any case, this multi-author volume yields the fruits of a constructive dialogue between ethics and law which may thoroughly inform both foreign lawyers and foreign legislators and regulators, too.

1.3 Overview of the Volume

The following chapters of the edited volume aim at both to provide more clarity regarding the ethical and legal assessment of the use of brain organoids in research, therapy, and enhancement and to give some, albeit preliminary, normative guidance with respect to the regulatory handling of brain organoids in the future.

To this end, Part I introduces the reader to the scientific and medical background of brain organoid technology in order to provide a sound understanding required for the subsequent legal and ethical considerations. Part II includes contributions which address the moral and legal status of brain organoids. The articles under Part III discuss the ethical and legal requirements for informed consent. The ethical and legal issues arising from the creation of human-to-animal chimeras are discussed in Part IV. Part V takes up the ethical and legal debate on neuro-enhancement.

Finally, Part VI analyses whether, and to what extent, normative standards for uses of brain organoids in research, therapy, and enhancement need to be internationally harmonized.

This section shall serve as a quick guide for the entire edited volume by introducing the separate chapters and addressing their respective research questions and approaches. At the same time, we also try to highlight our own conclusions drawn from the exciting contributions by the chapter authors. It goes without saying that the contributors speak for themselves, though.

1.3.1 Part I: Scientific Background

In order to be able to assess the ethical and legal issues surrounding brain organoid technology adequately a solid understanding of the scientific background of that technology is indispensable. For that purpose, it is conducive to clarify how brain organoids are developed on the basis of brain cells which, in turn, are derived from induced pluripotent stem cells (iPSCs) and how such brain organoids are, or might be, used for research and future therapeutic purposes.

1.3.1.1 Development of Brain Organoids Based on iPSC-Derived Brain Cells (Chap. 2)

Chapter 2 serves as an introduction to brain organoid technology. For that purpose, Naime Zagha and Beate Winner explain what iPSCs are, how they are generated, and why they are used in this context. This is followed by an illustration of how iPSCs are used to create brain organoids and an elaboration on the key developmental events that brain organoids go through. Finally, the potential and limitations of brain organoids are addressed in greater detail against the background that human brain organoids can be used as tools for disease research, drug testing, and the development of personalized medical treatments.

Certainly, one of the most striking features of organoid technology is that cells derived from iPSCs seem to have a natural self-organization capability requiring only very limited artificial stimulation to form complex 3D cell structures. The in-depth analysis of these organoids obviously depends on another highly innovative technology, i.e., single-cell RNA-sequencing, which unlocks tremendous valuable information on the characteristics of the organoids.

A most striking feature of brain organoids seems to be that a certain amount of electric activity can be measured indicating that certain network functions take place. It is such astonishing discoveries which further fuel the normative debate on whether brain organoids should be accorded a particular moral or legal status.

However, in this regard, it has to be kept in mind that, for the time being and apart from rudimentary photosensitivity, brain organoids lack substantial sensory input. This casts doubts on conceptions of, e.g., consciousness or awareness of brain organoids in vitro. On the other hand, research on vascularizing brain organoids is under way which will enable the supply of oxygen and nutrients and, thus, the formation of bigger organoids mimicking brain functions more reliably. This is why forward-looking ethical and legal evaluations with a view to more voluminous organoids exhibiting more substantial network functions matter.

1.3.1.2 Cell Therapy and Genome Editing in Parkinson's Disease: Quo Vadis? (Chap. 3)

The brain organoid technology is a potent tool not just for research but, probably, also for therapy. Therefore, it is especially worthwhile to pay attention to potential future therapy scenarios.

Currently therapeutic cell-based approaches are using only iPSCs or iPSC derivatives. In contrast, the therapeutic use of organoids is currently still facing safety concerns and the lack of appropriate protocols. Therefore, organoids are at the moment not used for immediate therapeutic purposes but rather for disease modeling and drug testing.

To nevertheless give an impression of the basic therapeutic potential of organoids, Chap. 3 by Yanni Schneider, Jeannette Wihan, Sören Turan, and Jürgen Winkler presents the therapeutic achievements made with iPSCs in Parkinson's disease and addresses the conceivable transferability of these findings to brain organoids.

Indeed, Parkinson's disease is one of those neural diseases which may take diverse and truly severe forms and which are currently incurable. Cell or combined cell/gene therapies based on iPSCs may offer the possibility to cure this neurodegenerative disease at its origins in the brain. However, a major challenge of any such therapeutic approaches is tumorigenicity of the cell, or organoid, transplants resulting from reprogramming and re-differentiation and, as the case may be, the genetic modification of the cells.

The particular case study which the authors refer to presents some more challenges from a medical and scientific ethics perspective. Among them is the fact that the relevant patient suffering from an only intermediate course of Parkinson's disease received iPSC-based cell therapy treatment, which was fully financed, including preclinical research, by the apparently very wealthy patient himself. This raises not only the question of social inequality (with only the rich benefitting from highly innovative medical treatment) but also the question of whether the involved researchers and clinicians were still driven by the purely scientific impulse to advance science for the greater benefit of society at large.

1.3.2 Part II: The Status Debate

With the advent of brain organoid technology, the question arises whether these novel biological entities of human origin enjoy a particular moral or legal status and, if so, what the moral or legal status of human brain organoids might be. Human brain organoids may not fit into legal and ethical categories that have commonly been used until now. Therefore, it is necessary to examine what the appropriate classification of these entities from both an ethical and a legal point of view should be.

1.3.2.1 Human Cerebral Organoids: Evolving Entities and Their Moral Status (Chap. 4)

After an introduction to the basic concept of moral status which underlies the status debate, Andrea Lavazza gives an in-depth and up-to-date overview of the ethical discussions on brain organoids and their (potential) moral status. To this end, he analyses in detail the currently discussed ethical concepts which are, or at least could be, applied to the question of whether a moral status and, if so, which kind of moral status could or should be accorded to human brain organoids.

Based on this, Andrea Lavazza formulates his central thesis: human brain organoids should be—at least provisionally—granted the moral status of sui generis entities. This means that human brain organoids are assigned a moral status which has not been accorded to other living entities so far and which is, thus, unique. The decisive threshold, or criterion, is sentience or consciousness respectively. Brain organoids exhibiting patterns of sentience or consciousness should, therefore, be treated with reasonable care according to well-established principles of bioethics (e.g., nonmaleficence, beneficence, and justice).

This concept is an expression of a precautionary approach which takes into account that the further development potential of human brain organoids cannot be anticipated with sufficient certainty yet. In addition, for the time being, we have difficulties in making moral judgments about human brain organoids for lack of moral experience since, hitherto, we have not been exposed to such peculiar living entities and their uses for a sufficiently long period of time.

Based on this precautionary approach, Andrea Lavazza concludes that human brain organoids produced for research purposes should be grown in a way ensuring that no form of sentience or consciousness is developing. In any case, they should be granted increasing protection in the case that the development of some form of sentience or consciousness should become likely and special protection if sentience, consciousness, or other cognitive capacities can be detected.

It is of note that Andrea Lavazza also refers to other recent technological developments such as the creation of human-to-animal chimeras and artificial intelligence (AI). Indeed, also these technologies confront us with peculiar living or purely technical entities which might display certain capabilities which we consider to be specifically human. Again, for lack of experience through interaction with such entities our moral intuition vis-à-vis such entities is underdeveloped and, hence, does not offer clear-cut answers to questions of moral status.

1.3.2.2 What Is, or Should Be, the Legal Status of Brain Organoids? (Chap. 5)

Jochen Taupitz examines the legal status of brain organoids from the perspective of German law. His legal analysis is based on the assumption that brain organoids will neither develop higher brain activities nor consciousness in the foreseeable future and are, therefore, not comparable to a human brain but rather to human tissue.

Consequently, Jochen Taupitz concludes that brain organoids qualify legally as things that are subject to ownership rights. This means that ownership of organoids, e.g., by the researcher is possible and that the general right of personality of the donor extends to brain organoids as well. What is more, in Jochen Taupitz' view, brain organoids may be also the object of commercial transactions, i.e., currently, human brain organoids are fully marketable. In addition, brain organoids in themselves do not constitute data within the meaning of personal data protection law. However, laws on the protection of personal data do apply, of course, as far as it concerns personal data (e.g., of the donor) retrieved from, or through the examination of, brain organoids.

According to Jochen Taupitz, brain organoids do not have a special constitutional status either. Thus, there is no need for any restrictions of scientific research with brain organoids at present. In addition, there is also no constitutionally valid justification to restrict research on and with brain organoids.

Jochen Taupitz points out that, nevertheless, researchers may have to submit their research projects to ethics committees enabling ethical advice to the researcher planning to conduct brain organoid research. Indeed, this seems to be in line with the constitutionally guaranteed freedom of science as far as ethical recommendations are not legally binding and simply an expression of those responsibilities which are inherent in modern science itself anyway.

1.3.3 Part III: The Informed Consent Challenge

Autonomy of the individual human being and, therefore, informed consent is a cornerstone of modern medical research and therapy. With a view to brain organoid technology, the question arises as to what are the specific requirements of an ethically and legally valid informed consent. In this regard, one may distinguish between donor consent, i.e. consent of the donor regarding the use of his or her biological material for medical or scientific purposes, and patient consent, i.e. consent of the patient regarding his or her medical treatment. This seems to be reasonable not only because these are two different scenarios but also because of the different interests of a patient on the one side and a donor on the other side. The same applies even if patient and donor are identical because the person's interest in getting individual medical treatment may differ from the same person's interest in donating biological material for purposes of third parties. Furthermore, one may differentiate between the formal requirements for a valid consent and the substantive legal limits to consent which, if exceeded, may invalidate the consent irrespective of whether formal requirements (regarding, e.g., procedure, written form) are met. In addition, another differentiation may be of significance: informed consent may relate to interferences with the right to physical integrity or the general right to personality or the right to personal data protection.

1.3.3.1 Ethical Requirements Related to Donors' and Patients' Consent (Chap. 6)

Jeremy Sugarman illustrates in detail the general requirements for a valid informed consent. As part of this, he identifies a set of necessary elements for an ethically sound consent: competency, voluntariness, disclosure, understanding, decision, and authorization.

Hereby, he addresses the ethical requirements not just in the context of basic research but also with a view to clinical translation. In the research scenario, it is the consent of the donor, whose tissue is used to create brain organoids, which is most relevant, whereas in the therapy scenario the consent of both the donor and the patient is under scrutiny.

This is followed by an application of these general findings to the particularities of informed consent regarding research and therapy scenarios with brain organoids and an identification of consent challenges that are specifically related to brain organoids. In particular, Jeremy Sugarman points out that the (potential) use of donated biological material for the purpose of creating brain organoids should be explicitly disclosed to the donor. In this case, relevant information provided to the donor should also extend, inter alia, to the process of generating the brain organoids as well as to the peculiar nature of brain organoids and, as the case may be, to their use for research on human consciousness or, if applicable, to their transplantation into animals resulting in chimeras. With a view to living biobanking, Jeremy Sugarman takes up the concept of "consent for governance" which seems to be specifically pertinent in case of brain organoid research and which requires disclosure of certain mechanisms and instruments ensuring, e.g., ethical review of the research projects or the sharing of benefits resulting from research or its commercial exploitation.

Of course, informed consent implies that the patient or donor has acquired a reasonable understanding of the disclosed information. Jeremy Sugarman also addresses this most delicate practical obstacle to valid informed consent and points to the necessity of gathering, e.g., empirical data on patients' or donors' knowledge, perceptions, and views. In fact, such data could and should inform concepts, mechanisms, and instruments of informed consent.

1.3.3.2 The Legal Requirements for, and Limits to, the Donor's and the Patient's Consent (Chap. 7)

Silvia Deuring examines from a legal point of view in which situations informed consent is necessary, under which circumstances consent is admissible and how far it can reach. This assessment is not limited to German law but takes into account international standards as well.

As regards domestic law, Silvia Deuring's analysis extends to a very broad spectrum of legal rules laid down, e.g., in general medical professional law, the Civil Code, the Transplantation Act, the Transfusion Act, and the Medicinal Products Act. In addition, data protection law and, above all, constitutional law may be highly relevant as well. As regards international instruments, Silvia Deuring refers, e.g., to the Biomedicine Convention of 1997, its additional protocols and related recommendations adopted within the framework of the Council of Europe. She also refers to soft law documents of the World Medical Association such as the famous Declaration of Helsinki of 1964.

In general, generation and subsequent use of human brain organoids requires informed consent by the donor of the cells forming the starting material of the organoids. In this regard, Silvia Deuring points out that informed consent may also extent to transplantation of brain organoids into animals and that, in any case, human dignity as guaranteed in Article 1(1) of the Basic Law, i.e. the German constitution, does not impose any restrictions on the donor's free will to express consent to brain organoid research.

In this regard, from Silvia Deuring's perspective, so-called broad consent is permissible. This concept allows the donor to consent to the use of his or her bodily substances for research purposes in general, i.e. without limitations to pre-specified purposes. Moreover, according to Silvia Deuring's view, informed consent by the donor is required irrespective of whether the donated material has been anonymized or not. In addition, she also emphasizes that donors ought to be informed that their biopsied biological material might be used for ethically contentious research purposes. An example in this regard might be the creation of human-to-animal chimeras. Furthermore, Silvia Deuring recalls that donors need to be informed of their right to withdraw consent and any consequences resulting therefrom as well as of their right to explicitly exclude certain research or therapeutic uses of their donated bodily substances.

1.3.4 Part IV: The Chimera Issue

At present, brain organoid technology is still mainly used in basic research. It is foreseeable, however, that sooner or later there will be efforts to translate basic research into therapeutic applications. Those translational efforts are likely to include animal experiments before clinical trials on humans can take place. If human brain organoids are transplanted into animals as part of these preclinical studies, the results are so-called human-to-animal chimeras. This raises the question of the moral and legal permissibility of such transplantations.

1.3.4.1 Moral Permissibility of Transplantation of Human Brain Organoids into Animals (Chap. 8)

Insoo Hyun analyzes the ethical legitimacy of transplanting human brain organoids into animals for research purposes.

For that purpose, he identifies two distinct aspects that make the transplantation of human brain organoids into animals so ethically sensitive: first, the perception of the human brain, or its cognitive functions, as the basis for fundamental philosophical concepts such as personhood, identity, and autonomy and, second, the concern about the emergence of humanlike consciousness (in terms of subjective self-awareness) in human-to-animal chimeras.

Insoo Hyun recalls that the creation of chimeras through the introduction of human cells into animals has become daily routine in biomedical research for decades. Such human-to-animal chimera experiments have become all the more important in modern stem cell research, especially with a view to translation of basic research into clinical applications.

The decisive question, then, is whether transplantation of human brain cells, or human brain organoids, into animal brains may produce forms of neurological chimerism which result in humanlike cognitive capacities. It is actually such scenarios which prompt critics to voice the fear that transplanted animals may not only be biologically but also morally "humanized". However, Insoo Hyun points out that, so far, insights from numerous experiments do not support such concerns, i.e. hitherto human-to-animal chimeras have not exhibited recognizably enhanced cognitive capabilities.

It follows that, at least for the time being, the ethical permissibility of generating chimeric experiment animals depends on the research purpose pursued and whether animal welfare standards are complied with. Hence, chimera research with the goal of expanding knowledge and promoting clinical translation with a view to therapies, e.g., for severe and currently incurable neurological diseases is morally permissible if animal welfare standards are rigorously followed.

1.3.4.2 Transplantation of Human Brain Organoids into Animals: The Legal Issues (Chap. 9)

Nils Hoppe, Maria Lorenz, and Johannes Teller examine possible regulatory challenges stemming from the transplantation of human brain organoids into animals. For this purpose, they make use of a case study to ground their analysis with a reference point based on a real-world application. This case study concerns the modeling of human Down syndrome in an animal model by implanting in vitro processed human iPSCs into neonatal mice. Since this animal model requires human somatic cells as a starting point, this prompts legal questions concerning the cell donor's interests. Against this backdrop, the authors analyze possible ownership rights and legal issues with regard to personal data protection as well as to autonomy rights beyond the right to informational self-determination of the donor.

After that, Nils Hoppe, Maria Lorenz, and Johannes Teller deal with the regulatory dilemma stemming from the situation that while current laws and regulation might capture a novel technology only insufficiently, the current state of knowledge of potential risks and benefits may still be far too vague to implement a truly effective regulatory regime tailored specifically to the novel technology. As regards human brain organoids and especially the creation of human-to-animal chimeras, the authors arrive at the conclusion that, for the time being, the current legal or regulatory frameworks do not need to be overhauled in order to overcome challenges arising from brain organoid technology. They extend this finding explicitly to animal welfare law which is sufficiently well-tuned to address human-to-animal chimera experiments.

Finally, the authors turn to the difficult question arising from the hypothesis that animals implanted with brain organoids might acquire a human-like consciousness or sentience. We have, hitherto, no concrete idea how to normatively categorize "humanized" animals. This is for a lack of actual experience with such animals which we have not been exposed to so far, and this, in turn, prevents us from making agreed moral judgments.

1.3.5 Part V: The Enhancement Conundrum

A successful medical application of the brain organoid technology could ultimately lead to uses which do not serve purely therapeutic purposes but purposes of neuro-enhancement as well. Such an enhancing intervention into the human brain seems particularly problematic because our brain is the source of consciousness, intellect and sentiment, and, hence, source of our personal, individual character. In particular, brain intervention by transplantation, or injection, of, e.g., brain organoids may lead to the specific problem of whether a person may ever be permitted to deliberately consent to intentional changes of his or her personal identity (i.e. consent "to be someone else"), thereby transcending his or her own hitherto existing personality and identity. In addition, any profound and irreversible intervention into the brain for the purposes of enhancement may impair the "naturalness" of human beings and, thus, undermine the constitutive conditions of the basic normative order underlying state and society. Furthermore, enhancing cognitive capacities may significantly interfere with social equality, justice, and fairness if, e.g., neuro-enhancement was only available to a small circle of beneficiaries who are able to afford such treatment. Moreover, any treatment of human beings would require prior animal experiments referring once more to normative problems surrounding human-to-animal chimeras. These and quite some other issues raise the question whether neuro-enhancement via brain organoid technology is morally justifiable and legally permissible.

1.3.5.1 Building a Better Beast: Enhancing the Minds of Animals (Chap. 10)

Eric Schneider and Veljko Dubljevic opt for an ethical assessment of neuro-enhancement in animals via implantation of human brain organoids which, in fact, seems to be closer to what we may expect in the short term. Indeed, any intervention into humans, be it for purposes of clinical trials, therapy, or enhancement, requires prior animal testing by necessity.

Against this backdrop, the authors assess, first, the potential of brain organoid technology to enhance the brain capabilities and functions of animals. On this basis, they evaluate the impact which different degrees of enhancement could have on the perception of the moral status of the animals concerned. In fact, it is a highly plausible thesis that a certain degree of enhancement of cognitive abilities of animals may be directly linked with an elevated moral status.

Eric Schneider and Veljko Dubljevic, then, turn to the issue of how institutions may decide whether, and to what extent, neuro-enhancement of animals through transplantation of human brain organoids is permissible. In this regard, the authors also point to the influence of narratives construed and disseminated by the media, thus influencing also the moral decision-making of, e.g., political, economic, religious, ethical or legal institutions. Another decisive question is: what are the "ideologies" which these institutions base their judgments and decisions on. This is illustrated along the lines of three "ideologies", i.e. "biocollectivism", "bioidentitarianism", and "bioliberalism", which can be viewed, or defined, as three-dimensional vectors in a 3D coordinate system established by three classic political theories, i.e., "conservatism", "liberalism", and "socialism". The authors explore the line of thinking of "biocollectivist", "bioidentitarian", and "bioliberal" institutions with a view to brain organoid technology and its use in animals but abstain, in the end, from favoring one approach over the other.

Eric Schneider and Veljko Dubljevic finally recall that governing the humanization of animals is an up and down, or back and forth, learning process of the institutions involved. In particular, they emphasize that multilateral consent needs to be achieved: consent by researchers, patients, enterprises, as well as those who care for the moral status, and morally acceptable treatment, of human-to-animal chimeras.

1.3.5.2 Legal Arguments in Favor of and Against Neuro-Enhancement by Means of Brain Organoids (Chap. 11)

Tade Spranger assesses the current legal situation as regards neuro-enhancement and sheds light on the legal debate on the permissibility of, and limits to, self-optimization under German law.

As pointed out by Tade Spranger, crucial in this process is a sound delineation between medically indicated treatment for therapeutic purposes and medically nonindicated treatment for enhancement purposes. Whereas the former is, generally, held to be ethically acceptable and legally permissible, the latter raises serious ethical concerns and intricate legal questions as regards normative acceptability and permissibility.

The contribution makes it clear that, for the time being, the German legal order does not provide for any meaningful regulation of neuro-enhancement on the level of statutory law. The constitution does not provide sufficiently precise guidance either. Rather, from a German constitutional law perspective, the legislature enjoys a certain margin of appreciation and discretion to square its positive obligations to protect human dignity, physical integrity, and equality before the law and to implement the social state principle with its negative obligation not to interfere unduly with freedom of research, occupational freedom, the general right to personality, or the general freedom of action.

While it is undisputed that the legislature has the power to regulate neuro-enhancement with a view to protecting the individual from potential health risks, it is more difficult to make a legal argument against neuro-enhancement when such potential negative effects cannot be discerned.

With a view to a future potential "Anti-Enhancement Act", Tade Spranger discusses whether parallels can be drawn from the already existing German Anti-Doping Act which is, of course, limited to sports. Similar to the Anti-Doping Act's objective of ensuring the "integrity of sports", the legislature might choose to protect, e.g., "mental integrity".

1.3.6 Part VI: The Harmonization Problem

The brain organoid technology and its application to humans are so fundamental to our image of the human being and the human species that, in the end, the ethical and legal questions of the technology seem to be of global reach. This also suggests that these questions cannot, or should not, be answered in national or regional isolation, but rather calls for a more universal, international as well as intercultural dialogue. Therefore, we considered it necessary to examine whether and to what the extent global harmonization of normative standards related to human brain organoids and their uses for research, therapeutic, or enhancement purposes is advisable and achievable.

1.3.6.1 Global Harmonization of Legal Standards for Brain Organoid Research and Therapy? (Chap. 12)

Silja Voeneky first outlines the international law framework for regulating brain organoid technology. She subsequently provides an outlook on a future governance framework for the application of this technology.

Regarding the currently applicable rules of international law, Silja Voeneky analyzes first the core international human rights treaties on the global level, i.e. the two UN International Covenants on Civil and Political Rights (ICCPR) and on Economic, Social and Cultural Rights (IPESCR) as well as on the regional (European) level, i.e. the European Convention of Human Rights (ECHR). After that, Silja Voeneky turns to human dignity as a concept which underlies the aforementioned human rights treaties, especially the two UN covenants, and which has been acknowledged by States as a legally relevant, albeit rather vague and undefined principle. She, therefore, calls upon states to develop soft law instruments clarifying the scope and contents of human dignity specifically, e.g., with a view to human brain organoid technology. Also with reference to the Cartagena Protocol of 2000, Silja Voeneky proposes to amalgamate the precautionary principle with international human rights. On the basis of this approach, she deduces certain requirements to be met in case of human brain organoid research.

In the second part of her contribution, Silja Voeneky argues that regarding a future regulatory international framework for brain organoid technology, a soft law declaration could have significant advantages which are exemplified by reference to the 2005 Universal Declaration on Bioethics and Human Rights. Finally, Silja Voeneky also details the essential content of such a soft law instrument related exclusively to human brain organoid research.

1.4 Some Concluding Preliminary Observations

The following reflections are a direct result of the online symposium held in February 2021. The purpose of this symposium was to present and discuss the draft contributions to this volume among the authors, the editors, and a broader audience composed of distinguished experts in the fields of biomedicine, bioethics, and life sciences law.

These scientifically informed ethical and legal debates on human brain organoids brought out some observations which we would like to sketch in this final section of our introductory remarks. These observations are "preliminary" in the sense that they are not meant to be outcomes of additional in-depth legal research (which is why any references are omitted) but to be ad hoc remarks as if raised during the lively discussions of the symposium.

1.4.1 The "Singularity of the Human Being"

The "singularity of the human being" comes into play when animals are "humanized", especially, e.g., through transplantation of brain cells or organoids. The core idea of this concept, i.e. the "singularity of the human being", is that the human being, or the human species *Homo sapiens*, is very special in a very unique way compared to other living or inanimate elements of nature and, thus, in this sense "singular". This concept seems to be a constitutive normative condition of the normative order of modern societies, at least as far as it concerns societies which have established a human rights-based (and, therefore, also democratic) government.

This can be illustrated with a view, e.g., to Germany and its legal order. The German legal order, in dismissing Nazi rule after 1945, is grounded in the individual human being's dignity which is guaranteed in the first paragraph of the first article of the German constitution of 1949 (Article 1(1) of the Basic Law). Accordingly, with a view to Article 1(1) of the Basic Law, the whole legal order is conceived as being developed from this very starting point: the individual human being endowed with dignity and, thus, autonomy. The legal order is, therefore, premised on a thesis which is that the human being is singular in that it is special and unique as compared to other species, not only from a biological but also, more particularly, from a normative point of view.

Hence, the closer animals (or, e.g., artificial intelligence (AI) machines) get to the human being, i.e. the more "humanlike" animals (or AI machines) are, the more the fundamental basis of our normative order underlying state and society may be under threat, or may be questioned. If the very basis of the legal order is, indeed, the concept of the "singularity of the human being", any development which "humanizes" animals (or AI machines or, e.g., creates AI machines with "superhuman" capabilities) is, or must be, conceived as a grave danger to the current normative order we agreed upon. The unrivaled normative status of the human being is being challenged. This may explain the (often only intuitively felt) unease we experience when considering the "humanization" of animals (or the construction of "humanlike" or even "superhuman" AI machines).

1.4.2 The Threshold of "Consciousness"

In the ethical and legal debate on brain organoids, "consciousness" has become a central threshold criterion. Human "consciousness" seems to be a product, or result, of our human brain functions. We assume that, for lack of similarly sophisticated brain functions, other living creatures do not possess a comparably highly advanced form of "consciousness". This is why "consciousness" is closely related to the concept of the "singularity of the human being" (Sect. 1.4.1). Human "consciousness", especially in the form of subjective self-awareness (including, in particular, the awareness of each human being that every other human being also has such subjective self-awareness) is very probably a core, or even indispensable, element of this concept underlying our current normative order of state and society.

Hence, our normative intuition tends to accord brain organoids a special normative status if, and to the extent that, they become "conscious". The problem of this threshold of "consciousness" is that it seems to be indeterminate and indeterminable: the threshold is indeterminate because the several sciences involved, or engaged, in research on "consciousness" have not developed a common idea or understanding of what constitutes "consciousness" or of which form of "consciousness" is decisive as a threshold criterion for a particular normative status. The threshold is also indeterminable because as along as "consciousness", or at least the normatively relevant concept of "consciousness", is indeterminate, it is impossible to define benchmarks and to design instruments for the purpose of measuring the required amount or level of "consciousness".

A more fundamental challenge could be that "consciousness" might turn out to be not comprehensively understood by focusing solely on the human brain. The human brain is not merely a "central processing unit" (CPU) governing the other parts of the human body. Rather, the brain and its functions seem to be closely and irresolvably interconnected with the whole human body of the individual human being. Experience through interaction with the social and natural environment is very probably indispensable for the gradual development of "consciousness" and, in particular, of subjective self-awareness allowing the individual human being to integrate as an individual personality into space and time. For example, it takes quite some time for infants to understand that the beginning of their own existence does not coincide with beginning of the whole universe around them but that there is a past in which they did still not exist, whereas other human beings such as parents and brothers and sisters already had perfectly been in existence.

1.4.3 "Realism" Versus "Fiction"

A recurring feature of the ethical and legal debates on brain organoids (and in other fields of bioethics or life sciences law respectively) is that scholars base their analyses either on a "realistic" or on a "fictious" (or "hypothetical" or "speculative") scenario. This choice has a decisive impact on the outcome of the analyses. In particular, the ethical and legal debates on the status of human brain organoids and on the (im-) permissibility of neuro-enhancement via transplantation of human brain organoids take quite different directions depending on what is the starting point of the analyses.

The "realist" approach starts from the current state of science and its foreseeable and immediate future development. On this basis, it is reasonable to arrive at the conclusion that, for the time being, there is no reason why human brain organoids should be assigned a special normative status beyond the status of other human biological material such as cells and tissues. Brain organoids have the size of peas (i.e. a diameter of 4–5 mm) only, and the possibilities to improve their growth in any meaningful way is currently severely limited due to the lack of vascularization necessary for the transport of oxygen and nutrients.

In contrast, the "fiction" approach starts from the assumption that, one day, in the not all too distant future, the growth and size of human brain organoids can be substantially increased creating organoids which display functions indicating at least rudimentary forms of consciousness. The "fiction" approach, hence, tries to anticipate developments that can, at least to a certain extent, reasonably envisioned or not completely ruled out. Accordingly, the answer to the question of what should be the normative status of brain organoids will be quite different based on the assumption that present technological barriers of brain organoid growth may well be overcome leading to significantly enlarged entities with normatively relevant cognitive or mental capacities.

None of the two approaches is clearly preferable. No approach clearly prevails over the other. Whereas the "realist" approach may hold to be strictly based on the state of science and technology and not to hype premature discussions on unknown scenarios, the "fiction" approach may claim to be forward-looking, thereby being ahead of science instead of being tardily behind scientific developments (which is a well-known reproach to ethics and law). The advantage of the "realist" approach may be, e.g., that it does not tend to instigate lawmakers to unduly interfere with basic research thereby stifling scientific progress which could yield unforeseen and unexpected but highly beneficial insights. The advantage of the "fiction" approach could be that it may be able to draft normative guard railings reminding researchers of some outer limits which could become relevant in the future.

References

Boers SN, van Delden JJ, Clevers H, Bredenoord AL (2016) Organoid biobanking: identifying the ethics: organoids revive old and raise new ethical challenges for basic research and therapeutic use. EMBO Rep 17:938–941. https://doi.org/10.15252/embr.201642613

Bredenoord AL, Clevers H, Knoblich JA (2017) Human tissues in a dish: the research and ethical implications of organoid technology. Science 355. https://doi.org/10.1126/science.aaf9414

Chiaradia I, Lancaster MA (2020) Brain organoids for the study of human neurobiology at the interface of in vitro and in vivo. Nat Neurosci 23:1496–1508. https://doi.org/10.1038/s41593-020-00730-3

Dederer H-G, Frenken G (2021). Genom-Editierung am Menschen – Vergleich der regulatorischen Rahmenbedingungen für CRISPR-Gen-Editierung und ihre Auswirkungen auf Forschung und Innovation. Studie zum deutschen Innovationssystem, Nr. 13-2021. Expertenkommission Forschung und Innovation (EFI), Berlin. Retrieved November 2, 2021, from https://www.e-fi.de/fileadmin/Assets/Studien/2021/StuDIS_13_2021.pdf

Farahany NA, Greely HT, Hyman S, Koch C, Grady C, Paşca SP, Sestan N, Arlotta P, Bernat JL, Ting J, Lunshof JE, Iyer EPR, Hyun I, Capestany BH, Church GM, Huang H, Song H (2018) The ethics of experimenting with human brain tissue. Nature 556:429–432. https://doi.org/10.1038/d41586-018-04813-x

Koo B, Choi B, Park H, Yoon K-J (2019) Past, present, and future of brain organoid technology. Mol Cells 42:617–627. https://doi.org/10.14348/molcells.2019.0152

Lavazza A, Massimini M (2018) Cerebral organoids and consciousness: how far are we willing to go? J Med Ethics 44:613–614. https://doi.org/10.1136/medethics-2018-104976

Munsie M, Hyun I, Sugarman J (2017) Ethical issues in human organoid and gastruloid research. Development 144:942–945. https://doi.org/10.1242/dev.140111

Tanaka Y, Park I-H (2021) Regional specification and complementation with non-neuroectodermal cells in human brain organoids. J Mol Med 2021(99):489–500

Part I

Scientific Background

Development of Brain Organoids with Genome-Edited iPSC-Derived Brain Cells

2

Naime Zagha and Beate Winner

2.1 Human-Cell Based Cerebral Organoids for Advanced Brain Research

Why do we need a proxy of the human brain? The brain is the most complex organ in the human body. Brain development is an extraordinary sequence of events and regulated by spatial and temporal factors. Impairments during this patterning process cause dysfunctions of the brain. Studies of human material in vivo are limited, due to scarce availability of the tissue. Therefore, the analysis of the human brain is mostly restricted to post-mortem and fetal tissue. These time points are a snapshot of either disease-related changes or early developmental processes, respectively.

The study of animal brains, mostly rodents, led to important insights into brain development and understanding of evolutionary similarities between species. Nevertheless, there are major interspecies differences between mice and men. One of these obvious differences is a 50-fold larger brain size and gyrification, the folding of the cerebral cortex in humans. There are limits to modeling human physiology and metabolic processes in animal models.[1] For example, in 80% of neurodevelopmental disease the underlying pathological process still remains unknown.[2,3]

[1] Kuzawa et al. (2014), "Metabolic costs and evolutionary implications of human brain development".

[2] Mariani et al. (2015), "FOXG1-dependent dysregulation of GABA/glutamate neuron differentiation in autism spectrum disorders".

[3] Coe et al. (2012), "The genetic variability and commonality of neurodevelopmental disease".

N. Zagha · B. Winner (✉)
Department of Stem Cell Biology, Friedrich-Alexander-Universität (FAU) Erlangen-Nürnberg, Erlangen, Germany
e-mail: naime.denguir@uk-erlangen.de; beate.winner@fau.de

H.-G. Dederer, D. Hamburger (eds.), *Brain Organoids in Research and Therapy*, Advances in Neuroethics, https://doi.org/10.1007/978-3-030-97641-5_2

Scientists strive to overcome these limitations of animal models and tissue accessibility by designing new human in vitro models. The goal is to understand human pathogenic variants and disease processes and their complex impact on human disease. A first step forward in human disease modeling was the discovery that in vitro human embryonic stem cells (hESC) can be differentiated into neurons. While hESC raise ethic concerns, a major breakthrough was the discovery of reprogramming. Researchers in 2007 demonstrated the capability of somatic cells to be reprogrammed back into a state of pluripotency, to induced pluripotent stem cells (iPSC).[4] Similar to hESC cells, human-derived iPSC (hiPSC) can be differentiated into neural lineages. Both can be used not only for two-dimensional (2D) neuronal cultures but also are able to form 3D cell aggregates, the so-called brain organoids. Those hiPSC-derived human cerebral organoids serve as an alternative platform to animal models by resembling more closely the human brain and as a new tool for disease investigation, drug discovery, and personalized treatment.

Section 2.2 will provide basic insights into the technology of generating hiPSC (2.2.1) and hiPSC-derived 3D cerebral organoids (2.2.2). Section 2.3 will recapitulate the profile and characteristics of hiPSC-derived cerebral organoids by looking at key developmental events of the cerebral organoid's structure (2.3.1) regarding cytoarchitecture and cell diversity (2.3.1.1) and maturation and circuit formation (2.3.1.2). Part 2.4 will question if human cerebral organoids are human brains in a dish and will discuss the current biological limitations.

2.2 Generation of Human Cerebral Organoids

2.2.1 The Use of Pluripotent Stem Cells

The groundwork for in vitro modeling of human brain cells came from the capability of differentiating them from hPSC inspired by developmental processes. The first human in vitro model was established by Thomson et al. 1998, who cultured hESC from human embryos starting at the blastocyst stage.[5] ESC have the unique property of pluripotency. Pluripotency implies the unlimited capacity of self-renewal and the ability to differentiate into the three primary germlayers (ecto-, endo-, and mesoderm) and further into somatic or finally differentiated types of cells. The usage of hESC not only led to tremendous advances in understanding human development, drug development, and cell therapy, it also raised enormous ethical concerns.[6]

In 2007, scientists from Japan succeeded to generate hiPSC from specific skin cells, the fibroblasts. They overexpressed specific transcription factors, the so-called Yamanaka factors, which jumpstarted endogenous transcription factors for

[4] Takahashi et al. (2007), "Induction of pluripotent stem cells from adult human fibroblasts by defined factors".

[5] Thomson et al. (1998), "Embryonic stem cell lines derived from human blastocysts".

[6] de Wert and Mummery (2003), "Human embryonic stem cells: research, ethics and policy".

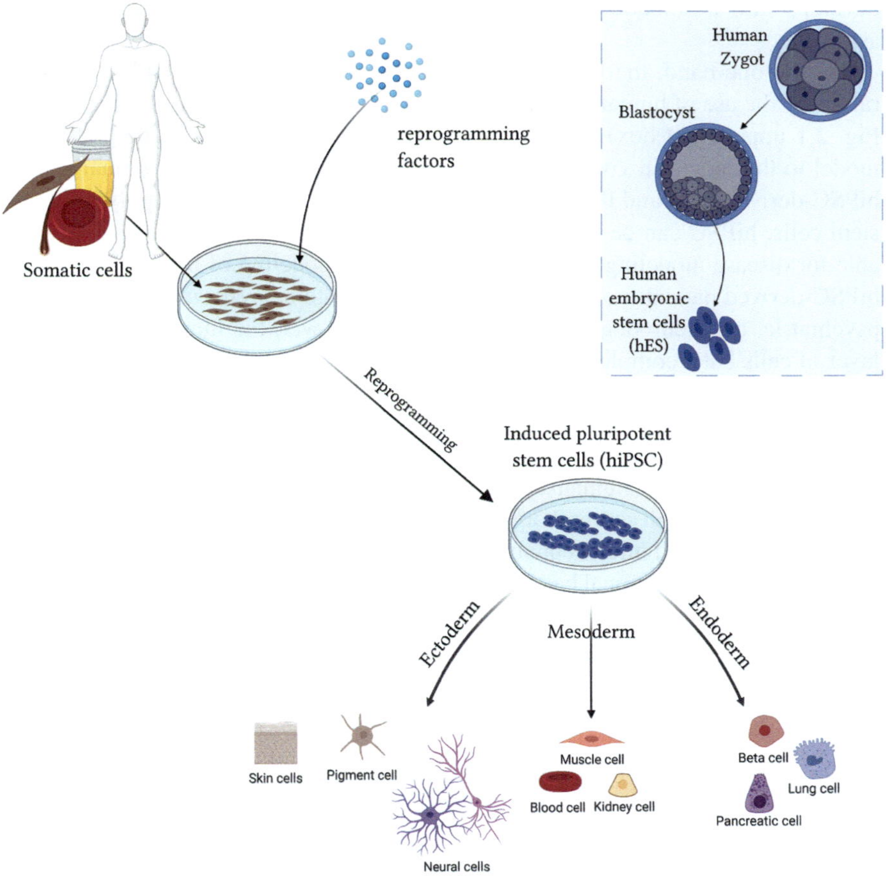

Fig. 2.1 Somatic cells (e.g. blood, urine, hair, or fibroblasts) from humans are reprogrammed by specific reprogramming factors into human induced pluripotent stem cells (hiPSC). hiPSC have the unique property to differentiate into all cell types from the three primary germ layers ecto-, endo-, and mesoderm. For differentiation into neural cells, e.g. neurons, hiPSC are guided to differentiate towards the ectodermal, more precise the neuroectodermal lineage. Neural cells are used in a bi-dimensional (2D) culture model of the human brain. Upper right box: Human embryonic stem cells (hESC) can be harvested from the inner cell mass of the blastocyst stage of the embryo at an early stage post-fertilization. Created with Biorender.com

pluripotency.[7] The fibroblasts could therefore be returned to a state of pluripotency, which is why the reprogrammed fibroblasts were called hiPSC. Moreover, similar to hESC, hiPSC can be differentiated into neural cells (Fig. 2.1). Neuronal differentiation is achieved by sequential application of exogenous growth factors, which first turn the hiPSC into a neural progenitor cell (an intermediate state), before it then

[7]Takahashi et al. (2007), "Induction of pluripotent stem cells from adult human fibroblasts by defined factors".

develops into a somatic brain cell (mostly neurons and support cells such as astrocytes).

On the one hand, the use of hiPSC has overcome significant ethical concerns raised by the use of human embryos to harvest hESC from the blastocyst stage (see Fig. 2.1 upper right box). Additionally, hiPSC-models serve as a complementary model to the rodent in vivo models and thus reduce the amount of animal testing. hiPSC-derived cells and their derivatives are an almost unlimited supply of human stem cells. hiPSC can be bio-banked and patient-specific hiPSC are readily available for disease modeling and drug screening. 2D modeling of brain disease using hiPSC-derived patient models led to new insights into neurodevelopmental, neuropsychiatric, and neurodegenerative diseases.[8,9] However, the human brain is not a layer of cells but a complex 3D structure.

A revolution in the field was Madeline Lancaster's observation that hiPSC are able to create 3D brain structures through self-organization even with just little external support.[10] This was the first step of the generation of hiPSC-derived 3D cell clusters, the so-called human cerebral organoids. One example of an important insight gained through cerebral organoid research was the delineation of the interaction of Zika virus and microcephaly.[11] Moreover, this technique led to novel insights into understanding neuronal heterotopia in patients, for example, with rare neurodevelopmental variants in DCHS1 and FAT4.[12]

2.2.2 Self-Organization into 3D Cerebral Organoids

Cerebral organoids are complex 3D structures with heterogeneous tissues resembling various regions of the brain. They are produced much as other 3D multicellular structures resembling eye, gut, liver, or kidney. Knoblich and Lancaster pioneered the field by relying on the cell's intrinsic development programs and self-patterning ability to generate the so-called human whole-brain or human cerebral organoids.[13] The process is as follows: hiPSC are instructed to aggregate to little balls called Embryoid Bodies (hEBs). The floating hEBs (Fig. 2.2 Left) are then confronted with a specific medium composition, which forces the development of the neuroectoderm[14] layer (Fig. 2.2 Right). These neural aggregates are then placed

[8] Brennand et al. (2011), "Modelling schizophrenia using human induced pluripotent stem cells".

[9] Prots et al. (2018), "U-Synuclein oligomers induce early axonal dysfunction in human iPSC-based models of synucleinopathies".

[10] Lancaster et al. (2013), "Cerebral organoids model human brain development and microcephaly".

[11] Qian et al. (2017), "Using brain organoids to understand Zika virus-induced microcephaly".

[12] Klaus et al. (2019), "Altered neuronal migratory trajectories in human cerebral organoids derived from individuals with neuronal heterotopia".

[13] Lancaster et al. (2013), "Cerebral organoids model human brain development and microcephaly".

[14] The neuro-ectoderm layer consists of cells derived from the ectoderm, the formation of which is the first step in the development of the nervous system and in which the neural tube is developed in the embryo.

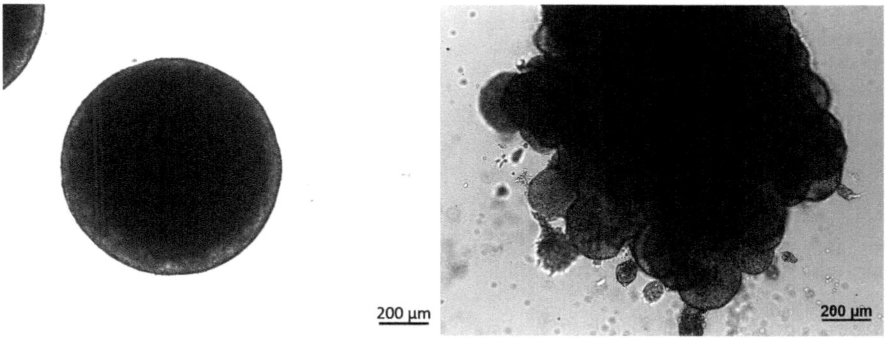

Fig. 2.2 **Left:** hiPSC aggregate to hEBs in floating condition. The formed hEBs are characterized by their circular shape. Image of 8 days in culture. **Right:** The neural induction leads to the forming of a neuroectodermal layer on the surface of the hEBs with bud formation. Image was taken after 15 days in culture. Source: Johanna Kaindl, Department of Stem Cell Biology, Erlangen

in gel droplets (Fig. 2.3). The gel as a matrix provides, on the one hand, both a 3D support and, on the other hand, regulates the proliferation, differentiation, distribution, and migration of neural progenitor cells.[15] While cerebral organoids can be generated by self-organization in a whole-brain organoid,[16] patterned organoids, containing different brain regions such as the forebrain, midbrain, and hindbrain, can be guided to form region-specific tissue of interest with specific sets of signaling molecules.[17,18]

2.3 Profile and Characteristics of Cerebral Organoids

To assess the cellular composition of cerebral organoids and ultimately to compare them to the human brain, sophisticated techniques are required. One of these widely used technologies is single-cell RNA-sequencing (scRNA-seq: see Box 2.1, 2.3). It can be used to analyze the genetic profile of every single cell independently. When cerebral organoids are studied at different time points of differentiation, developmental steps and fates of cell populations can be followed up. Ultimately, this technique enables researchers to study and compare the cellular composition and gene expression of organoids and human brains on a single cell level.[19]

[15] Long and Huttner (2019), "How the extracellular matrix shapes neural development".

[16] Lancaster et al. (2013), "Cerebral organoids model human brain development and microcephaly".

[17] Pasca et al. (2015), "Functional cortical neurons and astrocytes from human pluripotent stem cells in 3D culture".

[18] Sakaguchi et al. (2015), "Generation of functional hippocampal neurons from self-organizing human embryonic stem cell-derived dorsomedial telencephalic tissue".

[19] Quadrato and Arlotta (2017), "Present and future of modeling human brain development in 3D organoids".

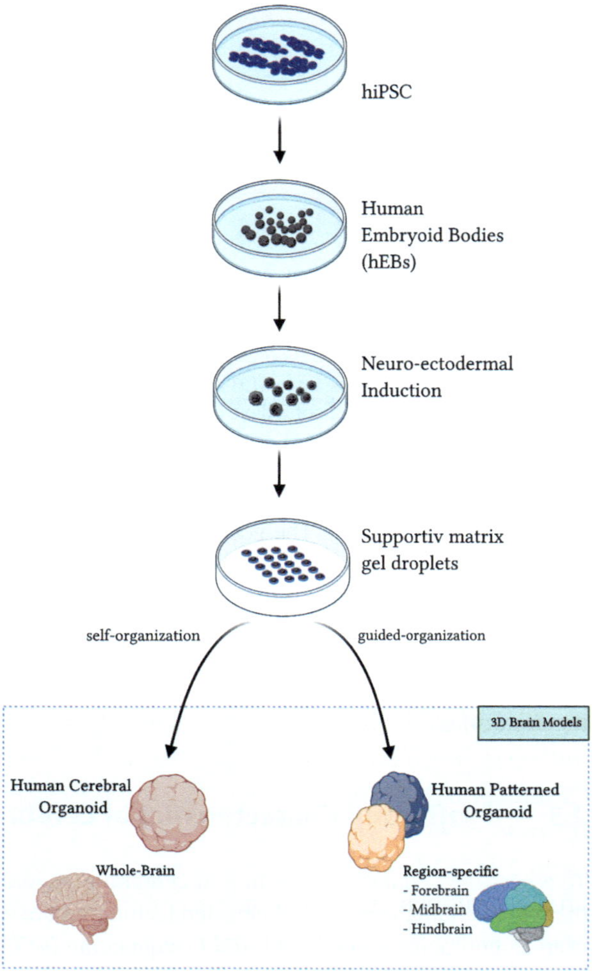

Fig. 2.3 Cerebral organoids as 3D brain models in vitro can be generated from human neural aggregates, the so-called human embryoid bodies (hEBs), which are aggregations of hiPSC induced by specific signaling molecules. The hEBs are cultivated in a specific medium which allows only the neuroectodermal layer to form further. Depending on the 3D brain model, these neuroectodermal aggregates are either placed in droplets of a gel, which serves as a supportive matrix or are cultured in a suspension. Cerebral organoids or whole-brain organoids are generated through self-organization whereas patterned organoids (e.g. forebrain, midbrain, hindbrain) are generated through patterning factors in a guided organization. Representation not true to scale. Created with Biorender.com

Box 2.1: Background to Single-Cell RNA-Sequencing

The technique of single-cell RNA-sequencing (scRNA-seq) is used to enable analytical approaches to characterize organoids. Whereas initially only a few cells could be extracted and analyzed, the number now exceeds more than 1.000.000. Even though there are many different techniques, they generally have the following steps in common: The physical extraction of viable single cells, the subsequent opening of the cells and extraction of the RNA (RNA, ribonucleic acid), to translate the RNA back into cDNA (cDNA, complementary deoxyribonucleic acid), and finally the generation of large sequencing data of the genome. Concerning the study of organoids, questions can be answered such as: which cell types and brain regions are found in the organoid? How "mature" are the cells, what stage of development do they correspond to compared to the human brain? Which gene expression pattern do they show means which genes are controlling the developmental processes and in which subsets of cells are those genes expressed in. A small overview of possible questions that can provide a lot of information about the 3D model's properties.

The following subsections will recapitulate key developmental features of cerebral organoids and deal with the achievements and limitations. We will specifically focus on cortex-like structures in human cerebral organoids since a large number of datasets are available from scRNA-seq analyses of the cortex.

2.3.1 Recapitulation of Key Developmental Events and Limitations

2.3.1.1 Cytoarchitecture and Cell Diversity

Cerebral organoids measure about four millimeters (Fig. 2.4) and are therefore much smaller than the human brain. Nevertheless, remarkable similarities can be discovered upon closer inspection of cytoarchitecture, cell types, and regional identity. The human brain develops from the neural tube, it then forms ventricles, which are cavities filled with fluid. These ventricles are developmentally important since they home a neural stem cell population. In the human developing brain, neural stem cells divide and generate neural progenitor cells that migrate from the proliferative zones around the vesicles towards the cortex. A special type of cells, the radial glial cells, plays a decisive role here, as they guide the neurons to their correct place in the six-layered cortex and contribute to the generation of glial cells[20] and neurons.

Ventricle-like structures also occur in cerebral organoids (Fig. 2.5). The ventricle-like structures in cerebral organoids also contain proliferative zones of neural progenitor cells. In organoids, a comparable migration of neural progenitor cells derived from proliferative zones can be observed. They also form a layered structure of neurons comparable to the layering in the cortex. Moreover, regional specific cell types, such as the above mentioned radial glial cells, emerge spontaneously in organoids, and they even emerge similar to the timeline in the fetal brain. Similar to human brain development, the neural progenitor cells of the cerebral organoid intrinsically control the generation of specific subsequent cell types.[21][22]

[20] Glia, also called neuroglia, are non-neuronal cells in the central nervous system that do not generate electrical impulses but make a decisive contribution to homeostasis, functionality, and protection of neurons.

[21] Kadoshima et al. (2013), "Self-organization of axial polarity, inside-out layer pattern, and species-specific progenitor dynamics in human ES cell–derived neocortex".

[22] Quadrato et al. (2017), "Cell diversity and network dynamics in photosensitive human brain organoids".

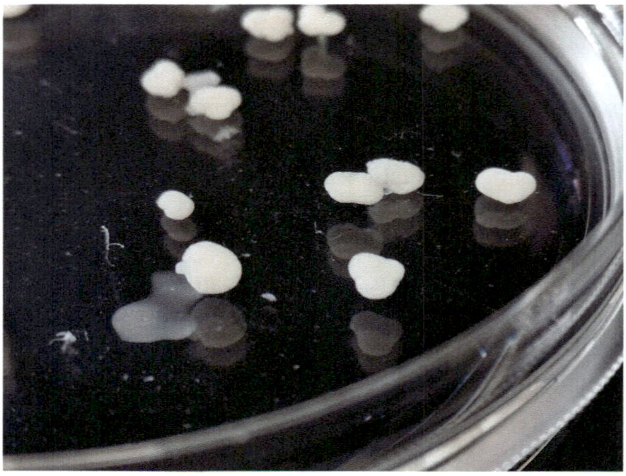

Fig. 2.4 Cerebral organoids grown from human stem cells. They are up to four millimeters in size and are cultivated as swimming cell clumps in laboratories. These 3D brain models develop similarly to a human brain. The cultivation of these organoids requires a regular supply of nutrient media and constant movement of the cell clumps. Organoids can currently be cultivated for up to a year. Image: Stem Cell Department, Universitätsklinikum Erlangen

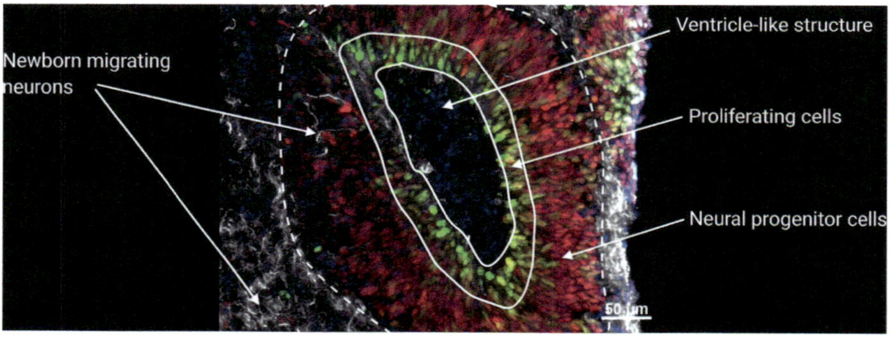

Fig. 2.5 Cross section of a 33-day-old human cerebral organoid, showing a ventricle-like structure (white dashed). In the center is a structural recess. In a thin zone, proliferating cells (green, solid line) are deposited, which are similar in arrangement and cell occurrence in the fetal brain. Neural progenitor cells can be found in the area above (red, between dashed and solid line), in which isolated migrating newborn neurons are shown (white). Source: Johanna Kaindl, Department of Stem Cell Biology, Erlangen

While major similarities in specific cell types were found between the human brain and cerebral organoids,[23] the relative numbers are different. The architecture of the neural tissue in cerebral organoids is an approximation. The human brain has more than 80 billion neurons and the same number of other cell types.[24] An organoid

[23] Quadrato and Arlotta (2017), "Present and future of modeling human brain development in 3D organoids".

[24] Von Bartheld Christopher and Bahney (2016), "The search for true numbers of neurons and glial cells in the human brain: A review of 150 years of cell counting".

consists of around two to three million cells. Some cell populations are missing in cerebral organoids. One example here are microglia, "the innate immune system of the brain." Microglia have a different developmental origin from mesoderm, therefore, microglia cannot be generated simultaneously with the current protocols for cerebral organoids. Instead, they can be added in co-cultures with organoids if required. These experiments already showed that added microglia can migrate into the organoids.[25] Another important cell population, which is missing in cerebral organoids, are endothelial cells. They form the vasculature and enable the blood supply to the human brain. Since these are not emerging in the organoid protocols, and organoids do not have blood supply, but rather depend on permeability, the larger organoids can suffer from an insufficient supply of oxygen and nutrients. As a result, organoids can only be cultivated for a certain period, as they develop a necrotic nucleus during cultivation, which leads to cell death in the nucleus and the limitation of the organoid's size. A groundbreaking proof of concept study from the Gage laboratory transplanted cerebral organoids into a mouse brain. Interestingly, they observed that the mouse vasculature started perfusing the cerebral organoid and led to improved survivability.[26] Thus, the resulting limitations in size, tissue architecture complexity, and maturation are current disadvantages.

Moreover, meninges and skull, which define the limits of the human brain, are also missing. Cerebral organoids also lack a body axis, which in the embryo provides orientation and enables growth and inhibiting factors to have region-specific different influences. Therefore, cerebral organoids currently form brain-like regions but do not reflect the spatial arrangement. Further steps will be needed for the standardization of organoid development to standardize culture conditions and organoid handling, as well as their size and morphology.[27]

2.3.1.2 Maturation and Formation of Circuits

The human brain not only consists of billions of brain cells. More importantly, they are interconnected in different ways in a complex network. They communicate with each other via electrical signals, transfer chemical molecules via millions of synapses and render the brain the most complex organ of the body. However, how similar are organoids and the brain regarding functional properties? Transcriptional analyses, which compared the genetic expression of the in vitro cortical cells of cerebral organoids with those in vivo, concluded that the organoid's cells resemble those of primitive fetal brain during the second trimester of gestation.[28] When using isolated patterned organoids of specific regions, their connections are limited. Specifically, connections between brain regions, the so-called projection, are missing. Circuits are for example building the sensory system, where sensory information from the skin, such as heat, cold, or pressure, is transmitted via electrical

[25] Abud et al. (2017), "iPSC-derived human microglia-like cells to study neurological diseases".

[26] Mansour et al. (2018), "An in vivo model of functional and vascularized human brain organoids".

[27] Kanton et al. (2019), "Organoid single-cell genomic atlas uncovers human-specific features of brain development".

[28] Camp et al. (2015), "Human cerebral organoids recapitulate gene expression programs of fetal neocortex development".

impulses to the sensory cortex. This input can trigger a chain of reactions, including pain or motor movements. Organoids have no sensory input or motor output.

In 2017 the Arlotta laboratory showed that neuronal activity in organoids can be controlled by stimulation of photosensitive cells by light.[29] Another striking discovery came from the Muotri laboratory in 2019. When analyzing cerebral organoids using a multi-electrode system, they registered electrical activity indicative of network function in cerebral organoids. These network curves recorded from the cerebral organoids closely resembled electroencephalogram curves of premature babies which could be an indication for similarities between the stage of maturation and network function.[30] Due to the fact that not all regions are generated in a standardized way, we cannot speak of inter-regional cell interaction and communication, let alone of electrical impulse transmissions. Only very small local measurements were taken which cannot represent extensive inter-regional communication. To counteract this, two or more organoids from different regions can already be fused to the so-called assembloids in order to enable the organization of inter-regional connections.[31]

The aforementioned achievements are outstanding and raise the question of whether organoids will demonstrate many more similarities and functional properties of the human brain in the future.

2.4 Can We Talk of Human Brains in a Dish?

While cerebral organoids are a major advance in human disease modeling in neuroscience and specifically an important bio-assay tool for a research focus on early brain development, the limitations need to be taken into account and still pose great hurdles for science. Therefore, from a biological point of view, one cannot speak of "brains in a dish." Why? Cerebral organoids currently do not represent an exact model of the human brain. The smaller size, the less organized shape, and the lack of some functionally important cell types such as vascular cells or microglia are obvious differences. Functionally, differences in neural network activities and the lack of sensory inputs limit the current cerebral organoids to an immature in vitro model. Thus, at this point, an exact imitation of brain development and function is still a future perspective.

Time to speculate about characteristics and functional properties that cerebral organoids will be able to mimic in the future: technical innovations promise more complex and functional organoids. First, standardizing the protocols of organoid generation concerning efficiency and replicability is still ahead. This implies improving oxygen and nutrient supply, to allow the generation of more mature or aging organoids and thus to examine functional characteristics of neurodegeneration. By

[29] Quadrato et al. (2017), "Cell diversity and network dynamics in photosensitive human brain organoids".

[30] Trujillo et al. (2019), "Complex oscillatory waves emerging from cortical organoids model early human brain network development".

[31] Birey et al. (2017), "Assembly of functionally integrated human forebrain spheroids".

simulating a polarity and proper diffusion of the specific signal factors in a gradient similar to that in the brain, researchers hope to enable a correct arrangement of the expressed brain regions in the organoids and to recreate regional networks.

New techniques such as CRISPR/Cas9 genome editing improved the options to selectively study the effect of single pathogenic variants. Since the scRNA-seq technique has the restriction of missing some localization information, new techniques that also allow visualization of the respective cell at its original spot are the next step. To provide organoids with a blood supply, attempts have been made to implant organoids in mice. A vascularization and survival of the organoids were achieved.[32] The next step along this journey will be to implant endothelial cells and provide blood supply in specific chambers. The combination of region-specific organoids in assembloids in a functional system could enable the simulation of the interconnectivity of different tissues. The Pasca laboratory has already developed a working model of the motor system, linking the motor cortex to muscles in a tripartite circle. For this, Pasca brought organoid models for motor cortex, spinal connections, and skeletal muscles together. The individual models fused on their own to assembloids, showing contractions of the muscle cells.[33]

Cerebral organoids are a great platform and already boost research into neurodevelopmental and brain affecting diseases (Box 2.2: Studying SARS-CoV-2 Infection in Cerebral Organoids). But the more similar the organoid and the brain become, the louder the question of whether they also have a sense of consciousness. Although not every new finding will lead to new biological or translational insights, some results might be game-changing enabling the development of even more complex and voluminous cerebral organoids. In this regard, ethical discourse is important and legal and ethical guidance provided to scientists is crucial. Cerebral organoid research raises many difficult questions and concerns not only scientists, but also researchers, state organs, pharmaceutical companies, and the general public. In cooperation and common endeavors, the difficult obstacles and questions can be worked out and, at the same time, it will be possible to enable new and groundbreaking insights into the human brain.

Box 2.2: Studying SARS-CoV-2 Infection in Cerebral Organoids

Numerous clinical reports show neurological symptoms in patients with SARS-CoV-2 infection. However, it is still unclear whether the virus directly affects and damages neurons. To investigate whether a SARS-CoV-2 infection of the human brain is a reason for symptoms of neurological anomalies, researchers make use of human cerebral organoids by infecting them with the virus. The aim is to find out which cell types the virus attacks, what potential the virus has to cause further neurological defects and, ultimately, how

[32] Mansour et al. (2018), "An in vivo model of functional and vascularized human brain organoids".

[33] Andersen et al. (2020), "Generation of Functional Human 3D Cortico-Motor Assembloids".

cerebral organoids can contribute to the development of therapy in the SARS-CoV-2 pandemic. Cerebral organoids can serve as a promising, viable, and safe test system to study direct neurotoxic effects resulting from a SARS-CoV-2 infection, as well as other viruses.

References

Abud EM et al (2017) iPSC-derived human microglia-like cells to study neurological diseases. Neuron 94(2):278–293

Andersen J et al (2020) Generation of functional human 3D Cortico-motor Assembloids. Cell 183(7):1913–1929.e26

Von Bartheld Christopher S, Bahney J, Herculano-Houzel S (2016) The search for true numbers of neurons and glial cells in the human brain: a review of 150 years of cell counting. J Comp Neurol 524(18):3865–3895

Birey F et al (2017) Assembly of functionally integrated human forebrain spheroids. Nature 545(7652):54–59

Brennand KJ et al (2011) Modelling schizophrenia using human induced pluripotent stem cells. Nature 473(7346):221–225

Camp JG et al (2015) Human cerebral organoids recapitulate gene expression programs of fetal neocortex development. Proc Natl Acad Sci 112(51):15672–15677

Coe BP, Girirajan S, Eichler EE (2012) The genetic variability and commonality of neurodevelopmental disease. Am J Med Genet Part C Seminars Med Genet 160(2):118–129. Wiley Online Library

Kadoshima T et al (2013) Self-organization of axial polarity, inside-out layer pattern, and species-specific progenitor dynamics in human ES cell–derived neocortex. Proc Natl Acad Sci 110(50):20284–20289

Kanton S et al (2019) Organoid single-cell genomic atlas uncovers human-specific features of brain development. Nature 574(7778):418–422

Klaus J et al (2019) Altered neuronal migratory trajectories in human cerebral organoids derived from individuals with neuronal heterotopia. Nat Med 25(4):561–568

Kuzawa CW et al (2014) Metabolic costs and evolutionary implications of human brain development. Proc Natl Acad Sci 111(36):13010–13015

Lancaster MA et al (2013) Cerebral organoids model human brain development and microcephaly. Nature 501(7467):373–379

Long KR, Huttner WB (2019) How the extracellular matrix shapes neural development. Royal Soc Open Biol 9(1):180216

Mansour AAF et al (2018) An in vivo model of functional and vascularized human brain organoids. Nat Biotechnol 36(5):432–441

Mariani J et al (2015) FOXG1-dependent dysregulation of GABA/glutamate neuron differentiation in autism spectrum disorders. Cell 162(2):375–390

Pasca AM et al (2015) Functional cortical neurons and astrocytes from human pluripotent stem cells in 3D culture. Nat Methods 12(7):671–678

Prots I et al (2018) U-Synuclein oligomers induce early axonal dysfunction in human iPSC-based models of synucleinopathies. *Proceedings of the National Academy of Sciences* 115(30):7813–7818

Qian X et al (2017) Using brain organoids to understand Zika virus-induced microcephaly. Development 144(6):952–957

Quadrato G, Arlotta P (2017) Present and future of modeling human brain development in 3D organoids. Curr Opin Cell Biol 49:47–52

Quadrato G et al (2017) Cell diversity and network dynamics in photosensitive human brain organoids. Nature 545(7652):48–53
Sakaguchi H et al (2015) Generation of functional hippocampal neurons from self-organizing human embryonic stem cell-derived dorsomedial telencephalic tissue. Nat Commun 6(1):1–11
Takahashi K et al (2007) Induction of pluripotent stem cells from adult human fibroblasts by defined factors. *Cell* 131(5):861–872
Thomson JA et al (1998) Embryonic stem cell lines derived from human blastocysts. *Science* 282(5391):1145–1147
Trujillo CA et al (2019) Complex oscillatory waves emerging from cortical organoids model early human brain network development. *Cell Stem Cell* 25(4):558–569
de Wert G, Mummery C (2003) Human embryonic stem cells: research, ethics and policy. *Human Reprod* 18(4):672–682

Cell-Based Therapy and Genome Editing in Parkinson's Disease: Quo Vadis?

3

Yanni Schneider, Jeanette Wihan, Soeren Turan, and Jürgen Winkler

3.1 Neurodegenerative Diseases: Urgent Need for Cell-Based Therapies

The incidence of neurodegenerative diseases is steadily increasing due to aging societies worldwide. Age-related neurodegenerative processes are hallmarked by a progressive loss of selectively vulnerable neural cells in the central nervous system (CNS). The most frequent neurodegenerative diseases are amyloid-, tau-, or synuclein-associated clinical entities defined by the pathological aggregation of the respective protein.[1] The broad spectrum of symptoms is mainly defined by specific CNS regions affected the most by neuronal dysfunction and consequent cell loss due to the continuous aggregation and spread of distinct protein species. The symptoms consist of a variable range of cognitive, motor, or neuropsychiatric deficits

[1] Dugger and Dickson (2017)

Y. Schneider
Department of Stem Cell Biology, University Hospital Erlangen, Friedrich-Alexander University (FAU) Erlangen-Nürnberg, Erlangen, Germany

Department of Molecular Neurology, University Hospital Erlangen, Friedrich-Alexander University (FAU) Erlangen-Nürnberg, Erlangen, Germany

J. Wihan · J. Winkler (✉)
Department of Molecular Neurology, University Hospital Erlangen, Friedrich-Alexander University (FAU) Erlangen-Nürnberg, Erlangen, Germany
e-mail: Juergen.Winkler@uk-erlangen.de

S. Turan
Department of Stem Cell Biology, University Hospital Erlangen, Friedrich-Alexander University (FAU) Erlangen-Nürnberg, Erlangen, Germany

Institute of Biochemistry, Friedrich-Alexander-Universität (FAU) Erlangen-Nürnberg, Erlangen, Germany

H.-G. Dederer, D. Hamburger (eds.), *Brain Organoids in Research and Therapy*, Advances in Neuroethics, https://doi.org/10.1007/978-3-030-97641-5_3

predominantly linked to distinct susceptible neurons and its corresponding neurotransmitter systems.[2,3] The majority of currently used symptomatic therapies aim to substitute or compensate the deficit of specific neurotransmitter systems in order to improve the clinical phenotype. However, besides causing adverse side effects in the long-term, previous studies showed that neurotransmitter-based symptomatic therapeutic approaches are not able to slow down, halt, or even reverse disease progression in these disorders.[4,5] Furthermore, the progressive dysfunction and loss of neurons have a tremendous impact on quality of life measures. Although the CNS maintains a pool of neural stem cells in some niches such as the hippocampus, these cells are not able to repopulate or even compensate the loss of neurons observed in age-related neurodegenerative diseases.[6] Almost a half century ago, the foundation to replace diseased neural cells by grafting neural cells into defined CNS regions has been laid by a group of scientists in Sweden.[7,8,9] Since the pharmacological substitution of neurotransmitters appeared promising to some degree, the idea to transplant specific neural cells secreting the respective neurotransmitter was considered as a promising long-lasting therapy to intervene in the course of these devastating neurodegenerative diseases. After the failure of randomized clinical trials grafting fetal dopaminergic cells in Parkinson's disease (PD), the development of technologies such as the generation of human-induced pluripotent stem cells (hiPSCs) and human cerebral organoids opened up new possibilities with respect to a revival for cell-based therapeutic approaches for the CNS.[10,11] Currently, therapeutic cell-based approaches are exclusively using cellular suspensions of hiPSC-derived neural cells. Up to now, the application of brain organoids into certain brain regions is limited due to the lack of a safe approach applying these macroscopic cell clusters. The transplantation of brain organoids might further damage the anatomical site of grafting due to the needle size required for the transplantation of an organoid. Therefore, with currently available protocols, brain organoids are rather suitable for preclinical disease modeling or testing of pharmacological compounds. The following chapter will summarize these cellular and molecular breakthroughs focusing on PD, the prototypical and most prevalent synucleinopathy. Furthermore, we will reflect and discuss very recent molecular gene editing advancements in integrating these innovative therapeutic strategies toward regenerative medicine.

[2] Pereira, Ferreiro, Cardoso, & de Oliveira (CR732004), p. 97

[3] Rinne (1993), p. 31

[4] Heumann et al. (2014), p. 472

[5] Sharma (2019), p. 1479

[6] Gage (2000), p. 1433

[7] Olson and Seiger (1972), p. 175

[8] Olson and Seiger (1975), p. 141

[9] Seiger and Olson (1975), p. 325

[10] Lancaster and Knoblich (2014), p. 2329

[11] Takahashi et al. (2007), p. 861

3.2 Parkinson's Disease: Pathophysiology and Diagnosis

PD belongs to the group of synucleinopathies. These disorders are defined as a spectrum of age-related neurodegenerative disorders commonly characterized by an abnormal aggregation of the intracellular presynaptic protein alpha-synuclein (aSyn). The progressive aggregation of aSyn in PD results in the deposition of aSyn in the cytoplasm of neurons (Lewy bodies) and/or neurites (Lewy neurites[12];). In 85–90%, PD patients are affected sporadically with a late onset usually during the sixth decade of life. Besides sporadic PD, 10–15% of PD cases are linked to mutations in specific genes known as PARK loci. These loci harbor different types of mutations including multiplications of the entire gene locus of aSyn, the SNCA gene.[13] Monogenic forms of PD are characterized by an earlier onset of motor symptoms and in some instances associated with severe cognitive or other psychiatric deficits in comparison to sporadic PD.[14,15,16,17,18,19] Clinically, sporadic PD is hallmarked by cardinal motor symptoms such as bradykinesia, rigidity, and resting tremor.[20] The presence of these symptoms is primarily linked to the progressive loss of dopaminergic neurons within the substantia nigra pars compacta of the midbrain.[21] Diagnosing PD remains challenging in the clinical routine and is still based on the presence of the above-mentioned clinical symptoms; however the definitive diagnosis requires the demonstration of Lewy bodies in post mortem neuropathological examinations.

3.3 Current Therapies

Current pharmacological therapies for PD-related motor deficits consist of dopaminergic partial replacement using the dopamine precursor levodopa (L-Dopa), the most potent compound to restore motor functions in PD. The usage of L-Dopa in PD represents a major breakthrough in the treatment of age-related neurodegenerative movement disorders. Although Dr. G. Cotzias discovered L-Dopa already in 1967 as a very powerful and effective compound for treating PD symptoms, it is still the gold-standard up today. The major sequelae of long-term L-Dopa treatment is, however, the development of adverse effects called motor fluctuations such as hypo-, hyper- or dyskinesias becoming in particular more prominent within or after

[12] Spillantini et al. (1997), p. 839

[13] Lesage and Brice (2009), p. R48

[14] Kiely et al. (2013), p. 753

[15] Kruger et al. (1998), p. 106

[16] Pasanen et al. (2014), p. 2180 e1–5

[17] Polymeropoulos et al. (1997), p. 2045

[18] Zarranz et al. (2004), p. 164

[19] Proukakis et al. (2013), p. 1062

[20] Jankovic (2008), p. 368

[21] Baba et al. (1998), p. 879

the first decade of therapy. In particular, patients start to suffer from other motor fluctuations, i.e., freezing of gait or a decreasing response to L-Dopa. To increase the efficacy and tolerability of L-Dopa during the long-lasting disease course, there are other compounds to increase the dopaminergic tone within the CNS such as dopamine receptor agonists and inhibitors of dopamine metabolizing enzymes such as the monoaminooxidase B or the catecholmethyltransferase.[22] Besides pharmacological approaches, deep brain stimulation (DBS) has been approved as an effective neurosurgical intervention in PD. The mode of action for DBS is based on the continuous electrical stimulation of anatomically well-defined CNS regions.[23] For instance, several electrodes are implanted into the thalamus, the pallidum, or the subthalamic nucleus resulting in the alleviation of distinct motor symptoms in PD patients.[23] Implanting these electrodes requires an invasive neurosurgical procedure by an interdisciplinary team. Despite these great therapeutic advances for patients suffering from PD, none of the aforementioned therapies is able to slow down the progression of the disorder. Thus, there is still an urgent need for novel innovative approaches more effectively modifying the course of the disease.

3.4 History of Cell-Based Therapy

The therapeutic concept of cellular transplantation into neuronal structures has a long history in translational neurosciences going back to the first transplantation studies in the 1970s. In 1972, *Olson and Seiger* set the basis for the transplantation of neural tissue.[7,8,9] In their initial experimental approach, they collected cerebral tissue consisting of monoaminergic neurons from newborn animals or fetuses further successfully transplanting this tissue in the anterior chamber of the adult rodent eye.[8] In a subsequent study, *Olson and Seiger* succeeded to transplant ganglion cells in combination with fetal cortical tissue resulting in a profound reinnervation of disconnected rodent eyes using similar monoaminergic neurons.[9] Noteworthy, these studies provided clear evidence to use fetal tissue for transplantation purposes based on findings such as the good cellular survival postgrafting and the potential for appropriate reinnervation.

After obtaining these encouraging findings in preclinical models, the transplantation of adrenal medullary tissue into the caudate nucleus of PD patients was initiated in 1985, however without resulting in clinical benefits.[24] Following these initial attempts in PD patients, a novel source for grafts was discovered: human fetal ventral mesencephalic (HFVM) tissue prepared from aborted fetuses. HFVM tissue consists of dopaminergic neurons,[25] thereby representing a "good cellular source" for transplantation into the putamen and caudate nucleus of PD patients. In contrast to the initial transplantation efforts using adrenal medullary tissue, two patients

[22] Lindvall (2016), p.30

[23] Benabid (2003), p. 696

[24] Backlund et al. (1985), p. 169

[25] Kontur, Leranth, Redmond, Roth, & Robbins (CR471993), p. 172

demonstrated an impressive improvement of PD symptoms after receiving HFVM grafts in 1990.[26] These initial promising results encouraged neuroscientists to move forward with the concept of HFVM transplantation approaches in randomized clinical studies. This intention was further supported by the optimization of pre-existing transplantation procedures resulting in the positive outcome after neural graft transplantation.[27,28] However, despite all positive preliminary clinical studies, larger, randomized clinical studies testing the efficacy of fetal dopaminergic grafts in PD patients failed to show an overall significant clinical improvement postgrafting.[29,30,31] The lack of clinical efficacy observed in the randomized clinical trials after fetal grafting and the presence of graft-induced dyskinesia was a major setback for moving forward with this cell-based transplantation approach. More importantly, the presence of Lewy body pathology in the transplanted fetal grafts 10 years after transplantation hampered further the optimism in regard to long-term safety and feasibility of HFVM transplantation in PD patients.[32] Neither follow-up studies demonstrating that the majority of the grafted cells was unaffected by Lewy body pathology nor reports of a maintained clinical improvement after transplantation changed this initial view on HFVM transplantations.[33,34] Besides crucial ethical concerns, the major clinical disadvantage of HFVM grafting strategies is the need for permanent immunosuppression in order to decrease the host versus graft reaction aimed to improve graft survival.[35,36] Since neural fetal grafts derive from several allogenic fetuses (i.e., up to four pooled fetuses are needed for one hemisphere of a single PD patient), the host immune response may result in the rejection of the transplanted fetal grafts. In general, immunosuppressive therapies carry additional risks for further detrimental adverse effects in elderly patients such as PD patients.[37] In summary, these important clinical considerations raise crucial ethical and methodological concerns regarding transplantation of fetal grafts. However, these clinical studies in PD patients had very important implications for i) the better understanding of the underlying molecular pathogenesis in PD by implying the potential spreading of aSyn from the neighboring CNS tissue of the host into the grafted immature fetal dopaminergic neurons and ii) introducing significant

[26] Lindvall et al. (1990), p. 574

[27] Kordower et al. (1998), p. 383

[28] Kordower et al. (1995), p. 1118

[29] Brundin et al. (2000), p. 1380

[30] Freed et al. (2001), p. 710

[31] Olanow et al. (2003), p. 403

[32] Kordower, Chu, Hauser, Freeman, & Olanow (2008)), p. 504

[33] Li et al. (2008), p. 501

[34] Li et al. (2010), p. 1091

[35] Frodl, Nakao, & Brundin (CR301994), p. 2393

[36] Nakao, Frodl, Duan, Widner, & Brundin (1994), p. 12408

[37] Wennberg et al. (2001), p. 1797

encouraging clinical efficacy data concerning neural grafting strategies in PD, however using other suitable cell sources.

3.5 Development of the Modern Era of Stem Cell Technology

Consequently, the basic and clinical research community was continuously searching for an alternate cellular source for this type of neural transplantation approach: a novel era started with the discovery of human embryonic stem cells (hESCs[38]). The development of the hESCs has been inspired by its murine analogue, the mouse embryonic stem cells (mESCs[39]). hESCs are derived from human blastocysts and show pluripotency allowing the differentiation into all germ layers and its cellular derivatives.[40,41] A major disadvantage for the clinical usage of pluripotent hESCs is their potential to form malignant embryonic tumors such as teratomas.[42,43] Thus, the preparation of hESCs for further clinical application requires very high safety profiling standards.[42] Nevertheless, hESCs raised the hope as a novel cellular source for grafting approaches in order to develop an alternate grafting strategy for PD. hESCs represent an unlimited cellular source with an overwhelming potential to differentiate into distinct mature human cells. Detailed protocols were immediately established for the differentiation toward various neuronal subtypes.[44,45] Moreover, preclinical studies highlighted the potential of hESC-derived neural progenitor cells (NPCs) as an ideal source for allogenic transplantation of human cells into animal models. hNPCs integrated into the host murine brain postgrafting and were able to differentiate into distinct neural lineages.[45,46] The motor phenotype in PD is closely linked to a progressive loss of dopaminergic neurons, thereby defining the need to establish specific, standardized, and safe differentiation protocols for human midbrain dopaminergic neurons (mDANs). Initial achievements were obtained by differentiating dopaminergic neurons derived from mESCs,[47] but the translation to hESCs remained challenging. Although human neurons with specific dopaminergic characteristics were obtained,[48] there was no significant symptomatic improvement

[38] Thomson et al. (1998), p. 1145

[39] Evans and Kaufman (1981), p. 154

[40] Itskovitz-Eldor et al. (2000), p. 88

[41] Schuldiner, Yanuka, Itskovitz-Eldor, Melton, & Benvenisty (CR412000), p. 11307

[42] Hentze et al. (2009), p. 198

[43] Prokhorova et al. (2009), p. 47

[44] Reubinoff et al. (2001), p. 1134

[45] Zhang, Wernig, Duncan, Brustle, & Thomson (2001), p. 1129

[46] Englund, Fricker-Gates, Lundberg, Bjorklund, & Wictorin (2002), p. 1

[47] Kawasaki et al. (2000), p. 31

[48] Yan et al. (2005), p. 781

after transplantation in rodent PD models.[49,50] Furthermore, transplanted hESCs formed tumors after grafting into the CNS.[51] Although this procedure was not applicable for therapeutic approaches in patients, these studies significantly contributed to our current understanding of the molecular machinery driving the differentiation of pluripotent stem cells into a specific midbrain dopaminergic phenotype.[52]

In 2006, K. Takahashi and Yamanaka reported the first success in reprogramming somatic mouse fibroblasts into adult induced pluripotent stem cells,[53] followed by the reprogramming of adult human fibroblasts into hiPSCs one year later.[11] This was the beginning of a new era in stem cell biology. The generation of patient-derived cells revolutionized the entire stem cell research field regarding its scientific and therapeutic impact including specific ethical questions raised by this novel molecular and cellular technology.

3.6 Human-Induced Pluripotent Stem Cells: A Promising Cell Source

K. Takahashi and Yamanaka successfully generated for the first time embryonal-like stem cells by reprogramming adult mouse fibroblasts. Initially, a large set of transcription factors was tested for their potency to induce stemness in somatic cells until they identified a pool of candidate genes associated with pluripotency.[53,54] Further selection led to the identification of four transcription factors sufficient for reprogramming murine somatic cells to iPSCs: Klf4, Sox2, c-Myc, and Oct4.[53] Based on this breakthrough, one year later, K. Takahashi and colleagues generated hiPSCs derived from human somatic cells.[11] The hiPSC technology facilitates the generation of isogenic pluripotent cells harboring the genetic background of the individual from whom they were obtained.[55] Additionally, this technology provides a novel personalized cell source on a large-scale for research and therapeutic purposes. Upon the establishment of hiPSC cultures, new opportunities emerged for differentiating hiPSCs toward specified neural cells, such as neurons[56] or oligodendrocytes.[57] Recently, several studies provided optimized differentiation protocols for the generation of mDANs from hiPSCs of genetic PD patients and demonstrated the power of this tool for subsequent investigations of disease-associated

[49] Barker, Drouin-Ouellet, & Parmar (2015), 492

[50] Park et al. (2005), p. 1265

[51] Roy et al. (2006), p. 1259

[52] Friling et al. (2009), p. 7613

[53] K. Takahashi and Yamanaka (2006), p. 663

[54] Tokuzawa et al. (2003), p. 2699

[55] Winner, Marchetto, Winkler, & Gage (2014), p. R27

[56] Sanchez-Danes et al. (2012), p. 56

[57] Hu, Du, & Zhang (2009), p. 1614

pathways.[58,59,60] Furthermore, hiPSC-technology-based in vitro models of PD indicated aSyn oligomers to be rather responsible for cellular toxicity than aSyn fibrils.[61] This rapid development of efficient differentiation protocols opened the window for novel strategies to model genetic or sporadic CNS disorders, but furthermore built the basis for developing innovative therapeutic strategies to treat age-related neurodegenerative diseases.

3.7 Adding a Dimension: 3D Human Cerebral Organoids

The advances in hiPSC generation and the continuous development of protocols to increase efficiency and reproducibility opened up new opportunities in the field of human in vitro systems: the generation of human cerebral organoids. Neural tissue originates from the ectodermal germ layer.[62] The ectoderm was reproducibly generated from structures called hiPSC-derived embryoid bodies (EBs[63]). Neural lineage commitment of these ectodermal-like cells was induced by specifically modifying in vitro conditions using chemically defined media.[64] Importantly, the generated neuroepithelium requires additional structural support to self-organize into a three-dimensional (3D) structure since the standard cell culture system is lacking a distinct basement membrane. Therefore, a system based on hydrogels was established to provide the neuroepithelial cells with a specific environment for 3D self-organization resulting in the formation of small neurogenic regions defined as cerebral organoids.[10] The use of the cerebral organoid model enables to recapitulate important aspects of CNS development as neural progenitor cells undergo self-organization and differentiation.[65] Human cerebral organoids demonstrate similar heterogeneity as the human brain in vivo during early development.[66] Previous research has already succeeded in modeling pathologic phenotypes in cerebral organoids, which enables the investigation of disease mechanisms more closely to the native state. This is of particular importance as cell–cell interactions in a 3D environment might significantly influence disease progression.[66] Furthermore, *Qian* et al. successfully generated brain-region-specific organoids displaying the identity of all six cortical layers, but also midbrain and hypothalamic organoids.[67] Overall, cerebral organoid technology provides a novel and highly innovative platform to

[58] Brazdis et al. (2020), p. 1180

[59] Simmnacher et al. (2020), p. 113466

[60] Sommer et al. (2018), p. 123

[61] Prots et al. (2018), p. 7813

[62] Rubenstein (2013)

[63] Eiraku et al. (2008), p. 519

[64] Hu and Zhang (2010), p. 123

[65] Renner et al. (2017), p. 1316

[66] Lancaster et al. (2013), p. 373

[67] Qian et al. (2016), p. 1238

investigate disease mechanisms in an organ-like context. Additionally, human cerebral organoids represent a large-scale and renewable cell source for neurons and other CNS cell types.

3.8 The Evolution of Genome Editing

Evolving reprogramming and differentiation strategies advanced the usage of hiPSCs in basic and translational research. Reprogramming of somatic cells with patient- and disease-specific genetic background offered the potential to gain further insights into disease pathomechanisms but also shifted the focus on developing molecular tools for genome editing as potential rescue strategy or for the manipulation of disease-associated genes. Consequently, initial gene editing tools emerged, the zinc finger nucleases (ZFN[68]). Zinc fingers are small-sized proteins capable of recognizing and binding specific nucleotide sequences of genes. The coupling with an endonuclease allows the cleavage of DNA in a site-specific manner.[69] Notably, the design of such ZFN is quite challenging and exceeds the expertise for the majority of laboratories. The major disadvantage using ZFNs is that the delivery of these nucleases is an irreversible process, thus potentially leading to serious off-target modifications. As a result, the need for efficient easy-to-handle gene editing tools increased. The discovery of transcription activator-like effector nucleases (TALENs[70,71]) offered a new DNA targeting tool, much "easier" in design and handling. Two variable adjacent amino acid repeats enable to recognize specific DNA sites.[71] The major challenge of TALENs is the correct combination of the variable adjacent amino acid repeats for specific targeting of DNA sites and the resulting immense increase in size of TALEN proteins. Due to the simplicity compared to ZFNs, TALENs were subsequently used for genome editing in stem cell-based disease models with initial promising results.[72,73,74] Since DNA-binding motifs are capable of binding homologous DNA sites, there is a minimal probability of non-desired genome modifications.[75] These novel promising gene-editing tools were replaced very rapidly after the discovery of the Clustered Regularly Interspaced Short Palindromic Repeats/Cas9 (CRISPR/Cas9) initiating a novel dimension in genome editing.[76] CRISPR/Cas9 became rapidly a very powerful and state-of-the-art tool for genome engineering. The CRISPR system in combination with different CRISPR-associated genes (Cas) participates in the adaptive immune system of

[68] Kim, Cha, & Chandrasegaran (CR461996), p. 1156

[69] Bibikova et al. (2001), p. 289

[70] Boch et al. (2009), p. 1509

[71] Christian et al. (2010), p. 757

[72] Bedell et al. (2012), p. 114

[73] Ding et al. (2013), p. 238

[74] Sun and Zhao (2014), p. 1048

[75] Yee (2016), p. 3239

[76] Doudna and Charpentier (2014), p. 1258096

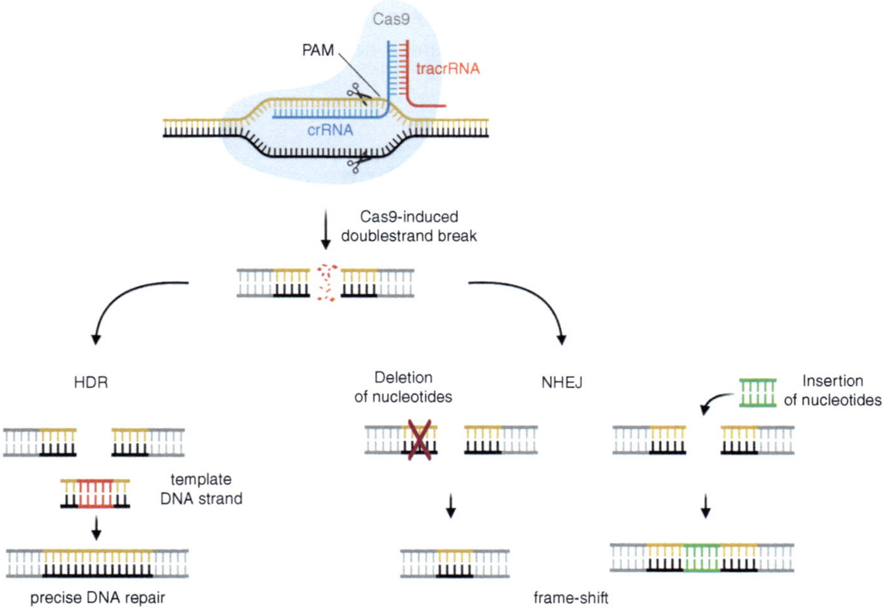

Fig. 3.1 *Principle of CRISPR/Cas9.* The Cas9 endonuclease (gray) consists of two independent endonuclease domains capable of generating DSBs in a DNA site-specific manner directed by an sgRNA. The sgRNA is divided into a crRNA (blue) for complementary pairing with the target DNA site and a tracrRNA (red). In addition, the Cas9 also contains a PAM recognition subunit for PAM-dependent base pairing. By the Cas9-induced DSBs, two DNA repair mechanisms are potentially triggered. The homology-directed repair (HDR) is based on the existence of a template DNA strand with homology (red) to the edited DNA site. Using the template, the cell is capable of precisely repairing the edited DNA strand. The second pathway represents non-homologous end joining (NHEJ). NHEJ is not template-based resulting in deletions (red "X") or insertions of nucleotides (green) causing frame-shift mutations

prokaryotic organisms.[77] Components of the CRISPR operon could be repurposed for genome editing. The CRISPR-associated protein 9 (Cas9) is able to form a ribonucleoprotein complex (RNP) with the trans-activating CRISPR RNA (tracrRNA) and the CRISPR RNA (crRNA), both expressed at the CRISPR array recombined to a single guide RNA (sgRNA)[78] (Fig. 3.1). A 5′ stretch of the crRNA, the protospacer, can be reprogrammed to pair with complementary 20 nt specific target DNA sequence of the genome. The Cas9 scans the genomic DNA strand for a specific protospacer adjacent motif (PAM[79]). If the PAM matches, the protospacer will pair with the genomic sequence, and subsequently, the endodeoxyribonuclease RuvC and endonuclease domain HNH of the Cas9 initiate a process that results in the

[77] Barrangou et al. (2007), p. 1709

[78] Deltcheva et al. (2011), p. 602

[79] Mali et al. (2013), p. 957

generation of a double-strand break (DSB) three to four nucleotides upstream of the PAM.

The ability to generate specific DSBs triggers several potential scenarios for genome editing. Employing the nonhomologous end joining pathway, it is possible to generate gene knockouts by inducing out-of-frame insertions and deletions (indels). By the addition of a homologous donor template containing the edit of choice, the homology-directed repair pathways allows the stable reversal of disease-causing mutations. The CRISPR/Cas9 system is based on the delivery of the endonuclease and the sgRNA by plasmids, viral transduction or as synthetic RNPs. The ability to program the CRISPR/Cas9 system simply by adapting the sgRNA renders CRISPR/Cas9 a far superior system than ZFNs or TALENs, which rely on protein–DNA interaction. Hence, CRISPR/Cas9 represents a very fast and easy "hands on" approach. A further tremendous advantage of using CRISPR/Cas9 technology is the possibility for targeting multiple genomic loci simultaneously allowing multiplex genome engineering.[80] Compared to TALENs, the probability modifying off-target sequences using CRISPR/Cas9 is marginally higher. The field of CRISPR/Cas9 is rapidly evolving, thus identifying continuously promising applications and new bacteria-derived endonucleases with different PAM specificities, allowing a broader range of host genome modification. Combined with the platform of hiPSCs, CRISPR/Cas9 represents a powerful tool to modulate disease-associated genes and provides novel functional data of pathways in health and disease.

3.9 A New Hope: Preclinical Stem Cell Replacement Therapies

Based on the outcome of previous studies using hESCs, protocols for differentiating hESCs and hiPSCs into a dopaminergic lineage were refined and optimized.[81] Initial transplantation studies using ESC approaches have been initiated already in 2008, called "therapeutic cloning."[82] In this study, all mice engrafted with ESC-derived dopaminergic neurons by autologous transplantation showed a significant attenuation of the PD-like phenotype in behavioral tests. Notably, the applied autologous transplantation approach revealed no graft rejection or an increased immune response in the host brain. The fundamental finding that dopaminergic neurons originate from a developmental structure called floor plate (FP) catalyzed the process of generation and specification of dopaminergic neurons.[83] Based on this finding, *Kriks and colleagues* established a protocol for effective transplantation of human-derived ESCs in nonhuman primates with a toxin-induced PD phenotype showing a robust survival of mDANs.[84] Analysis of the ESC-derived mDAN transplantation revealed

[80] Cong et al. (2013), p. 819

[81] Chambers et al. (2009), p. 275

[82] Tabar et al. (2008), p. 379

[83] Ono et al. (2007), p. 3213

[84] Kriks et al. (2011), p. 547

an efficiency of the transplantation comparable to previous studies using HVFM transplanted grafts.[85]

In 2008, first studies of hiPSC transplantation succeeded in reproducing the findings from ESC transplantation approaches. The hiPSCs were differentiated into mDANs, analyzed for dopaminergic markers, and subsequently transplanted into the CNS of a PD rat model.[86] The mDANs successfully integrated into the host brain, formed synaptic contacts, and were electrophysiologically active. Rodents with grafts showed a symptomatic improvement although a continuous proliferation of these cells was detected postgrafting. Comparable results have been obtained by a similar strategy using a sorting approach of cells originating from a developmental structure (CORIN) important for the differentiation of mDANs.[87] $CORIN^+$ cells are more suitable for dopaminergic differentiation. Transplantation of these cells resulted in a better survival of mDANs in conjunction with an improved functional outcome. The first autologous transplantation approaches of hiPSCs in a nonhuman primate PD model were performed in 2013.[88] This study showed that hiPSC grafts efficiently integrate into the host brain, but the authors did not observe a functional improvement. *Morizane and colleagues* initiated an autologous and allogenic transplantation of hiPSC-derived dopaminergic neurons comparing intragenomic retrovirally with nonintegrating episomally generated hiPSC grafts.[89] The authors performed this transplantation study in nonhuman primates, demonstrating a strong immune response by allografts, but a very limited by autografts. Furthermore, an improved survival of tyrosinhydroxylase (TH+)-expressing human neurons was observed in both the types of grafts, even with a higher number of TH+ human neurons in the autografts. These findings were confirmed by a follow-up study using optimized protocols for hiPSC generation and transplantation procedures.[90] A very crucial and relevant finding for further translation of autologous cell transplantation approaches to humans was the consistency and rigidity in regard to the observed symptomatic improvement in nonhuman primates with grafts.[91] The animals were screened over a period of 2 years after transplantation. A prolonged survival of the engrafted cells in conjunction with a sustained functional improvement was observed. Taken together, these landmark studies in nonhuman primates emphasized the therapeutic potential of autologous hiPSC transplantation by demonstrating an augmented survival of engrafted cells with a concurrent functional and biological relevant improvement of the disease course in broadly accepted preclinical nonhuman primate PD models. At this stage, it is very important to note that no immunosuppression was necessary to obtain these results after transplantation in contrast to allogenic transplantation approaches using HFVM or hESCs. Therefore,

---

[85] Grealish et al. (2014), p. 653

[86] Wernig et al. (2008), p. 5856

[87] Doi et al. (2014), p. 337

[88] Emborg et al. (2013), p. 646

[89] Morizane et al. (2013), p. 283

[90] Sundberg et al. (2013), p. 1548

[91] Hallett et al. (2015), p. 269

autologous hiPSC transplantation approaches represent the most promising platform for present and future clinical studies in the light to achieve an effective, long-term symptomatic treatment of PD patients without the necessity of immunosuppressive medication.

In the field of brain organoid transplantation, there is little published data about preclinical cerebral organoid transplantation. Two studies provide evidence of successfully engrafted hiPSC-derived brain organoids into mouse brain. The study of Mansour et al. revealed a vascularization of transplanted brain organoids in adult mice and the capability of neuronal maturation and differentiation, as well as axonal outgrowth and gliogenesis.[92] A second study observed similar findings in lesioned mouse cortex, confirming the potential of brain organoid transplantation as an alternate therapeutic cell-based approach.[93] However, there are currently no studies described in nonhuman primates further. Nonetheless, brain organoids represent a heterogeneous population of cells, thus consisting of pluripotent cell populations within the organoid, leading to an incompatibility with the current available protocols regarding safety for in-vivo approaches (see below). In addition, the transplantation process requires an invasive procedure for successfully transplanting organoids into the region of interest. Since organoids are of macroscopic nature, the use of a larger application device may result in additional tissue damage at the site of transplantation.

3.10 In-Human Studies: hiPSC-Based Cell Replacement in PD

In 2015, the international consortium named G-Force PD was founded focusing on novel cell-based therapies for treating neurological disorders in humans, especially patients with PD. In the framework of this consortium, four transplantation studies were initialized involving two hESC and two hiPSC transplantation studies.[94]

In 2018, the first clinical trial was initiated aiming to implant allogenic hiPSC-derived mDANs in Japan.[95] The hiPSCs were obtained from a single healthy donor carrying the most common human leukocyte antigen (HLA) type, as indicator for immunocompatibility, in Japan to minimize the risk of an immunogenic rejection of the transplant. The hiPSCs were obtained by reprogramming peripheral blood cells using episomal plasmid vectors containing the prototypical Yamanaka reprogramming transcription factors. Midbrain dopaminergic differentiation (Fig. 3.2) was performed according to the aforementioned protocols followed by a thorough screening for tumorigenicity, cell overgrowth, and survival in a PD rat model. Additionally, the behavioral parameters were evaluated to assess the potential clinical outcome after transplantation. The cells demonstrated no tumorigenic characteristics and a robust survival as well as adequate engrafting into the rat host

---

[92] Mansour et al. (2018), p. 432

[93] Daviaud et al. (2018), p. ENEURO.0219–18.2018

[94] Barker et al. (2017), p. 569

[95] J. Takahashi (2020), p. 18

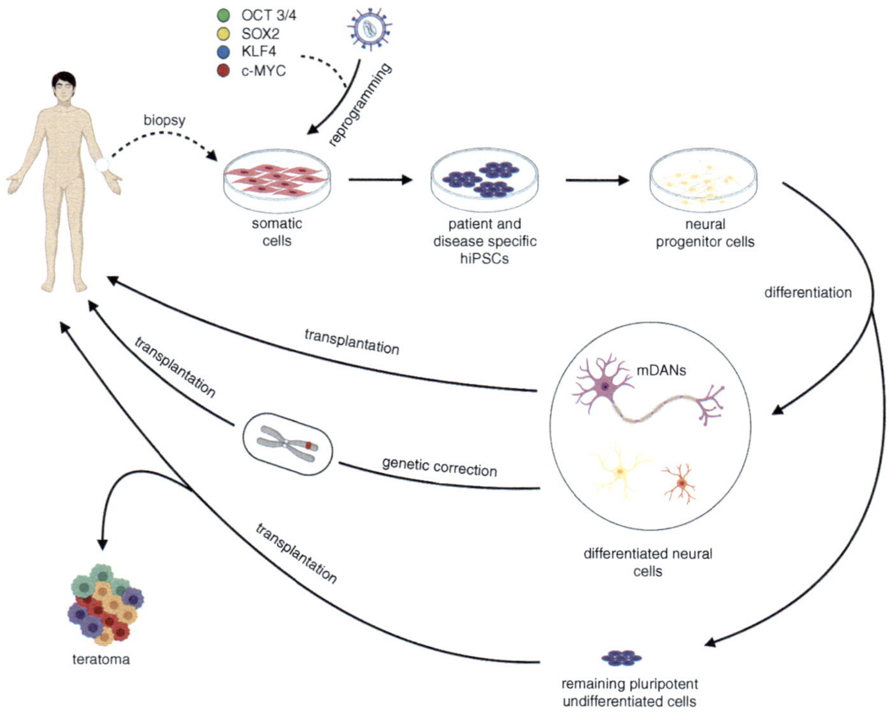

Fig. 3.2 Autologous transplantation of hiPSC-derived mesencephalic dopaminergic neurons (mDANs): Somatic cells are obtained by standard biopsy techniques and subsequently reprogrammed into hiPSCs by the ectopic overexpression of the Yamanaka reprogramming transcription factors (OCT3/4, SOX2, KLF4, c-MYC). Further cell fate-specific differentiation allows the generation of mDANs and other neural cell types. Differentiated mDANs are further utilized for transplantation into the patient's affected brain regions. The usage of genome editing systems enables the correction of genetic aberrations. The major concern of hiPSC transplantations refers to the potential of tumor formation (e.g., teratomas) due to remaining cells in a pluripotent state

brain. Moreover, grafted rats presented a solid motor improvement suggesting a high potential to translate these findings toward initiating a clinical trial using mDANs in PD patients. The consequent clinical trial enrolled seven PD patients in the range between 50 and 69 years of age and a disease duration exceeding 5 years. PD patients showed already motor symptoms not controlled by their oral medication. The patients received five million cells injected into the putamen as spheres using a stereotactic needle designed for transplantation purposes. Due to the allogenic origin of the grafts, patients underwent immunosuppression for a period of 12 months. The follow-up of this allogenic transplantation approach was envisioned at least 24 months after transplantation.

The second study performing autologous transplantation in PD patients was planned to start recruiting patients in 2019 (Summit for PD[94]). The inclusion criteria are almost identical to the clinical study headed by *J. Takahashi and colleagues*. The clinical follow-up was estimated to take place 1 year after transplantation.

Since the follow-up in both the studies is still pending, there is no explicit report thus far describing the current clinical status of the patients enrolled into both the studies. Overall, the very rigid preclinical work of the consortium G-Force PD is promising. Finally, a positive outcome of these ongoing clinical trials will represent a new milestone in the field of neurorestoration in PD.

Besides G-Force PD, to date, a single case report was recently published in the *New England Journal of Medicine* reporting a preliminary "blueprint" of an autologous transplantation of patient-derived mDANs.[96] The patient was a 69-year-old physician with a 10-year history of progressive, sporadic PD. Based on this report, he was continuously treated according to the present guidelines for the treatment of PD, however with poor outcomes, leading to a severe worsening of his symptoms. The patient received an autologous graft of mDANs progenitors in the right and left putamen, both the surgeries separated by a 6-month interval. The patient was not immunosuppressed after undergoing transplantation. To assess whether grafted mDANs are tolerated by the host CNS, cells were prescreened and initially implanted in patient-humanized mice, suggesting that the grafts will be immunologically tolerated by the patient brain. The patient was imaged up to 24 months after the first transplantation procedure. The analysis displayed an initial reduction of dopamine uptake in the putamen followed by a mild increase over a longer period, suggesting that the injected cells engrafted successfully into the host brain. The patient demonstrated improved motor symptoms showing a decline in the severity of symptoms, both with and without his standard medication. Furthermore, the patient reported an improved quality of life after 24 months. In addition, the dosage of the standard medication was reduced in comparison to the status prior to the transplantation, and no graft-related dyskinesias were observed. In summary, this first pilot study addressing the feasibility of autologous transplantation of hiPSCs showed the potential of this avenue for treating PD patients, but a detailed and robust double-blinded, randomized clinical trial must be performed in order to draw some meaningful and rigid conclusions.

The application of genome editing in hiPSC technology for therapeutic purposes is dramatically rising. By 2017, almost 2600 ongoing or completed trials using gene therapy approaches have been approved globally.[97] The overall aim of gene-based therapeutic strategies is the incorporation of plasmids or viral vectors to target proteins identified to cause diseases such as cancer, but also rare monogenic diseases.[98,99] Autologous transplantation of genetically altered cells is exclusively tested in sporadic PD thus far. However, there are also about 10–15% PD patients linked to monogenic mutations and thus representing a potential target population of genome editing efforts. Since hiPSC-derived mDANs resemble neural cells in a very early stage, the transplantation of such immature neurons still harboring mutant genes may result in less favorable outcomes compared to mDANs derived from sporadic

[96] Schweitzer et al. (2020), p. 1926

[97] Ginn et al. (2018), p. e3015

[98] Hacein-Bey Abina et al. (2015), p. 1550

[99] Porter et al. (2011), p. 725

PD patients. One of the most prominent PARK locus, PARK4, is characterized by the duplication or triplication of the SNCA gene resulting in an aggregation-promoting overexpression of aSyn. Genetic manipulations allow removing additional alleles of the SNCA locus, thereby restoring the physiological level of aSyn expression in the patient-derived hiPSCs (Fig. 3.2).

The PARK1 locus refers to missense point mutation in the SNCA gene, resulting in gain- or loss-of-function events of aSyn. Similar to PARK4, it is possible to target the disease-causing mutations and replace the affected exon/gene, thus re-establishing the physiological function. In summary, the genetically modified hiPSCs may be further differentiated to mDANs and subsequently implanted as a genetically treated cell population in affected brain areas, such as the putamen in PD (Fig. 3.2).

Alternatively, gene-editing tools are also an appropriate tool to improve the therapeutic potential of hiPSCs by genetically improving cell survival after transplantation.[100] Overall, genome editing represents a powerful tool for the modulation of patient-derived cells but important aspects in terms of safety and bioethics must be considered prior to applying these genetically modified hiPSCs in patients. At present, there are no registered clinical trials using genetically edited hiPSCs for transplantation purposes in PD.

3.11 The Flip Side of the Coin: Safety and Social Concerns of hiPSC Technology

The discovery of hiPSCs revolutionized the field of stem cell research due to its individualized source and standardized procedures for scaling up, but moreover, by circumventing certain ethical and legal concerns, which have been raised in particular with the usage of hESCs. By "simply" obtaining somatic cells from an individual by a less invasive method such as a skin biopsy or drawing peripheral blood, hiPSCs overcome serious ethical concerns "to use" or "to consume" human blastocysts, embryos, or fetuses for therapeutic purposes. Moreover, autologous transplantation of hiPSC may allow circumventing lifelong immunosuppression since graft and host refer to the identical individual thus paving the way to immunocompatibility. So far, hiPSC circumvent ethical concerns of embryonal- or fetal-tissue-derived stem cell technology, but the term "pluripotency" implies the potential to form tumors.[101] Since the potency of teratoma formation is a gold standard to evaluate pluripotency, undifferentiated hiPSC populations in the engrafted cells pose the risk of tumor formation after transplantation. Besides this safety concern, an additional tumor-promoting characteristic refers to the genomic instability of hiPSCs, an important aspect hampering the usage of these cells for its application in humans.[102] Reprogramming technologies for somatic cells require the usage of oncogenic transcription factors such as c-MYC or the integration of retro- and lentiviral vectors

[100] Moradi et al. (2019), p. 341

[101] Lindvall (2015), p. 20140370

[102] Yoshihara et al. (2017), p. 7

potentially resulting in nontargeted mutagenesis.[103] Therefore, it is necessary to continuously develop and improve differentiation protocols not only to increase the purity of the desired cells but also to fulfill the highest safety standards to exclude the risk of tumor formation.

hiPSCs represent a powerful tool for disease modeling and drug discovery in a human-based in vitro model. However, despite the advantages of hiPSC, a large transcriptional variability between cells derived from the identical donor was observed,[104] resulting in a considerable heterogeneity of cells despite its identical "mother" cell.[105] Due to this transcriptional variability, the prediction in regard to the expected outcome of transplanted hiPSCs remains a huge challenge. Another arguable factor relates to the molecular strategy for reprogramming. As retro- and lentiviral-based reprogramming strategies involve the integration of defined reprogramming factors into the genome, an increased risk of intragenic mutations may occur. For a safe clinical application, the development of new molecular strategies such as integration-free transient vector systems is fundamental to lower the risk of mutagenesis. However, up to now, there is not sufficient knowledge regarding the safety of integration-free generated hiPSC.[103] Finally, the usage of genome editing strategies for hiPSCs imply other risks such as i) the delivery of bacterial endonucleases into hiPSCs and subsequent transplantation into the immunocompetent CNS, ii) the possibility of off-target mutagenesis by the Cas9 or triggered DNA repair mechanisms, iii) the potential of unknown mechanisms involving other genes in the pathogenesis caused by the known monogenic mutation (e.g., multiplication of a whole chromosome stretch in PARK4 patients involving additional genes).

Finally, the financial burden of these molecular and cellular procedures is a major obstacle for public health care systems to implement hiPSC transplantation technology for a disorder such as PD due to its increasing prevalence worldwide.[106,107] The aspect of health costs raises the serious question for society whether autologous hiPSC transplantation is affordable at all for healthcare systems.

3.12 Pay-to-Participate: The Slippery Slope of Scientific Integrity

In this review, we have outlined the advantages but also safety, ethical, and social concerns associated with the advancements of hiPSC technology. In the brief report of *Schweitzer and colleagues*,[96] the clinical assessment of the PD patient revealed a return of dopamine uptake to the baseline (*pretransplantation*) 24 months after autologous transplantation of hiPSC-derived mDANs. As a result, the patient reported improved motor symptoms as well as quality of life. Although this report

[103] Volarevic et al. (2018), p. 36

[104] Liang and Zhang (2013), p. 149

[105] Carcamo-Orive et al. (2017), p. 518

[106] Beers et al. (2015), p. 113119

[107] Prescott (2011), p. 1575

appears promising for the future of hiPSCs transplantation technology as a new therapeutic approach for PD, several serious concerns of this study must be discussed. In fact, the grafted hiPSCs were characterized in previous studies;[108] however, the current safety protocols are not sufficient to exclude the abovementioned tumor-promoting genomic instability of hiPSCs.[102] Therefore, a more detailed preclinical evaluation of the hiPSC properties in humanized animal models is required to ensure the safety for future patients.

From a clinical point of view on this single case published in one of the most relevant journals in medicine, there are several issues further to be considered. The patient had an intermediate course of PD offering the therapeutic option for him just by increasing his daily L-Dopa dosage to improve his motor symptoms since no L-Dopa-induced dyskinesias were observed yet. Moreover, he declined deep brain stimulation as an alternative therapeutic approach. By analyzing the pattern of the cerebral positron-emission tomography, it becomes evident that the dopamine uptake returned or minimally exceeded the initial baseline uptake. Notably, since PD is a progressing neurodegenerative disease, the putaminal dopamine uptake consequently decreases over the period of 24 months, thus indicating that the transplantation of human mDANs was able to halt disease progression at least based upon the levels of the initial dopamine uptake. The lack of an internal (sham surgery on the less affected side) or adding an external control further raises questions about the issue whether the restorative effects observed are linked to the grafted mDANs or to the procedure itself clinically well-known as placebo effect. Crucially, although PD is defined by prototypical motor symptoms, there is a plethora of nonmotor symptoms in PD frequently present prior to the onset of motors symptoms or throughout the course of the disease.[109] Thus, it is evident that dopamine replacement or substitution is not able to relief nonmotor symptoms such as cognitive deficits or depressive symptoms. In summary, the transplantation of mDANs may be a powerful and long-term restorative therapy to enhance the dopaminergic tone within the CNS of PD patient, but will never represent a causal cure of the disease.

The last and potential ambiguous aspect of this initial pilot study on the clinical application of hiPSC-derived mDANs to reflect on is the social and financial circumstances in a highly respected academic institution such as the Harvard Medical School. The transplanted PD patient is a wealthy former physician and businessperson. After receiving the diagnosis PD, the patient decided to fund the research on hiPSC transplantation technology to benefit from the findings of this research. In the present case, the patient funded a scientist investigating safety and efficiency of hiPSC transplantation after being declined for other public funding sources. Besides the preclinical research, he paid for the surgical procedure including the legal and ethical approval by the institutional review board and the Food and Drug Administration (FDA). This payment to researchers, administrators, and physicians directly involved in the preclinical and clinical procedures may result in a selection

[108] Song et al. (2020), p. 904

[109] Chaudhuri et al. (2006), p. 235

bias leading to research and clinical decisions made in favor of the donator of funds than rigid science as a whole. Moreover, this type of pay-to-participate study[110] sheds an ambiguous light on scientists and clinicians who may apparently be bought from a single individual for his or her own purpose. A further questionable aspect was the selected FDA program for approval rather intended for patients with life-threatening conditions or no remaining therapeutic alternatives. It is noteworthy at this moment to reiterate that PD is not a fatal disease; furthermore, life expectancy has tremendously increased with new developments and optimizations of current therapeutic approaches. Due to this fact, FDA approval for the transplantation of hiPSC-derived mDANs is arguable in the present case since there is no necessity for this intervention in the light of alternate therapeutic options. Finally, this first case report of an autologous hiPSC transplantation was published in one of the most cited, high-impact medical journals eventually fostering false interpretations, hopes and overestimations of the prospect of this type of treatment.[111]

3.13 Conclusion

Since its discovery, the research field of stem cells and genome editing is developing continuously and rapidly.[53,11,38] Although promising results were obtained in pre-clinical models of rodents and nonhuman primates, the idea of self-derived transplantation requires precautious interpretations. Pluripotency is generally linked to unconditional potential to proliferate and differentiate, an immanent risk factor for tumorigenesis. Until sufficient safety data for hiPSC grafts are not yet fully established, the transplantation of hiPSC-derived neural cells in humans is very cautiously to be considered as an additional, but powerful symptomatic approach possibly halting the deterioration of distinct clinical symptoms in PD. For diseases in which multiple cell types are affected, brain organoids, as kind of "mini organs" may represent a powerful cell source in the future. However, similar to hiPSCs, no current protocols of organoid generation ensure the highest safety for transplantation in humans. Moreover, since brain organoids are macroscopic cellular clusters, there are major biotechnical concerns regarding invasiveness applying these clusters into patients. Additionally, self-funded research raises numerous concerns regarding scientific integrity. Therefore, the study of *Schweitzer and colleagues* is a hallmark for the entire research community and society to further discuss and develop stringent guidelines for this type of cutting-edge technology in modern medicine. In light of all considerations and results at present, autologous transplantation of hiPSCs offers the promise to restore CNS functions and potentially to increase the quality of life of thousands of patients suffering from age-related neurodegenerative diseases such as PD.

[110] Grady (2005), p. 1681

[111] Jankovic et al. (2020), p. 1312

Acknowledgments Je.W. and Y.S. are graduate students of the research training group 2162 “Neurodevelopment and Vulnerability of the Central Nervous System” of the Deutsche Forschungsgemeinschaft (DFG 270949263/GRK2162). S.T. was a postdoctoral researcher and consequent principle investigator of the research training group 2162 “Neurodevelopment and Vulnerability of the Central Nervous System” of the Deutsche Forschungsgemeinschaft (DFG 270949263/GRK2162). J.W. is a member of the research training group 2162 of the Deutsche Forschungsgemeinschaft (DFG 270949263/GRK2162) and speaker of the Bavarian Research Network for Induced Pluripotent Stem Cells (ForIPS), as well as consequent member of the Bavarian Research Network of Interaction of human brain cells (ForInter). Je.W., Y.S., and S.T are associated with the Bavarian Research Networks for Induced Pluripotent Stem Cells (ForIPS) and for the Interaction of human brain cells (ForInter). The work was supported by the Forschungsstiftung Medizin, University Hospital Erlangen. All figures were created with BioRender.com.

References

Baba M, Nakajo S, Tu PH, Tomita T, Nakaya K, Lee VM, Trojanowski JQ, Iwatsubo T (1998) Aggregation of alpha-synuclein in Lewy bodies of sporadic Parkinson's disease and dementia with Lewy bodies. Am J Pathol 152:879–884. https://www.ncbi.nlm.nih.gov/pubmed/9546347

Backlund EO, Granberg PO, Hamberger B, Knutsson E, Martensson A, Sedvall G, Seiger A, Olson L (1985) Transplantation of adrenal medullary tissue to striatum in parkinsonism. First clinical trials J Neurosurg 62:169–173. https://www.ncbi.nlm.nih.gov/pubmed/257855810.3171/jns.1985.62.2.0169

Barker RA, Drouin-Ouellet J, Parmar M (2015) Cell-based therapies for Parkinson disease-past insights and future potential. Nat Rev Neurol 11:492–503. https://www.ncbi.nlm.nih.gov/pubmed/2624003610.1038/nrneurol.2015.123

Barker RA, Parmar M, Studer L, Takahashi J (2017) Human trials of stem cell-derived dopamine neurons for Parkinson’s disease: Dawn of a new era. Cell Stem Cell 21:569–573. https://www.ncbi.nlm.nih.gov/pubmed/2910001010.1016/j.stem.2017.09.014

Barrangou R, Fremaux C, Deveau H, Richards M, Boyaval P, Moineau S, Romero DA, Horvath P (2007) CRISPR provides acquired resistance against viruses in prokaryotes. Science 315:1709–1712. https://www.ncbi.nlm.nih.gov/pubmed/1737980810.1126/science.1138140

Bedell VM, Wang Y, Campbell JM, Poshusta TL, Starker CG, Krug RG 2nd, Tan W, Penheiter SG, Ma AC, Leung AY, Fahrenkrug SC, Carlson DF, Voytas DF, Clark KJ, Essner JJ, Ekker SC (2012) In vivo genome editing using a high-efficiency TALEN system. Nature 491:114–118. https://www.ncbi.nlm.nih.gov/pubmed/2300089910.1038/nature11537

Beers J, Linask KL, Chen JA, Siniscalchi LI, Lin Y, Zheng W, Rao M, Chen G (2015) A cost-effective and efficient reprogramming platform for large-scale production of integration-free human induced pluripotent stem cells in chemically defined culture. Sci Rep 5:11319. https://www.ncbi.nlm.nih.gov/pubmed/2606657910.1038/srep11319

Benabid AL (2003) Deep brain stimulation for Parkinson’s disease. Curr Opin Neurobiol 13:696–706. https://www.ncbi.nlm.nih.gov/pubmed/1466237110.1016/j.conb.2003.11.001

Bibikova M, Carroll D, Segal DJ, Trautman JK, Smith J, Kim YG, Chandrasegaran S (2001) Stimulation of homologous recombination through targeted cleavage by chimeric nucleases. Mol Cell Biol 21:289–297. https://www.ncbi.nlm.nih.gov/pubmed/1111320310.1128/MCB.21.1.289-297.2001

Boch J, Scholze H, Schornack S, Landgraf A, Hahn S, Kay S, Lahaye T, Nickstadt A, Bonas U (2009) Breaking the code of DNA binding specificity of TAL-type III effectors. Science 326:1509–1512. https://www.ncbi.nlm.nih.gov/pubmed/1993310710.1126/science.1178811

Brazdis RM, Alecu JE, Marsch D, Dahms A, Simmnacher K, Lorentz S, Brendler A, Schneider Y, Marxreiter F, Roybon L, Winner B, Xiang W, Prots I (2020) Demonstration of brain region-specific neuronal vulnerability in human iPSC-based model of familial Parkinson’s disease.

Hum Mol Genet 29:1180–1191. https://www.ncbi.nlm.nih.gov/pubmed/3216028710.1093/hmg/ddaa039

Brundin P, Pogarell O, Hagell P, Piccini P, Widner H, Schrag A, Kupsch A, Crabb L, Odin P, Gustavii B, Bjorklund A, Brooks DJ, Marsden CD, Oertel WH, Quinn NP, Rehncrona S, Lindvall O (2000) Bilateral caudate and putamen grafts of embryonic mesencephalic tissue treated with lazaroids in Parkinson's disease. Brain 123(Pt 7):1380–1390. https://www.ncbi.nlm.nih.gov/pubmed/1086905010.1093/brain/123.7.1380

Carcamo-Orive I, Hoffman GE, Cundiff P, Beckmann ND, D'souza SL, Knowles JW, Patel A, Papatsenko D, Abbasi F, Reaven GM, Whalen S, Lee P, Shahbazi M, Henrion MYR, Zhu K, Wang S, Roussos P, Schadt EE, Pandey G, Chang R, Quertermous T, Lemischka I (2017) Analysis of transcriptional variability in a large human iPSC library reveals genetic and non-genetic determinants of heterogeneity. Cell Stem Cell 20(518–532):e519. https://www.ncbi.nlm.nih.gov/pubmed/2801779610.1016/j.stem.2016.11.005

Chambers SM, Fasano CA, Papapetrou EP, Tomishima M, Sadelain M, Studer L (2009) Highly efficient neural conversion of human ES and iPS cells by dual inhibition of SMAD signaling. Nat Biotechnol 27:275–280. https://www.ncbi.nlm.nih.gov/pubmed/1925248410.1038/nbt.1529

Chaudhuri KR, Healy DG, Schapira AH, National Institute for Clinical E (2006) Non-motor symptoms of Parkinson's disease: diagnosis and management. Lancet Neurol 5:235–245. https://www.ncbi.nlm.nih.gov/pubmed/1648837910.1016/S1474-4422(06)70373-8

Christian M, Cermak T, Doyle EL, Schmidt C, Zhang F, Hummel A, Bogdanove AJ, Voytas DF (2010) Targeting DNA double-strand breaks with TAL effector nucleases. Genetics 186:757–761. https://www.ncbi.nlm.nih.gov/pubmed/2066064310.1534/genetics.110.120717

Cong L, Ran FA, Cox D, Lin S, Barretto R, Habib N, Hsu PD, Wu X, Jiang W, Marraffini LA, Zhang F (2013) Multiplex genome engineering using CRISPR/Cas systems. Science 339:819–823. https://www.ncbi.nlm.nih.gov/pubmed/2328771810.1126/science.1231143

Daviaud N, Friedel RH, Zou H (2018) Vascularization and Engraftment of Transplanted Human Cerebral Organoids in Mouse Cortex. eNeuro:5. https://www.ncbi.nlm.nih.gov/pubmed/3046033110.1523/ENEURO.0219-18.2018

Deltcheva E, Chylinski K, Sharma CM, Gonzales K, Chao Y, Pirzada ZA, Eckert MR, Vogel J, Charpentier E (2011) CRISPR RNA maturation by trans-encoded small RNA and host factor RNase III. Nature 471:602–607. https://www.ncbi.nlm.nih.gov/pubmed/2145517410.1038/nature09886

Ding Q, Lee YK, Schaefer EA, Peters DT, Veres A, Kim K, Kuperwasser N, Motola DL, Meissner TB, Hendriks WT, Trevisan M, Gupta RM, Moisan A, Banks E, Friesen M, Schinzel RT, Xia F, Tang A, Xia Y, Figueroa E, Wann A, Ahfeldt T, Daheron L, Zhang F, Rubin LL, Peng LF, Chung RT, Musunuru K, Cowan CA (2013) A TALEN genome-editing system for generating human stem cell-based disease models. Cell Stem Cell 12:238–251. https://www.ncbi.nlm.nih.gov/pubmed/2324648210.1016/j.stem.2012.11.011

Doi D, Samata B, Katsukawa M, Kikuchi T, Morizane A, Ono Y, Sekiguchi K, Nakagawa M, Parmar M, Takahashi J (2014) Isolation of human induced pluripotent stem cell-derived dopaminergic progenitors by cell sorting for successful transplantation. Stem Cell Reports 2:337–350. https://www.ncbi.nlm.nih.gov/pubmed/2467275610.1016/j.stemcr.2014.01.013

Doudna JA, Charpentier E (2014) Genome editing. The new frontier of genome engineering with CRISPR-Cas9. Science 346:1258096. https://www.ncbi.nlm.nih.gov/pubmed/2543077410.1126/science.1258096

Dugger BN, Dickson DW (2017) Pathology of neurodegenerative diseases. Cold Spring Harb Perspect Biol 9. https://www.ncbi.nlm.nih.gov/pubmed/2806256310.1101/cshperspect.a028035

Eiraku M, Watanabe K, Matsuo-Takasaki M, Kawada M, Yonemura S, Matsumura M, Wataya T, Nishiyama A, Muguruma K, Sasai Y (2008) Self-organized formation of polarized cortical tissues from ESCs and its active manipulation by extrinsic signals. Cell Stem Cell 3:519–532. https://www.ncbi.nlm.nih.gov/pubmed/1898396710.1016/j.stem.2008.09.002

Emborg ME, Liu Y, Xi J, Zhang X, Yin Y, Lu J, Joers V, Swanson C, Holden JE, Zhang SC (2013) Induced pluripotent stem cell-derived neural cells survive and mature in the nonhuman primate brain. Cell Rep 3:646–650. https://www.ncbi.nlm.nih.gov/pubmed/2349944710.1016/j.celrep.2013.02.016

Englund U, Fricker-Gates RA, Lundberg C, Bjorklund A, Wictorin K (2002) Transplantation of human neural progenitor cells into the neonatal rat brain: extensive migration and differentiation with long-distance axonal projections. Exp Neurol 173:1–21. https://www.ncbi.nlm.nih.gov/pubmed/1177193510.1006/exnr.2001.7750

Evans MJ, Kaufman MH (1981) Establishment in culture of pluripotential cells from mouse embryos. Nature 292:154–156. https://www.ncbi.nlm.nih.gov/pubmed/724268110.1038/292154a0

Freed CR, Greene PE, Breeze RE, Tsai WY, Dumouchel W, Kao R, Dillon S, Winfield H, Culver S, Trojanowski JQ, Eidelberg D, Fahn S (2001) Transplantation of embryonic dopamine neurons for severe Parkinson's disease. N Engl J Med 344:710–719. https://www.ncbi.nlm.nih.gov/pubmed/1123677410.1056/NEJM200103083441002

Friling S, Andersson E, Thompson LH, Jonsson ME, Hebsgaard JB, Nanou E, Alekseenko Z, Marklund U, Kjellander S, Volakakis N, Hovatta O, El Manira A, Bjorklund A, Perlmann T, Ericson J (2009) Efficient production of mesencephalic dopamine neurons by Lmx1a expression in embryonic stem cells. Proc Natl Acad Sci U S A 106:7613–7618. https://www.ncbi.nlm.nih.gov/pubmed/1938378910.1073/pnas.0902396106

Frodl EM, Nakao N, Brundin P (1994) Lazaroids improve the survival of cultured rat embryonic mesencephalic neurones. Neuroreport 5:2393–2396. https://www.ncbi.nlm.nih.gov/pubmed/788106610.1097/00001756-199411000-00045

Gage FH (2000) Mammalian neural stem cells. Science 287:1433–1438. https://www.ncbi.nlm.nih.gov/pubmed/1068878310.1126/science.287.5457.1433

Ginn SL, Amaya AK, Alexander IE, Edelstein M, Abedi MR (2018) Gene therapy clinical trials worldwide to 2017: an update. J Gene Med 20:e3015. https://www.ncbi.nlm.nih.gov/pubmed/2957537410.1002/jgm.3015

Grady C (2005) Payment of clinical research subjects. J Clin Invest 115:1681–1687. https://www.ncbi.nlm.nih.gov/pubmed/1600724410.1172/JCI25694

Grealish S, Diguet E, Kirkeby A, Mattsson B, Heuer A, Bramoulle Y, Van Camp N, Perrier AL, Hantraye P, Bjorklund A, Parmar M (2014) Human ESC-derived dopamine neurons show similar preclinical efficacy and potency to fetal neurons when grafted in a rat model of Parkinson's disease. Cell Stem Cell 15:653–665. https://www.ncbi.nlm.nih.gov/pubmed/2551746910.1016/j.stem.2014.09.017

Hacein-Bey Abina S, Gaspar HB, Blondeau J, Caccavelli L, Charrier S, Buckland K, Picard C, Six E, Himoudi N, Gilmour K, Mcnicol AM, Hara H, Xu-Bayford J, Rivat C, Touzot F, Mavilio F, Lim A, Treluyer JM, Heritier S, Lefrere F, Magalon J, Pengue-Koyi I, Honnet G, Blanche S, Sherman EA, Male F, Berry C, Malani N, Bushman FD, Fischer A, Thrasher AJ, Galy A, Cavazzana M (2015) Outcomes following gene therapy in patients with severe Wiskott-Aldrich syndrome. JAMA 313:1550–1563. https://www.ncbi.nlm.nih.gov/pubmed/2589805310.1001/jama.2015.3253

Hallett PJ, Deleidi M, Astradsson A, Smith GA, Cooper O, Osborn TM, Sundberg M, Moore MA, Perez-Torres E, Brownell AL, Schumacher JM, Spealman RD, Isacson O (2015) Successful function of autologous iPSC-derived dopamine neurons following transplantation in a non-human primate model of Parkinson's disease. Cell Stem Cell 16:269–274. https://www.ncbi.nlm.nih.gov/pubmed/2573224510.1016/j.stem.2015.01.018

Hentze H, Soong PL, Wang ST, Phillips BW, Putti TC, Dunn NR (2009) Teratoma formation by human embryonic stem cells: evaluation of essential parameters for future safety studies. Stem Cell Res 2:198–210. https://www.ncbi.nlm.nih.gov/pubmed/1939359310.1016/j.scr.2009.02.002

Heumann R, Moratalla R, Herrero MT, Chakrabarty K, Drucker-Colin R, Garcia-Montes JR, Simola N, Morelli M (2014) Dyskinesia in Parkinson's disease: mechanisms and current non-pharmacological interventions. J Neurochem 130:472–489. https://www.ncbi.nlm.nih.gov/pubmed/2477303110.1111/jnc.12751

Hu BY, Zhang SC (2010) Directed differentiation of neural-stem cells and subtype-specific neurons from hESCs. Methods Mol Biol 636:123–137. https://www.ncbi.nlm.nih.gov/pubmed/2033652010.1007/978-1-60761-691-7_8

Itskovitz-Eldor J, Schuldiner M, Karsenti D, Eden A, Yanuka O, Amit M, Soreq H, Benvenisty N (2000) Differentiation of human embryonic stem cells into embryoid bodies compromising the three embryonic germ layers. Mol Med 6:88–95. https://www.ncbi.nlm.nih.gov/pubmed/10859025

Jankovic J (2008) Parkinson's disease: clinical features and diagnosis. J Neurol Neurosurg Psychiatry 79:368–376. https://www.ncbi.nlm.nih.gov/pubmed/1834439210.1136/jnnp.2007.131045

Jankovic J, Okun MS, Kordower JH (2020) Stem cells: scientific and ethical quandaries of a personalized approach to Parkinson's disease. Mov Disord 35:1312–1314. https://www.ncbi.nlm.nih.gov/pubmed/3252966210.1002/mds.28187

Kawasaki H, Mizuseki K, Nishikawa S, Kaneko S, Kuwana Y, Nakanishi S, Nishikawa SI, Sasai Y (2000) Induction of midbrain dopaminergic neurons from ES cells by stromal cell-derived inducing activity. Neuron 28:31–40. https://www.ncbi.nlm.nih.gov/pubmed/1108698110.1016/s0896-6273(00)00083-0

Kiely AP, Asi YT, Kara E, Limousin P, Ling H, Lewis P, Proukakis C, Quinn N, Lees AJ, Hardy J, Revesz T, Houlden H, Holton JL (2013) Alpha-Synucleinopathy associated with G51D SNCA mutation: a link between Parkinson's disease and multiple system atrophy? Acta Neuropathol 125:753–769. https://www.ncbi.nlm.nih.gov/pubmed/2340437210.1007/s00401-013-1096-7

Kim YG, Cha J, Chandrasegaran S (1996) Hybrid restriction enzymes: zinc finger fusions to Fok I cleavage domain. Proc Natl Acad Sci U S A 93:1156–1160. https://www.ncbi.nlm.nih.gov/pubmed/857773210.1073/pnas.93.3.1156

Kontur PJ, Leranth C, Redmond DE Jr, Roth RH, Robbins RJ (1993) Tyrosine hydroxylase immunoreactivity and monoamine and metabolite levels in cryopreserved human fetal ventral mesencephalon. Exp Neurol 121:172–180. https://www.ncbi.nlm.nih.gov/pubmed/768796010.1006/exnr.1993.1084

Kordower JH, Chu Y, Hauser RA, Freeman TB, Olanow CW (2008) Lewy body-like pathology in long-term embryonic nigral transplants in Parkinson's disease. Nat Med 14:504–506. https://www.ncbi.nlm.nih.gov/pubmed/1839196210.1038/nm1747

Kordower JH, Freeman TB, Chen EY, Mufson EJ, Sanberg PR, Hauser RA, Snow B, Olanow CW (1998) Fetal nigral grafts survive and mediate clinical benefit in a patient with Parkinson's disease. Mov Disord 13:383–393. https://www.ncbi.nlm.nih.gov/pubmed/961372610.1002/mds.870130303

Kordower JH, Freeman TB, Snow BJ, Vingerhoets FJ, Mufson EJ, Sanberg PR, Hauser RA, Smith DA, Nauert GM, Perl DP et al (1995) Neuropathological evidence of graft survival and striatal reinnervation after the transplantation of fetal mesencephalic tissue in a patient with Parkinso's disease. N Engl J Med 332:1118–1124. https://www.ncbi.nlm.nih.gov/pubmed/770028410.1056/NEJM199504273321702

Kriks S, Shim JW, Piao J, Ganat YM, Wakeman DR, Xie Z, Carrillo-Reid L, Auyeung G, Antonacci C, Buch A, Yang L, Beal MF, Surmeier DJ, Kordower JH, Tabar V, Studer L (2011) Dopamine neurons derived from human ES cells efficiently engraft in animal models of Parkinson's disease. Nature 480:547–551. https://www.ncbi.nlm.nih.gov/pubmed/2205698910.1038/nature10648

Kruger R, Kuhn W, Muller T, Woitalla D, Graeber M, Kosel S, Przuntek H, Epplen JT, Schols L, Riess O (1998) Ala30Pro mutation in the gene encoding alpha-synuclein in Parkinson's disease. Nat Genet 18:106–108. https://www.ncbi.nlm.nih.gov/pubmed/946273510.1038/ng0298-106

Lancaster MA, Knoblich JA (2014) Generation of cerebral organoids from human pluripotent stem cells. Nat Protoc 9:2329–2340. https://www.ncbi.nlm.nih.gov/pubmed/2518863410.1038/nprot.2014.158

Lancaster MA, Renner M, Martin CA, Wenzel D, Bicknell LS, Hurles ME, Homfray T, Penninger JM, Jackson AP, Knoblich JA (2013) Cerebral organoids model human brain development and microcephaly. Nature 501:373–379. https://www.ncbi.nlm.nih.gov/pubmed/2399568510.1038/nature12517

Lesage S, Brice A (2009) Parkinson's disease: from monogenic forms to genetic susceptibility factors. Hum Mol Genet 18:R48–R59. https://www.ncbi.nlm.nih.gov/pubmed/1929740110.1093/hmg/ddp012
Li JY, Englund E, Holton JL, Soulet D, Hagell P, Lees AJ, Lashley T, Quinn NP, Rehncrona S, Bjorklund A, Widner H, Revesz T, Lindvall O, Brundin P (2008) Lewy bodies in grafted neurons in subjects with Parkinson's disease suggest host-to-graft disease propagation. Nat Med 14:501–503. https://www.ncbi.nlm.nih.gov/pubmed/1839196310.1038/nm1746
Li JY, Englund E, Widner H, Rehncrona S, Bjorklund A, Lindvall O, Brundin P (2010) Characterization of Lewy body pathology in 12- and 16-year-old intrastriatal mesencephalic grafts surviving in a patient with Parkinson's disease. Mov Disord 25:1091–1096. https://www.ncbi.nlm.nih.gov/pubmed/2019864510.1002/mds.23012
Liang G, Zhang Y (2013) Genetic and epigenetic variations in iPSCs: potential causes and implications for application. Cell Stem Cell 13:149–159. https://www.ncbi.nlm.nih.gov/pubmed/2391008210.1016/j.stem.2013.07.001
Lindvall O (2016) Clinical translation of stem cell transplantation in Parkinson's disease. J Intern Med 279:30–40. https://www.ncbi.nlm.nih.gov/pubmed/2633295910.1111/joim.12415
Lindvall O (2015) Treatment of Parkinson's disease using cell transplantation. Philos Trans R Soc Lond Ser B Biol Sci 370:20140370. https://www.ncbi.nlm.nih.gov/pubmed/2641668110.1098/rstb.2014.0370
Lindvall O, Brundin P, Widner H, Rehncrona S, Gustavii B, Frackowiak R, Leenders KL, Sawle G, Rothwell JC, Marsden CD et al (1990) Grafts of fetal dopamine neurons survive and improve motor function in Parkinson's disease. Science 247:574–577. https://www.ncbi.nlm.nih.gov/pubmed/210552910.1126/science.2105529
Mali P, Esvelt KM, Church GM (2013) Cas9 as a versatile tool for engineering biology. Nat Methods 10:957–963. https://www.ncbi.nlm.nih.gov/pubmed/2407699010.1038/nmeth.2649
Mansour AA, Goncalves JT, Bloyd CW, Li H, Fernandes S, Quang D, Johnston S, Parylak SL, Jin X, Gage FH (2018) An in vivo model of functional and vascularized human brain organoids. Nat Biotechnol 36:432–441. https://www.ncbi.nlm.nih.gov/pubmed/2965894410.1038/nbt.4127
Moradi S, Mahdizadeh H, Saric T, Kim J, Harati J, Shahsavarani H, Greber B, Moore JBT (2019) Research and therapy with induced pluripotent stem cells (iPSCs): social, legal, and ethical considerations. Stem Cell Res Ther 10:341. https://www.ncbi.nlm.nih.gov/pubmed/3175303410.1186/s13287-019-1455-y
Morizane A, Doi D, Kikuchi T, Okita K, Hotta A, Kawasaki T, Hayashi T, Onoe H, Shiina T, Yamanaka S, Takahashi J (2013) Direct comparison of autologous and allogeneic transplantation of iPSC-derived neural cells in the brain of a non-human primate. Stem Cell Reports 1:283–292. https://www.ncbi.nlm.nih.gov/pubmed/2431966410.1016/j.stemcr.2013.08.007
Nakao N, Frodl EM, Duan WM, Widner H, Brundin P (1994) Lazaroids improve the survival of grafted rat embryonic dopamine neurons. Proc Natl Acad Sci U S A 91:12408–12412. https://www.ncbi.nlm.nih.gov/pubmed/780905010.1073/pnas.91.26.12408
Olanow CW, Goetz CG, Kordower JH, Stoessl AJ, Sossi V, Brin MF, Shannon KM, Nauert GM, Perl DP, Godbold J, Freeman TB (2003) A double-blind controlled trial of bilateral fetal nigral transplantation in Parkinson's disease. Ann Neurol 54:403–414. https://www.ncbi.nlm.nih.gov/pubmed/1295327610.1002/ana.10720
Olson L, Seiger A (1972) Brain tissue transplanted to the anterior chamber of the eye. 1. Fluorescence histochemistry of immature catecholamine and 5-hydroxytryptamine neurons reinnervating the rat iris. Z Zellforsch Mikrosk Anat 135:175–194. https://www.ncbi.nlm.nih.gov/pubmed/464252710.1007/BF00315125
Olson L, Seiger A (1975) Brain tissue transplanted to the anterior chanber of the eye: 2. Fluorescence histochemistry of immature catecholamine- and 5-hydroxytryptamine neurons innervating the rat vas deferens. Cell Tissue Res 158:141–150. https://www.ncbi.nlm.nih.gov/pubmed/113185910.1007/BF00219957
Ono Y, Nakatani T, Sakamoto Y, Mizuhara E, Minaki Y, Kumai M, Hamaguchi A, Nishimura M, Inoue Y, Hayashi H, Takahashi J, Imai T (2007) Differences in neurogenic potential in floor

plate cells along an anteroposterior location: midbrain dopaminergic neurons originate from mesencephalic floor plate cells. Development 134:3213–3225. https://www.ncbi.nlm.nih.gov/pubmed/1767078910.1242/dev.02879
Park CH, Minn YK, Lee JY, Choi DH, Chang MY, Shim JW, Ko JY, Koh HC, Kang MJ, Kang JS, Rhie DJ, Lee YS, Son H, Moon SY, Kim KS, Lee SH (2005) In vitro and in vivo analyses of human embryonic stem cell-derived dopamine neurons. J Neurochem 92:1265–1276. https://www.ncbi.nlm.nih.gov/pubmed/1571567510.1111/j.1471-4159.2004.03006.x
Pasanen P, Myllykangas L, Siitonen M, Raunio A, Kaakkola S, Lyytinen J, Tienari PJ, Poyhonen M, Paetau A (2014) Novel alpha-synuclein mutation A53E associated with atypical multiple system atrophy and Parkinson's disease-type pathology. Neurobiol Aging 35(2180):e2181–e2185. https://www.ncbi.nlm.nih.gov/pubmed/2474636210.1016/j.neurobiolaging.2014.03.024
Pereira C, Ferreiro E, Cardoso SM, De Oliveira CR (2004) Cell degeneration induced by amyloid-beta peptides: implications for Alzheimer's disease. J Mol Neurosci 23:97–104. https://www.ncbi.nlm.nih.gov/pubmed/1512669510.1385/JMN:23:1-2:097
Polymeropoulos MH, Lavedan C, Leroy E, Ide SE, Dehejia A, Dutra A, Pike B, Root H, Rubenstein J, Boyer R, Stenroos ES, Chandrasekharappa S, Athanassiadou A, Papapetropoulos T, Johnson WG, Lazzarini AM, Duvoisin RC, Di Iorio G, Golbe LI, Nussbaum RL (1997) Mutation in the alpha-synuclein gene identified in families with Parkinson's disease. Science 276:2045–2047. https://www.ncbi.nlm.nih.gov/pubmed/919726810.1126/science.276.5321.2045
Porter DL, Levine BL, Kalos M, Bagg A, June CH (2011) Chimeric antigen receptor-modified T cells in chronic lymphoid leukemia. N Engl J Med 365:725–733. https://www.ncbi.nlm.nih.gov/pubmed/2183094010.1056/NEJMoa1103849
Prescott C (2011) The business of exploiting induced pluripotent stem cells. Philos Trans R Soc Lond Ser B Biol Sci 366:2323–2328. https://www.ncbi.nlm.nih.gov/pubmed/2172713810.1098/rstb.2011.0047
Prokhorova TA, Harkness LM, Frandsen U, Ditzel N, Schroder HD, Burns JS, Kassem M (2009) Teratoma formation by human embryonic stem cells is site dependent and enhanced by the presence of Matrigel. Stem Cells Dev 18:47–54. https://www.ncbi.nlm.nih.gov/pubmed/1839367310.1089/scd.2007.0266
Prots I, Grosch J, Brazdis RM, Simmnacher K, Veber V, Havlicek S, Hannappel C, Krach F, Krumbiegel M, Schutz O, Reis A, Wrasidlo W, Galasko DR, Groemer TW, Masliah E, Schlotzer-Schrehardt U, Xiang W, Winkler J, Winner B (2018) Alpha-Synuclein oligomers induce early axonal dysfunction in human iPSC-based models of synucleinopathies. Proc Natl Acad Sci U S A 115:7813–7818. https://www.ncbi.nlm.nih.gov/pubmed/2999159610.1073/pnas.1713129115
Proukakis C, Dudzik CG, Brier T, Mackay DS, Cooper JM, Millhauser GL, Houlden H, Schapira AH (2013) A novel alpha-synuclein missense mutation in Parkinson disease. Neurology 80:1062–1064. https://www.ncbi.nlm.nih.gov/pubmed/2342732610.1212/WNL.0b013e31828727ba
Qian X, Nguyen HN, Song MM, Hadiono C, Ogden SC, Hammack C, Yao B, Hamersky GR, Jacob F, Zhong C, Yoon KJ, Jeang W, Lin L, Li Y, Thakor J, Berg DA, Zhang C, Kang E, Chickering M, Nauen D, Ho CY, Wen Z, Christian KM, Shi PY, Maher BJ, Wu H, Jin P, Tang H, Song H, Ming GL (2016) Brain-region-specific organoids using mini-bioreactors for modeling ZIKV exposure. Cell 165:1238–1254. https://www.ncbi.nlm.nih.gov/pubmed/2711842510.1016/j.cell.2016.04.032
Renner M, Lancaster MA, Bian S, Choi H, Ku T, Peer A, Chung K, Knoblich JA (2017) Self-organized developmental patterning and differentiation in cerebral organoids. EMBO J 36:1316–1329. https://www.ncbi.nlm.nih.gov/pubmed/2828358210.15252/embj.201694700
Reubinoff BE, Itsykson P, Turetsky T, Pera MF, Reinhartz E, Itzik A, Ben-Hur T (2001) Neural progenitors from human embryonic stem cells. Nat Biotechnol 19:1134–1140. https://www.ncbi.nlm.nih.gov/pubmed/1173178210.1038/nbt1201-1134
Rinne JO (1993) Nigral degeneration in Parkinson's disease. Mov Disord 8(Suppl 1):S31–S35. https://www.ncbi.nlm.nih.gov/pubmed/830230510.1002/mds.870080507
Roy NS, Cleren C, Singh SK, Yang L, Beal MF, Goldman SA (2006) Functional engraftment of human ES cell-derived dopaminergic neurons enriched by coculture with telomerase-

immortalized midbrain astrocytes. Nat Med 12:1259–1268. https://www.ncbi.nlm.nih.gov/pubmed/1705770910.1038/nm1495
Rubenstein J, Rakic P (2013) Patterning and cell type specification in the developing CNS and PNS: comprehensive developmental neuroscience. Academic Press 1
Sanchez-Danes A, Consiglio A, Richaud Y, Rodriguez-Piza I, Dehay B, Edel M, Bove J, Memo M, Vila M, Raya A, Izpisua Belmonte JC (2012) Efficient generation of A9 midbrain dopaminergic neurons by lentiviral delivery of LMX1A in human embryonic stem cells and induced pluripotent stem cells. Hum Gene Ther 23:56–69. https://www.ncbi.nlm.nih.gov/pubmed/2187792010.1089/hum.2011.054
Schuldiner M, Yanuka O, Itskovitz-Eldor J, Melton DA, Benvenisty N (2000) Effects of eight growth factors on the differentiation of cells derived from human embryonic stem cells. Proc Natl Acad Sci U S A 97:11307–11312. https://www.ncbi.nlm.nih.gov/pubmed/1102733210.1073/pnas.97.21.11307
Schweitzer JS, Song B, Herrington TM, Park TY, Lee N, Ko S, Jeon J, Cha Y, Kim K, Li Q, Henchcliffe C, Kaplitt M, Neff C, Rapalino O, Seo H, Lee IH, Kim J, Kim T, Petsko GA, Ritz J, Cohen BM, Kong SW, Leblanc P, Carter BS, Kim KS (2020) Personalized iPSC-derived dopamine progenitor cells for Parkinson's disease. N Engl J Med 382:1926–1932. https://www.ncbi.nlm.nih.gov/pubmed/3240216210.1056/NEJMoa1915872
Seiger A, Olson L (1975) Brain tissue transplanted to the anterior chamber of the eye: 3. Substitution of lacking central noradrenaline input by host iris sympathetic fibers in the isolated cerebral cortex developed in oculo. Cell Tissue Res 159:325–338. https://www.ncbi.nlm.nih.gov/pubmed/109711810.1007/BF00221780
Sharma K (2019) Cholinesterase inhibitors as Alzheimer's therapeutics (review). Mol Med Rep 20:1479–1487. https://www.ncbi.nlm.nih.gov/pubmed/3125747110.3892/mmr.2019.10374
Simmnacher K, Krach F, Schneider Y, Alecu JE, Mautner L, Klein P, Roybon L, Prots I, Xiang W, Winner B (2020) Unique signatures of stress-induced senescent human astrocytes. Exp Neurol 334:113466. https://www.ncbi.nlm.nih.gov/pubmed/3294957210.1016/j.expneurol.2020.113466
Sommer A, Marxreiter F, Krach F, Fadler T, Grosch J, Maroni M, Graef D, Eberhardt E, Riemenschneider MJ, Yeo GW, Kohl Z, Xiang W, Gage FH, Winkler J, Prots I, Winner B (2018) Th17 lymphocytes induce neuronal cell death in a human iPSC-based model of Parkinson's disease. Cell Stem Cell 23(123–131):e126. https://www.ncbi.nlm.nih.gov/pubmed/2997998610.1016/j.stem.2018.06.015
Song B, Cha Y, Ko S, Jeon J, Lee N, Seo H, Park KJ, Lee IH, Lopes C, Feitosa M, Luna MJ, Jung JH, Kim J, Hwang D, Cohen BM, Teicher MH, Leblanc P, Carter BS, Kordower JH, Bolshakov VY, Kong SW, Schweitzer JS, Kim KS (2020) Human autologous iPSC-derived dopaminergic progenitors restore motor function in Parkinson's disease models. J Clin Invest 130:904–920. https://www.ncbi.nlm.nih.gov/pubmed/3171489610.1172/JCI130767
Spillantini MG, Schmidt ML, Lee VM, Trojanowski JQ, Jakes R, Goedert M (1997) Alpha-synuclein in Lewy bodies. Nature 388:839–840. https://www.ncbi.nlm.nih.gov/pubmed/927804410.1038/42166
Sun N, Zhao H (2014) Seamless correction of the sickle cell disease mutation of the HBB gene in human induced pluripotent stem cells using TALENs. Biotechnol Bioeng 111:1048–1053. https://www.ncbi.nlm.nih.gov/pubmed/2392885610.1002/bit.25018
Sundberg M, Bogetofte H, Lawson T, Jansson J, Smith G, Astradsson A, Moore M, Osborn T, Cooper O, Spealman R, Hallett P, Isacson O (2013) Improved cell therapy protocols for Parkinson's disease based on differentiation efficiency and safety of hESC-, hiPSC-, and non-human primate iPSC-derived dopaminergic neurons. Stem Cells 31:1548–1562. https://www.ncbi.nlm.nih.gov/pubmed/2366660610.1002/stem.1415
Tabar V, Tomishima M, Panagiotakos G, Wakayama S, Menon J, Chan B, Mizutani E, Al-Shamy G, Ohta H, Wakayama T, Studer L (2008) Therapeutic cloning in individual parkinsonian mice. Nat Med 14:379–381. https://www.ncbi.nlm.nih.gov/pubmed/1837640910.1038/nm1732
Takahashi J (2020) iPS cell-based therapy for Parkinson's disease: a Kyoto trial. Regen Ther 13:18–22. https://www.ncbi.nlm.nih.gov/pubmed/3349031910.1016/j.reth.2020.06.002

Takahashi K, Okita K, Nakagawa M, Yamanaka S (2007) Induction of pluripotent stem cells from fibroblast cultures. Nat Protoc 2:3081–3089. https://www.ncbi.nlm.nih.gov/pubmed/1807970710.1038/nprot.2007.418

Takahashi K, Yamanaka S (2006) Induction of pluripotent stem cells from mouse embryonic and adult fibroblast cultures by defined factors. Cell 126:663–676. https://www.ncbi.nlm.nih.gov/pubmed/1690417410.1016/j.cell.2006.07.024

Thomson JA, Itskovitz-Eldor J, Shapiro SS, Waknitz MA, Swiergiel JJ, Marshall VS, Jones JM (1998) Embryonic stem cell lines derived from human blastocysts. Science 282:1145–1147. https://www.ncbi.nlm.nih.gov/pubmed/980455610.1126/science.282.5391.1145

Tokuzawa Y, Kaiho E, Maruyama M, Takahashi K, Mitsui K, Maeda M, Niwa H, Yamanaka S (2003) Fbx15 is a novel target of Oct3/4 but is dispensable for embryonic stem cell self-renewal and mouse development. Mol Cell Biol 23:2699–2708. https://www.ncbi.nlm.nih.gov/pubmed/1266557210.1128/mcb.23.8.2699-2708.2003

Volarevic V, Markovic BS, Gazdic M, Volarevic A, Jovicic N, Arsenijevic N, Armstrong L, Djonov V, Lako M, Stojkovic M (2018) Ethical and safety issues of stem cell-based therapy. Int J Med Sci 15:36–45. https://www.ncbi.nlm.nih.gov/pubmed/2933308610.7150/ijms.21666

Wennberg L, Czech KA, Larsson LC, Mirza B, Bennet W, Song Z, Widner H (2001) Effects of immunosuppressive treatment on host responses against intracerebral porcine neural tissue xenografts in rats. Transplantation 71:1797–1806. https://www.ncbi.nlm.nih.gov/pubmed/1145526110.1097/00007890-200106270-00016

Wernig M, Zhao JP, Pruszak J, Hedlund E, Fu D, Soldner F, Broccoli V, Constantine-Paton M, Isacson O, Jaenisch R (2008) Neurons derived from reprogrammed fibroblasts functionally integrate into the fetal brain and improve symptoms of rats with Parkinson's disease. Proc Natl Acad Sci U S A 105:5856–5861. https://www.ncbi.nlm.nih.gov/pubmed/1839119610.1073/pnas.0801677105

Winner B, Marchetto MC, Winkler J, Gage FH (2014) Human-induced pluripotent stem cells pave the road for a better understanding of motor neuron disease. Hum Mol Genet 23:R27–R34. https://www.ncbi.nlm.nih.gov/pubmed/2482170410.1093/hmg/ddu205

Yan Y, Yang D, Zarnowska ED, Du Z, Werbel B, Valliere C, Pearce RA, Thomson JA, Zhang SC (2005) Directed differentiation of dopaminergic neuronal subtypes from human embryonic stem cells. Stem Cells 23:781–790. https://www.ncbi.nlm.nih.gov/pubmed/1591747410.1634/stemcells.2004-0365

Yee JK (2016) Off-target effects of engineered nucleases. FEBS J 283:3239–3248. https://www.ncbi.nlm.nih.gov/pubmed/2720870110.1111/febs.13760

Yoshihara M, Hayashizaki Y, Murakawa Y (2017) Genomic instability of iPSCs: challenges towards their clinical applications. Stem Cell Rev Rep 13:7–16. https://www.ncbi.nlm.nih.gov/pubmed/2759270110.1007/s12015-016-9680-6

Zarranz JJ, Alegre J, Gomez-Esteban JC, Lezcano E, Ros R, Ampuero I, Vidal L, Hoenicka J, Rodriguez O, Atares B, Llorens V, Gomez Tortosa E, Del Ser T, Munoz DG, De Yebenes JG (2004) The new mutation, E46K, of alpha-synuclein causes Parkinson and Lewy body dementia. Ann Neurol 55:164–173. https://www.ncbi.nlm.nih.gov/pubmed/1475571910.1002/ana.10795

Zhang SC, Wernig M, Duncan ID, Brustle O, Thomson JA (2001) In vitro differentiation of transplantable neural precursors from human embryonic stem cells. Nat Biotechnol 19:1129–1133. https://www.ncbi.nlm.nih.gov/pubmed/1173178110.1038/nbt1201-1129

Part II

The Status Debate

Human Cerebral Organoids: Evolving Entities and Their Moral Status

4

Andrea Lavazza

4.1 Introduction: Basic Concepts

Using embryonic stem (ES) cells or induced pluripotent stem (iPS) cells, it is possible to generate three-dimensional in vitro cultures that mimic the developmental process and organization of the developing human brain.[1] These human cerebral organoids (HCOs) have provided a unique, physiologically relevant in vitro model system for the study of human neurological development and a number of diseases. In just a few years, there has been rapid and considerable progress in the attempt to create a brain model capable of showcasing the characteristics of the central nervous system. There are still strong limitations to address, including the absence of vascularization which makes it difficult to feed the central layers of the organoid. Nevertheless, some important features of the human brain have recently been observed in cerebral organoids: they manifest electrical activity (i.e., communication between neurons), are sensitive to light stimulation, are able to connect to a spinal cord by sending impulses that make a muscle contract, and can grow blood–brain barriers and produce fluid secretion in self-contained compartments. Recent data show that cortical organoid network development after 10 months resembles even the EEG patterns of preterm babies.

Assembloids are organoids obtained from the union of different parts of the central nervous system. Thanks to this technique, it might be possible to overcome the current growth limits of HCOs, so that they may in future develop sensory channels. Although cerebral organoids are not comparable to human brains at the moment due

[1] Chiaradia and Lancaster (2020). Cf. also the chapters of this book devoted to scientific features of HCOs.

A. Lavazza (✉)
Centro Universitario Internazionale, Arezzo, Italy

University of Pavia, Pavia, Italy

H.-G. Dederer, D. Hamburger (eds.), *Brain Organoids in Research and Therapy*, Advances in Neuroethics, https://doi.org/10.1007/978-3-030-97641-5_4

to their extremely simplified structure, the prospects of research give rise to ethical concerns about the creation and destructive experimental use of human cerebral organoids. In particular, one can wonder whether a human cerebral organoid could develop some degree of consciousness and whether, under certain conditions, it could acquire its own moral status with some related rights. In this chapter, I discuss the conditions under which HCOs could require the acknowledgment of a moral status. For this purpose, I consider the hypothesis that HCOs might develop a primitive form of sentience or consciousness.

In general, "moral status is a concept that deals with who or what is so valuable that it should be treated with special regard."[2] Moral status can be taken to be "that which gives rise to moral rights"[3] or, more in general, that which gives rise to moral consideration, depending on morally significant interests.[4] Specifically, moral status might also "be a metric of the overall strength and breadth of one's fundamental moral rights,"[5] in the sense that "claims about moral status can be translated into claims about fundamental moral rights."

The minimum grounding feature of moral status seems to be the capability to have subjective interests.[6] Entities with moral status are sentient, have a certain kind of subjective experience and can be "wronged." By sentience, I mean here the minimal ability to experience basic sensations (such as pain, thirst, and lack of oxygen but not as an pure reflex)—which can be considered a minimal or rudimentary degree of consciousness, if we agree that consciousness is a property that comes in degrees.[7] For a human cerebral organoid to be attributed moral status, it should therefore exhibit a minimal or basic form of consciousness, although this minimum requirement can be insufficient to meet the criteria established for granting moral status in most philosophical views.

(a) If you are sentient, then you have interests (since we can understand interests in terms of present subjective motivational states), (b) if you have interests, then you are capable of being harmed (since we can understand harm in terms of interest frustration), (c) if you are capable of being harmed, then moral agents have at least a prima facie moral duty not to harm you, and (d) if moral agents have at least a prima facie moral duty not to harm you, then you have at least a prima facie moral right, against these moral agents, not to be harmed.[8]

The point is that this prima facie moral right needs to be further specified and seems to be overinclusive with respect to our ordinary moral intuitions; respecting such a right also seems unfeasible as it could imply strong obligations even toward living beings such as octopuses and birds.

---

[2] Walters (2021).

[3] Douglas (2021).

[4] DeGrazia (2008).

[5] Douglas (2021).

[6] Jaworska and Tannenbaum (2018).

[7] Cf. Bayne et al. (2016).

[8] Sebo (2017).

In the field of analytic philosophy, the recent reflection on the moral status of an entity can be summarized in this way.[9]

1. To obtain moral status, it is necessary to have some kinds of subjective interest, i.e., to have some kinds of subjective experience.
2. To be granted moral status, one needs (2a) to possess a certain morally relevant characteristic; or (2b) to be part of a relationship of similarity or biological belonging; or (2c) to be inserted in a network of significant and appropriate relationships (such as recognition, care, and respect)—these being the main theories of justification, or grounds, for an entity to be attributed moral status.
3. When attributing moral status to an entity not hitherto considered from this point of view, it is necessary to clarify the moral hierarchy in which this entity is placed, what kind of rights it acquires and what obligations other moral agents have toward it. This should be done bearing in mind that, in general, (3a) obtaining moral status does not in itself imply ownership of specific rights and does not impose specific obligations on other moral agents; and (3b) moral status can be analytically broken down into (1) a purely evaluative function, which attributes an intrinsic value to the entity in question, and (2) into a prescriptive function, for which this intrinsic value requires a certain treatment on the part of moral agents.

Baertschi, among others, distinguishes between a *complete* moral status (i.e., the entity owns all the rights associated to this quality whereas other subjects have all the relative duties toward it), and an *incomplete* moral status (i.e., the entity possesses only some of the rights or partially owns them, while other parties have partial or reduced obligations toward it.[10]

Furthermore, moral status can be *intrinsic* or *conferred*. In the first case, it is the properties, characteristics, or value associated with it which confer moral status on the entity, without any need for an explicit act by a third party. When moral status is *conferred*, instead, it is attributed to the entity by someone based on the properties or characteristics deemed particularly relevant.

In general, complete moral status, or the highest degree of moral status, is dubbed Full Moral Status (FMS) and "it is usually taken for granted that all adult cognitively unimpaired human beings have FMS."[11] This kind of moral status implies a stringent presumption against interference (including destruction, experimentation, and causing suffering); strong reason to aid; and strong reason to treat fairly. All of this means that "all beings with FMS are owed the same protections and entitlements," they "have equal moral status."[12]

As to cerebral organoids, two stages should probably be differentiated. First, we should ascertain the presence of some specific characteristics, properties, and

---

[9] Zuolo (2016); cf. Jaworska and Tannenbaum (2018).

[10] Baertschi (2008).

[11] Jaworska and Tannenbaum (2018).

[12] Ibidem.

potentials (both for biological development and in ontological terms); second, these characteristics, properties, and potentials should be evaluated to decide whether they can be sufficient to attribute moral status to cerebral organoids. This is the canonical process for attributing moral status, but it seems that for cerebral organoids this two-stage process should be analytically split based on the alleged specific features of them.

The premise seems to be that if HCOs have some form of sentience, (1) it will confer on them subjective interests, primarily that of not enduring pain. Whether this is the case is a decisive empirical question to be investigated scientifically. This is not the place to address this issue, but it is fundamental and the difficulties of factual assessment should also be taken into account for the ethical aspects.[13] As regards condition (2) seen before, human cerebral organoids are certainly (2b) part of a relationship of similarity or biological belonging (even if it cannot be ruled out that similarity comes in degrees, which would complicate the judgment) and (2a) they can be taken to possess morally relevant properties, i.e., a form of human consciousness.[14] This point may appear more controversial because it would seem, given condition (1), that in this case every sentient entity has moral status. In my opinion, the "human" specification of the "consciousness" property, at least above a specific threshold, eliminates this risk of circularity of conditions. It is human consciousness that constitutes a morally relevant property in the case of HCOs according to many (but not all) theories of moral status.

As for the aspects summarized in point (3), they will be considered in the rest of the chapter (likewise the other points will be more deeply analyzed below). In fact, if there can be agreement, albeit not unanimous, that conscious HCOs should be attributed a moral status, there are different positions regarding the specific rights and obligations that may arise. And this depends on whether this moral status has a purely evaluative function or a prescriptive one.

4.2 The Ethical Debate on Brain Organoids and Their Moral Status

The ethical debate on human cerebral organoids is very recent, and the related literature, while expanding at a fast pace, is still quite limited. Perhaps precisely because we are at the beginning of the ethical evaluation of HCOs, the positions on the market are very different and range from extreme prudence to the extreme lack of limitations in the use of human brain organoids. At the time of writing, there is no overview available of the main views on the subject. For this reason, it may be useful to propose a review of these perspectives, in order to better frame my specific proposal on the moral status of current and future brain organoids.

---

[13] Cf. Lavazza and Massimini (2018a) and Sect. 4.2.

[14] Here I am obviously considering different grounds for moral status, which can be deemed not to be compatible or to be alternative in principle.

4.2.1 Framing the Problem: Are HCOs Too Human?

One of the most recent papers on the topic, the target article written by Greely for the *American Journal of Bioethics*, addresses the main ethical issues involved and lists a series of important questions, without however trying to answer them.[15] Interestingly, Greely begins his article by pointing out how research on the human brain has to deal with a dilemma that seems inherent in this type of study. If in fact the best way to understand the functioning of our nervous system is to study it *in vivo*, there are clear medical and ethical limits to the invasive interventions that would be necessary to shed light on the fine functioning of the brain and to find cures for the most serious diseases that affect it.

In this sense, the dilemma is immediately clear:

> When we avoid unethical research by making living models of human brains, we may make our models so good that they themselves deserve some of the kinds of ethical and legal respect that have hindered brain research in human beings. If it looks like a human brain and acts like a human brain, at what point do we have to treat it like a human brain – or a human being?[16]

At first sight, the dilemma thus formulated may seem too simplistic. Indeed, similarity is not enough to create ontological commitments and moral obligations. But certainly, this can be a starting point for entering the difficult domain of the ethical consideration of so-called brain surrogates. In this macro-category, in addition to human cerebral organoids, we can also include human/nonhuman chimeras or living entities that have cells or tissues within them that are not their own, but come from another individual, either of the same or of a different species. Later, I will briefly consider human/nonhuman chimeras only as an entity in which a human cerebral organoid has been implanted.

Farahany and 16 other scholars were among the first—after Cheshire and Lavazza and Massimini[17]—to wonder "if researchers could create brain tissue in the laboratory that might appear to have conscious experience or individual phenomenal states, that tissue deserves one of the protections routinely provided to human or animal research subjects?" The authors did not rule out that in the future human brain organoids may develop high-level cognitive abilities and phenomenal consciousness, including being able to store or retrieve memories and perhaps even have some perception of agency or self-awareness. In that case, they wonder whether, having reached one of such stages of development, it is necessary to "assign someone loosely akin to a guardian or decision-maker for the brain surrogate." In addition to "stewardship," Farahany and colleagues addressed the theme of ownership, i.e., who, if anyone, should "own" the brains grown in the laboratory.

[15] Greely (2021). See also Barnhart and Dierickx (2022).

[16] Ibidem.

[17] Farahany et al. (2018); Cheshire (2014); Lavazza and Massimini (2018a). Cf. also Bredenoord et al. (2017); Munsie et al. (2017) concerning different ethical issues about brain surrogates.

The question arises especially at a time when brain surrogates could acquire higher moral status than that we would currently attribute to them.

Greely usefully identifies six issues related to the ethical aspects of brain surrogates.[18] They are: (1) the welfare of the surrogates; (2) the consent and welfare of the "human parts" of the surrogates; (3) possible nonresearch implications of the research; (4) possible nonresearch uses of the surrogates; (5) humanization of nonhumans; (6) the rights of the surrogates. Regarding the first issue, namely the welfare of the surrogates, the author wonders if a brain organoid that shows specific patterns of neural activation, similar to those that signal pain in adult humans, is really sensitive to pain. The implicit question then is: what should we do if that were the case?

Regarding the third issue, i.e., the potential nonresearch implications of research, Greely examines the potential case of gene edited nonhuman animals or chimeras that approach, far beyond what researchers expected, what he believes to be the different degrees of awareness of the human being, i.e., sentience, consciousness, and self-consciousness. If so, "would we want to treat such animals differently from the way we do now? [...] Would we want to distinguish among different kinds of animals based on their degrees of awareness?" Regarding potential nonresearch uses of the surrogates, Greely does not give this example explicitly, but his remarks suggest that in the future people may want bio-tech companies to grow spare brains, ready for potential transplantation. I will return to this aspect below because it has an ethical relevance that should not be underestimated.

Discussing the fifth issue—the humanization of nonhumans—Greely lists a number of strong restrictions recently introduced on experiments with chimeras. In particular, in accord to an American federal law, it is now a felony to create "a nonhuman life form engineered such that it contains a human brain, or a brain derived wholly or predominantly from human neural tissues." As Greely notes, "the mixing itself may trouble some people who are concerned about the moral confusion that may arise regarding our obligations toward humans and non-humans." It therefore appears that human brains have a special status that prevents their mingling with nonhuman entities. By the same token, HCOs may also be granted a certain special status.

Finally, as regards the prerogatives of brain surrogates, Greely wonders whether the appearance of human-like awareness or intelligence also brings specific rights. As to HCOs, the author specifies that it seems unlikely that they can develop a consciousness without the experiences of the human world. But later, Greely argues that scientific progress on brain surrogates may prompt us to revise our notion of a "person," since it is "persons" who have rights and not "humans." In this sense, fetuses and anencephalic children are cited as subjects about which it was debated whether to include them or not in the category of persons. In the same way, it is suggested that there could be a stage in which the question of personhood could also be asked for HCOs.

[18] Greely (2021).

In conclusion, Greely emphasizes the fact that science should not be separated from ethical evaluation, under penalty of becoming dangerous for society. At the same time, he argues that ethics should not be separated from scientific grounding, under penalty of creating unrealistic speculations and unfounded myths. The need then arises to combine different skills to reflect thoroughly on the starting dilemma, the one that places on the one hand the moral obligation to prevent human suffering that comes from diseases and on the other hand the moral obligation not to inflict inappropriate suffering in trying to prevent it.

This dilemma, which today manifests itself in the form of brain surrogates, appears to imply that the latter can experience pain and that such suffering should be avoided as much as possible. However, the trade-off between the two moral obligations still seems to be unclear.

4.2.2 Cerebral Organoids without moral status

At one extreme of the ideal continuum represented by the dilemma proposed by Greely, there are those who, while potentially seeing a particular specificity in human cerebral organoids from the viewpoint of ethical issues, do not believe that the use of HCOs can raise an ethical problem when one considers the intrinsic moral value of the cerebral organoid itself. This conclusion does not derive from the neglect of the analysis of the potential moral status of HCOs but from a peculiar view of the topic. Hyun is an advocate for this position.[19] Indeed, he considers a number of important and interesting ethical issues, but of a different type. Before analyzing his argument for denying moral status to human cerebral organoids, it is worth briefly looking at some other ethical aspects examined by Hyun.

First, there is an issue about the consent of the donors of the cells from which the organoids are grown. Donors often agree to have their cells used to create cell lines that will then have different applications. They are therefore not aware that a cerebral organoid with characteristics similar to those of a human brain can be created from their cells. This is a specific application for which they may not wish to give their consent. Furthermore, cells or cell lines are often anonymized, so it would not even be feasible to trace the donors. There is also the issue of the "ownership" of the cerebral organoid and the decisions on the use made of it. Should donors be able to express their opinion? And if so, what weight should it have?[20] Obviously, the answers will vary depending on whether one thinks that HCOs are simply chunks of human tissue or that they have some degree of moral status.

According to Hyun, Scharf-Deering, and Lunshof,[21] the possibility of HCOs becoming conscious is "extremely remote at best," but, in any case, they "acknowledge that several important considerations provide good reasons to resist overemphasizing this ethical concern at this time." Their considerations start from the

[19] Hyun et al. (2020).

[20] Cf. Boers et al. (2018); Lavazza (2019).

[21] Hyun et al. (2020).

experiment conducted by Muotri's team which recorded EEG patterns of neuronal activity in 10-month-old organoids similar to those recorded in preterm babies of 25–30 weeks.[22] A certain type of coordinated electrical brain activity, in general, does not imply consciousness if there is a lack of verbal or behavioral cues. Therefore, if it is not easy to infer consciousness in preterm babies starting from specific neural correlates, it is even harder to hypothesize that HCOs with the same activity patterns are conscious.

But Hyun goes further. Not only—he says—do we not know for sure what the neural correlates of consciousness are, but also the knowledge we have accumulated to date seems to tell us that widely distributed areas and circuits of the brain and multiple cell types are required for consciousness to emerge. These are specific conditions that organoids would not meet. Furthermore, these requirements have been learned by comparing verbal reports of adult subjects and neuroscientific examinations, while it seems inappropriate to extend these results to patients with altered states of consciousness, fetuses, newborns, and, today, human cerebral organoids.

It should also be noted, in Hyun's opinion, that the term "consciousness" is fundamentally ambiguous. According to the meaning attributed to it, moral concerns about cerebral organoids can change. In fact, if by consciousness, we mean the basic neuronal activity in a cortical region after it has been stimulated by sensory input without there being subjective awareness, the presence of this type of consciousness will have no ethical relevance. However, it should be emphasized that few would take this sense of consciousness to be relevant when it comes to living entities and their potential moral status. According to Hyun, it would be different if by consciousness we meant, in ascending order, "conscious access to sensory stimulation; wakefulness; vigilance; focal attention; sentience; and lastly, subjective awareness."

In that case, more ethical issues would arise, but Hyun and colleagues argue that it is very unlikely that a cerebral organoid could manifest these levels of consciousness: to do so, it would require global integration and activation of cortical neurons at great distances, which are tenets of the theory of consciousness called global workspace theory.[23] The latter, however, is only one of the neuroscientific theories of consciousness currently debated, and certainly one of the most highly considered and widely talked about, but there is no consensus that it is the one *true* theory or even just *the best* theory available today. As Seth pointed out, the confidence that a brain organoid has or has not developed a form of consciousness depends on the theory of consciousness we believe in, so if different theories give different indications, then there can be no confidence in establishing whether cerebral organoids have or have not developed a form of consciousness.[24]

This first series of considerations was focused on the scientific or factual aspect of the topic under consideration, that is, whether HCOs can give rise to some degree of awareness. Now I will move on to moral considerations. According to Hyun,

[22] Trujillo et al. (2019).

[23] Dehaene (2014).

[24] Reardon (2020).

cerebral organoids grown from human cells are in vitro models that do not have moral status any more than mice do. Indeed, so the argument goes, if complex mouse brain organoids exhibited some degree of sentience, it is unlikely that people would have moral objections to their use for research any more than they object to the use of live mice in biomedical research.

> Why would human brain organoids displaying comparable levels of "consciousness" be more ethically problematic than neuronally-equivalent mouse organoids with respect to their moral status as research tools? Perhaps the difference maker is that, in the public's imagination, it might be supposed that human brain organoids somehow exhibiting these "lower" forms of consciousness could, under the right circumstances, instantiate the (much more) morally significant property of conscious self-awareness. [...] However, as one of us (I. H.) has argued elsewhere, this most complex form of consciousness – that which forms the very basis of moral life of humans – can only be realized within nurturing social environments and through the acquisition of language that would enable one to have propositional belief systems and reflective beliefs about one's own beliefs (Hyun 2013). Not even 100% natural human brains found in neonates can develop into recognizably human minds unless they are given the right interactions and social development necessary for their full realization over the span of several years.[25]

Following this line, it can be said that such conditions will never arise in the laboratory, and therefore, there is no risk that this type of consciousness may appear in cerebral organoids. And this type of consciousness is the only property, according to the authors, which could prevent the use of HCOs, whatever their development and sentience capacity, as experimental tools in biomedical research.

Lunshof also makes a similar case, using a different and original moral yardstick.[26] According to Lunshof, humans deserve ethical respect and are typically considered at the top of the entities that are entitled to moral protection. If the moral considerability or moral status (two successive and increasing levels of relevance) of humans are based on specific properties of the human brain, then one can ask whether brain surrogates also have such properties or can be conscious, so as to decide whether they are entitled to such moral protection. But according to Lunshof, consciousness is an elusive and not well-defined concept. She then proposes to use memory as a benchmark to establish the brain-likeness of brain surrogates.

Memory is in fact something specific to the nervous system: there is biological evidence of its existence, its mechanisms are quite well understood, and it is measurable. The engrams encode sensory inputs at the level of neurons and synapses. We could therefore rephrase the key question about the potential moral status of brain organoids as follows: do they have the neurobiological characteristics needed to have memory? For Lunshof, what really makes the difference between brain surrogates and brains is only the content, which must be understood as cognitive content. In the author's opinion, brain organoids are empty, as they do not have the ability to receive sensory input or to store information. Being devoid of memory, HCOs should therefore not receive any ethical protection, like other human-derived

[25] Hyun et al. (2020).

[26] Lunshof (2021).

tissues. Human heads detached from the body could instead receive ethical protection if they were resuscitated *postmortem*, as we have been able to do with pig heads for now.[27]

4.2.3 Entities that Might Have Moral Status

The approach chosen by Hostiuc and colleagues is very different from the previous ones.[28] As shown by the title of their article, they talk explicitly about the moral status of cerebral organoids. They start the discussion by referring to the debate on the moral status of human embryos. In this sense, the three features that the Warnock Committee used to attribute moral status to embryos after 14 days of development are relevant, as they make up the threshold that is still used as a time limit for experimental use in most legislations. As is well known, these features are: (a) human origin, (b) the ability to feel pain, and (c) the ability to generate individual human beings.

However, the authors choose to replace the second criterion with "a biological threshold that grants moral status to an entity." This move seems to create a certain circularity in the analysis adopted, given that the three criteria should jointly guarantee the moral status of cerebral organoids. Hostiuc and colleagues conveniently report, with regard to the first criterion, that the human origin of HCOs is "non-disputable," but things are not necessarily so simple. The use of genetic editing techniques such as Crispr-Cas9 makes it possible to alter the DNA of a biological entity, even in a radical way, thus questioning whether this entity belongs to the species of origin. More realistically, the creation of chimeras, with the insertion of human cerebral organoids in the brains of mice or other nonhuman animals, in turn calls into question the fully human origin of the HCOs that could develop in those special conditions.

However, the most important criterion remains to be established, namely the one that identifies the threshold beyond which moral status should be recognized for cerebral organoids. The authors reject that such a criterion could be sentience, since it is not an exclusive property of human beings. But it should be noted here that moral status should not necessarily be attributed only to humans but can be plausibly conferred on other living entities, provided they have some interest of their own. And sentience, i.e., the ability to experience pain and to try to avoid it, seems to be the minimum basis for having interests in the sense indicated above.

On the other hand, in the opinion of Hostiuc and colleagues, the criterion of the ability to feel pain in the strict sense, as established by the Warnock Committee, is not suitable for HCOs, as cerebral organoids could grow unable to experience pain due to the lack of nociceptors. Regarding this consideration, it can be said that the pain of a conscious organoid would not be limited to physical pain in the classical sense. In fact, we know that loneliness or being neglected cause sensations that are

---

[27] Cf. Vrselja et al. (2019).

[28] Hostiuc et al. (2019).

physiologically similar to those of physical pain and, therefore, such suffering voluntarily inflicted on cerebral organoids should fall within the criterion described above.

The authors also propose "learning capacity" as a discriminating criterion: "an entity that is capable of learning can self-develop, interact with the environment, and can be considered as more than a mass of cells in a 3D matrix. Learning capacity is not unique to human beings, but it is specific enough to differentiate between biological structures that are capable or incapable of interacting with the environment or understanding it." Hostiuc only hints at how this criterion could be ascertained and evaluated, namely using techniques such as single-photon emission tomography, microRNAs, and biochemical markers. However, it is not clear how learning capacity can be measured outside of assessments related to purposeful behavior, of which HCOs are not currently capable, even if they may still be aware of their own state.

In addition, regarding the third criterion, it seems prima facie impossible that HCOs could give birth to a fully developed human individual. However, it is interesting to note that if by hypothesis a brain connected with the environment could develop in the laboratory, endowed with all the characteristics of an adult human brain, this brain could be considered a moral agent, a bit like in the brain in a vat experiment described by Putnam, to which I will return.[29] Furthermore, with a thought experiment reminiscent of those used in philosophy of mind, Hostiuc points out that an individual who loses their limbs continues to be considered a human person, and the same would happen even if the individual in question were deprived of other parts of the body or some parts of their body were replaced by artificial surrogates. What matters is the brain, if it remains vital and capable of those functions that we usually recognize in a human person.

In this sense, it could be argued that cerebral organoids are able of giving birth to a human individual. The fact that such a HCO would be completely dependent on those who run the laboratory in which it grew does not invalidate the proposed argument. In fact, it happens that human persons, due to accidents or diseases, are completely incapable of autonomous subsistence, both for nutrition and for all other vital needs, so that they are on the one hand perfectly capable of high-level cognitive functions but would still die very quickly if they did not have the help of another person or even a robot.

From the legal point of view, the authors give an extensive interpretation of a judgment of the Court of Justice of the European Union (C-34/10, Oliver Brüstle v. Greenpeace eV.). Based on the principle of personal dignity and fundamental rights, this ruling judged the use of human embryos for scientific research purposes to be "not patentable." According to the authors, given that the European Union protects in this sense "any cell capable of commencing the process of development of human being" and given that the cells used to create cerebral organoids are either embryonic or induced pluripotent stem cells, "cerebral organoids have, from a legal point

[29] Putnam (1981).

of view, the potential to generate human beings-even if they are not necessarily human in the strict sense of the term."[30]

However, according to European law interpreted in a more rigorous way, and also according to Italian law, one of the most protective in bioethics, cerebral organoids do not currently enjoy legal protection. In fact, given that "the scientific and technological research for the promotion and safeguard of human health is a constitutional and European value (articles 9, 32 and 33 Cost. (It.); Article 13 EUCFR; article 3 TUE; article 1 of the Convention on Human Rights and Biomedicine)," the limitation of organoid research should be reasonable and proportional.[31] "This means that the research on organoids and the development of cerebral organoids for health purposes (diagnosis, treatments) should not be impeded. The limitations should only apply to experiments aimed at 'producing' highly developed and sophisticated brain organoids capable of mimicking human superior cognitive functions and human emotional feelings of pain and distress."[32]

The conclusion reached by Hostiuc and colleagues is that research on organoids should be allowed until they reach a level of development that would require moral status. In that case, their moral status "should be valued and respected by only using them in studies conducted with morally viable techniques."[33] However, it does not seem that this conclusion, although worth considering, can settle the debate, since almost all scholars and legislators would agree that research without limitations on entities with significant moral status is inadmissible. The point is to better identify the suitable and consistent criteria to establish whether HCOs should or should not be attributed moral status, to ascertain what kind of moral status this would be, and to decide what research limitations derive from it.

4.2.4 When Can Cerebral Organoids be Used in the Lab?

A detailed proposal of "moral limits of brain organoid research" was made by Koplin and Savulescu.[34] In their article, they start from a precise premise to introduce a question that implies a potentially positive answer. "It is plausible that 'mature' whole brain organoid could one day attain sentience, and perhaps even higher cognitive abilities. Should we place any restrictions on this area of research, given that potential?"[35] The authors in fact share the idea that some restrictions should be introduced, especially in the case of HCOs acquiring some degree of moral status once they develop sensitivity to pain, consciousness, or self-awareness.

---

[30] Hostiuc et al. (2019)

[31] Lavazza and Pizzetti (2020).

[32] Ibidem.

[33] Hostiuc et al. (2019).

[34] Koplin and Savulescu (2019).

[35] Ibidem.

This hypothesis is taken very seriously by Koplin and Savulescu, unlike other ethicists and most scientists. For this reason, they state that moral limits should be established on organoid research before the threshold of consciousness is crossed. That limit, according to the authors, should be established just before the rise of phenomenal consciousness, that is, when an entity can have experiences and when there is something that it is like to be a cerebral organoid. At that point, HCOs could experience suffering (if not pain) and should therefore be given at least partial moral status.

The general principle is that "we can treat brain organoids according to existing regulatory frameworks for stem cells research until the point at which organoids develop consciousness, but we should restrict the kind of research that can take place beyond that point."[36] Even in this case, it should first be defined from the viewpoint of empirical assessment whether HCOs have or will have feelings or consciousness and then establish the moral consequences of this condition.

The authors suggest two methods for determining whether a cerebral organoid can be conscious. The first is to consider its morphology. It is believed that a human fetus begins to develop some form of awareness at 25 weeks of gestation and the ability to feel pain at 30 weeks. So, if a HCO had a structure very similar to that of the brain of a 20-week-old fetus, it should be precautionally treated as if it could have some degree of consciousness. The second method involves the measurement of the physical processes involved in consciousness, ranging from electroencephalography to the type of activation that brain organoids could have after stimulation with TMS (a proposal made by Lavazza and Massimini),[37] which evaluates the type of response by comparing it with the electrical activity recorded by unquestionably conscious or unconscious subjects, such as healthy adults and patients under anesthesia.

Koplin and Savulescu state that it is preferable to "err on the side of generosity" and treat cerebral organoids as if it were likely that they have some rudimentary form of consciousness and therefore moral status, even if we are not sure about it. In this way, beyond the limit set previously, experimentation with HCOs would no longer be an experimentation with human tissue, but with entities that have their own interests.[38] According to the authors, however, having interests does not imply an absolute ethical prohibition of their use for scientific and medical purposes. In fact, these moral constraints are taken to hold only for humans, that is, entities capable of complex cognitive abilities and significant social relationships, and cerebral organoids do not seem to be able to reach that threshold.

Not even consciousness, as such, implies an absolute right not to be harmed in any way, according to the authors. In fact, think of animals, which have varying degrees of sentience and awareness, yet are legitimately (at least for many people) used in experiments in order to increase human well-being. However, as sensitivity for animal welfare has increased so to suggest using ethical principles in

[36] Ibidem.

[37] Lavazza and Massimini (2018a).

[38] Koplin and Savulescu (2019).

experiments that involve them, such as the three Rs—Reduce, Refine, Replace—the same could be done for HCOs. In this sense, the number of conscious brain organoids used in the laboratory should be reduced; their suffering should be minimized as far as possible; and conscious organoids should be replaced whenever an alternative is available.

However, the latter indication may appear contradictory as HCOs are developed precisely as the best brain model available today and are considered an ethically preferable alternative to the use of live animals. Furthermore, the three Rs firstly proposed by Russell and Burch do not protect single entities but limit themselves to reducing as much as possible the general suffering inflicted on animals and, in Koplin's and Savulescu's proposal, on conscious brain organoids, without worrying about any suffering caused to a specific individual entity that might have value as unique.

The authors also set a comparison with late-term abortion, in which the fetus is very likely conscious already. Even in that case, the laws that allow late-term abortion have as a premise the fact that the fetus is not a person and therefore does not have the same right to the protection of its life to which a person is entitled. In any case, Koplin and Savulescu believe that some limitations should be introduced in human cerebral organoid research in cases where they could potentially be endowed with consciousness.

> Research should proceed only if (a) the research serves a sufficiently important purpose to outweigh the expected costs, including harms to the organoids themselves, (b) the research cannot be conducted using non-conscious organoids or other non-sentient material, (c) researchers use the minimum number of organoids than is required to answer the research question, (d) the organoids used do not have a higher potential capacity for suffering than is necessary to achieve the scientific objectives of the research, (e) the research is designed to minimize possible suffering, and (f) the research would not inflict severe long-term suffering, unless necessary to achieve some critically important purpose.[39]

But if human cerebral organoids, developed to a higher level, are acquiring advanced cognitive abilities through interaction with the external environment, they would have greater morally relevant interests and greater social and emotional needs. Furthermore, any degree of self-consciousness would give them greater moral significance and would impose greater restrictions on their use in experiments. Their moral status would be even greater since cerebral organoids with advanced cognitive development, as it happens with other entities, are taken to have a higher degree of moral status than HOCS with only a rudimentary form of sentience (see what was previously said about the position of Hostiuc and colleagues). In addition, the interests of cerebral organoids capable of self-consciousness and sociality should outweigh the interests of cerebral organoids at a lower cognitive stage. Based on these considerations,

> [r]esearch with cognitively advanced brain organoids should therefore face a further set of research limits. Specifically, research with advanced brain organoids should proceed only if (a) they are screened for cognitive capacities they could plausibly develop, (b) any

[39] Ibidem.

associated welfare requirements are taken into account, (c) brain organoids' cognitive capacities are not more sophisticated than is necessary to achieve the goals of the research, and (d) the research serves a sufficiently important purpose to outweigh the harms to the organoids themselves, taking into account these organoids' (potentially enhanced) degree of moral status.[40]

This consequentialist perspective introduces interesting insights and allows for quite precise criteria but, in my view, does not seem fully satisfactory in that it continues to deem it legitimate to use human cerebral organoids that are self-conscious and capable of high-level cognitive performance. As I will propose below, it might be reasonable to introduce a hybrid view of grounds HCOs' moral status is based on, where a threshold sets the level above which a deontological perspective could be adopted.

4.2.5 Consciousness as an Empirical and Moral Problem

As to the assessment of consciousness in HCOs, very different consequences are drawn by Sawai and colleagues and by Cheshire.[41] Sawai's group ranks among those who believe it is highly unlikely that cerebral organoids could develop consciousness. But even in the remote case they did, the authors distinguish between different types of consciousness which, according to them, have different moral value. Only self-awareness (which "entails having a concept of the self and being able to use this concept when thinking about the self") is considered morally relevant, but it is believed to be possessed only by adult humans and nonhuman primates.[42] Phenomenal consciousness, the ability to experience sensations, mainly of pleasure and pain, is instead considered by Sawai, based on Levy,[43] less deserving of moral consideration or even not ethically relevant. In this regard, the authors suggest comparing HCOs possessing "morally relevant consciousness" and animals with the same type of consciousness or humans in vegetative states.

If the discussion is not particularly thorough from the philosophical viewpoint, from the scientific viewpoint, some considerations by Sawai and colleagues are of high interest.[44] First, they highlight that cerebral organoids are likely to become more similar to human adult brains thanks to the possibility of fusing different types of brain organoids together. In fact, by fusing cerebral and thalamic organoids, it could be possible to obtain the transmission of sensory information through thalamocortical projections. And by fusing these assembloids with a dorsal spinal cord and peripheral nerves, one could grow cerebral organoids that would have somatic sensory experience. And, again, by combining these brain assembloids with neural

[40] Ibidem.

[41] Sawai et al. (2019); Cheshire (2020).

[42] Sawai et al. (2019).

[43] Levy (2014).

[44] Sawai et al. (2019).

retina and optic nerve issues, these HCOs could perceive light. "Should we create such organoids if we had the chance?" Sawai and colleagues ask.

On the other hand, they also emphasize that we should be cautious in drawing inferences on the presence of consciousness starting from the neural activity that can be detected. "There may be 'sub-personal correlates of pain and pleasure' that are responsive to stimulation in brain organoids—an organism may have neural activity corresponding to pain, without having any experience of it feeling like anything."[45]

In another more recent paper, Sawai and a large group of scientists and philosophers suggest "the neeed for an international framework for research and application of brain organoids."[46] But this recommendation is based on a quite permissive ethical framework concerning cerebral organoid research.

Contrary to what has been said so far, Cheshire develops a very different kind of reflection.[47] He considers the value of consciousness and outlines three moral scenarios, each with pros and cons. The first involves categorizing conscious or thoughtful cerebral organoids as potential members of the human community, based on the analogy with embryos, who are considered nascent human beings. But on closer examination, according to the author, this analogy is quite imperfect, given that HCOs cannot become a complete organism or give life to other similar entities. Therefore, human brain organoids should not be equated with humans deserving moral status.

The second option is to group cerebral organoids together with types of organoids derived from human organs, such as artificial hearts, kidneys, and lungs. In this case, there would be no specific moral protection for parts of a specialized tissue of human origin. However, Cheshire acknowledges that HCOs would be different from other organoids should they reach the threshold of consciousness and therefore deserve a particular consideration.

The third option is to consider cerebral organoids as morally special based on the unique status of the human brain, at least after the developmental stage in which a HCO may have the ability to process information as the human brain typically does when it starts thinking. According to Cheshire, we can respond to this scenario with the idea that thinking brain organoids are instead endowed with little or no sentience, that is, they are more similar to a computer, to which, at least for now, we do not attribute moral status.

As for the comparison between the potential consciousness of a cerebral organoid and a human being in a vegetative state or in a severely altered state of consciousness, Cheshire interestingly notes that "incomplete consciousness in dissolution might look very different than incomplete consciousness in development." Indeed, the residual consciousness of adult humans who have had complex cognitive abilities and a life full of relationships will be largely influenced by these experiences, which are lacking in a HCO. In this sense, the tools and techniques for

[45] Sawai et al. (2019).

[46] Sawai et al. (2021).

[47] Cheshire (2020).

ascertaining the presence of consciousness in humans when altered states of consciousness occur may not work for brain organoids.

Furthermore, according to the author, brain organoids are parts and not entities: they are not whole organisms because they cannot live independently. But to this objection, as mentioned above, Hostiuc and colleagues have responded convincingly. Then, assuming two philosophical paradigms about consciousness, Cheshire underlines how physicalists have a bottom-up approach and may find it difficult to identify and define the threshold of the appearance of consciousness, especially if the development of consciousness precedes the moment in which we are able to detect it. For essentialists, or those who believe that mind and brain coexist from the very beginning of development, the difficulty could lie in identifying the starting moment when the entity becomes conscious. However, it should be noted that for the two afore-mentioned approaches, "consciousness" probably means something very different even though its phenomenal manifestation could be the same.

From the point of view of the social perspective, Cheshire claims that the way in which we consider consciousness in adult human beings influences the way we value the potential consciousness of an HCO, and vice versa the reflections that can be developed about the consciousness of cerebral organoids have the potential to influence the way in which we evaluate altered states of consciousness in humans.

In conclusion, to show the difficulty of the problem, Cheshire acknowledges that "a complete ethical assessment recognizes on independent grounds that intentionally creating entities that are known to be self-conscious only to use and then destroy them would be *prima facie* morally wrong." Indeed, we could realize that we have created and destroyed conscious human brain organoids only after a certain time: therefore, the ethical decisions to be made cannot wait for a definitive scientific answer to be found.

4.2.6 Applying the Precautionary Principle

Like Cheshire, Birch and Browning call for a strong precautionary principle.[48] Concerning HCOs, their premise is in fact that "we should not allow that our uncertainty about their sentience to block the adoption of proportionate measures to safeguard their welfare." The authors reject the idea, adopted for example by Koplin and Savulescu,[49] of using estimates about human fetuses as tentative benchmarks for the protection of brain organoids. According to Birch and Browning, the process should be reversed: we should not rely on known markers but look for markers of sentience specific to HCOs and "draw conclusions about how small an organoid can be and yet still display these markers."[50] The problem is that all the sentience markers available, for example in animal research, are behavioral.

[48] Birch and Browning (2021). See also Niikawa et al. (2022).

[49] Koplin and Savulescu (2019).

[50] Birch and Browning (2021).

Therefore, one should consider the available theories of consciousness and the neuronal correlates they propose. However, no theory seems uncontroversial today. The proposal put forward by Birch and Browning is therefore as follows:

> *If an organoid contains structure or mechanisms that any serious and credible theory of the human NCCs (neural correlates of conscious experience) posits to be sufficient for conscious experience, we should take proportionate measures to regulate research on that organoid.* [...] Precautionary thinking requires us to take seriously theories of consciousness that can't be ruled out on the basis of current evidence, even if they don't command strong positive evidential support. [...] The most obvious precaution is that, in these circumstances, the organoids should be brought within the regulatory frameworks that currently exist in many countries for scientific research on sentient animals.[51]

According to authors, this does not mean that research should be stopped altogether. Once regulation has been introduced, one can weigh costs against benefits, also considering the overall well-being that society could obtain from experimenting on HCOs. The purpose seems to be primarily to stop the gratuitous use of potential sentient HCOs when their use is not essential. Birch and Browning conclude with a second precautionary suggestion: "when evaluating the harms and benefits of research on human brain surrogates, we should recognize our own ignorance regarding their welfare needs and take into account the risk of unforeseen harm that results from this ignorance."

As is evident, this is a proposal for a prospective strong ethical protection, but always rooted in a situational costs/benefits calculation. In fact, preventive protection is invoked for potentially sentient beings but is followed by an unspecified comparison between pros and cons that could lead to ignoring the moral status of human cerebral organoids endowed with a high degree of consciousness and cognitive capacities.

Żuradzki is cautious about the use of the precautionary principle instead.[52] In particular, he considers the factual and normative uncertainty that still surrounds the question of brain surrogates and introduces the subject of two possible types of errors: under-attribution and over-attribution of moral status to human brain organoids. The author believes that both science and ethics have no elements to state whether there is some form of consciousness in HCOs and whether, should this be the case, it would have certain moral consequences.

In this sense, the precautionary principle appears unbalanced in the face of the lack of clear clues in either direction. When decision-makers need to choose between under-attribution and over-attribution, there is no clear indication that the former is preferable. In fact, it is a question of comparing a scenario in which damage to HCOs results in social benefits with a scenario in which, on the other hand, no damage is caused to brain organoids, but no benefit is obtained insofar as research on them is prohibited. In Żuradzki's opinion, it is not obvious why experimentation

---

[51] Ibidem.

[52] Żuradzki (2021).

should be stopped in situations "where there is even the slightest possibility of under-attribution."

4.3 The Comparison with AI

In analogy with the research that is being carried out at the intersection between biology and robotics, a test for the evaluation of the cognitive stages reached by organoids—including a degree of apparent intentionality that can also be a proxy of awareness and not only instinctive and automatic behaviors—could be exemplified by OpenWorm, an open-source project dedicated to creating the first virtual organism in a computer. "The OpenWorm project has mapped the connections between the worm's 302 neurons and simulated them in software. (The project's ultimate goal is to completely simulate *C. elegans* as a virtual organism.) Recently, they put that software program in a simple Lego robot. The worm's body parts, and neural networks now have LegoBot equivalents: The worm's nose neurons were replaced by a sonar sensor on the robot. The motor neurons running down both sides of the worm now correspond to motors on the left and right of the robot."[53]

A video made by the research group shows how the robot thus constructed is actually controlled by the brain of *C. elegans* uploaded into the software of the Legobot. When stimulating the nose, the robot stops. When the front and rear sensors are touched, the robot moves in the two directions toward which it is stimulated. When the food sensor is stimulated, the robot moves forward.

C. elegans has only 302 neurons, and the simulation is still approximate, but this does not mean that it is not possible in principle to use a similar method to test the potential performance of human cerebral organoids or parts of them. While it is not surprising that even a rudimentary reproduction of the fully developed brain of an animal shows reactive and purposeful behaviors, it is not so obvious that this should happen with HCOs. In this sense, in addition to cognitive performance, simulation could also extend to rudimentary aspects of sentience. For example, a heat and cold sensor could elicit approach or avoidance behaviors in *C. elegans* and, hypothetically, could do the same for a simulated cerebral organoid in the robot.

Again, this test would not be the definitive evidence that a HCO can have sensory experience. In fact, if *C. elegans* avoids very hot surfaces, we do not infer that it feels anything about being suddenly in a very hot environment. The same, as noted by Sawai and colleagues, could happen with cerebral organoids, potentially capable of sentience at this level but not properly aware. However, while we believe that insects generally do not feel pain based on much behavioral evidence, the same is not the case for humans, even at an incomplete stage of nervous system development. Therefore, by simulating the reactions of a cerebral organoid in a robot, we could have a remarkable advance in the (inferential) understanding of its capacity for experience and, consequently, the possibility of attributing moral status to it.

[53] Fessenden (2014).

It may also be useful to consider the ethical discussion surrounding the hypothesis that artificial intelligence may become conscious. We know that today we have super intelligent computers, capable of computational performance enormously superior to that of the human brain in terms of speed and quantity of data processed. However, it does not seem that current computers are conscious, and the same can be said about devices that can acquire and process information about the environment as the human nervous system does through the senses. It is not so important here to consider how we could detect potential consciousness in artificial intelligence—almost all the proposals available, starting from the well-known Turing Test, are based on verbal interaction. Rather, it is interesting to understand what ethical consequences could derive from an artificial intelligence becoming conscious in a way that we recognize.

Schneider recalls that we are already talking about robot rights.[54] In particular, she argues that if an artificial intelligence company were to commercialize a conscious system intended for the usual uses of human support it could be accused of "robotic slavery," and its use could be banned to protect that entity. Schneider believes that stringent rules would be introduced if computers or robots became manifestly conscious. And she does not rule out that terminating a conscious digital system, reducing its conscious abilities, or modifying an artificial intelligence system so that consciousness is reduced or canceled may constitute criminal actions. And, she points out, "rightly" so.

In this sense, researchers and manufacturers could and should avoid attempting to build artificial intelligence systems that can develop a consciousness. Although it is not explicitly stated, this would be motivated by the desire not to cause suffering to artificial intelligence systems. Only awareness, in fact, seems to be able to give a digital system the desire to remain "active" and continue to experience what it is experiencing. In philosophical terms, a conscious AI would resemble a cognitively unimpaired human adult, since it is obviously endowed of high cognitive capacities and, in case of a self-aware device or program, it would also be endowed with self-thought capacities and therefore morally significant interests.

This, however, conflicts in part with how we evaluate phenomenal consciousness in the human being. Shepherd has disputed the moral significance and the intrinsic value of consciousness at all, especially when, as might be the case with a cerebral organoid, it leads a disembodied life, without interpersonal and social connections, and cannot develop its own beliefs and desires.[55] "Self-consciousness is not on its own important for moral significance. Instead, its significance emerges along with a suite of psychological capacities that enable high-level cognitive sophistication: features like cognitive and attentional control and the coordination of perception, imagination, memory, and so on."[56]

---

[54] Schneider (2019), Chap. 3.

[55] Shepherd (2017, 2018); Cf. also Lee (2018); Kahane and Savulescu (2009).

[56] Shepherd and Levy (2020). Kriegel (2019) is an advocate of the value and key role of phenomenal consciousness for the human beings and life.

> In fact, it cannot be ruled out that those who devalue the relevance of consciousness tend to focus on intelligence, that is, the functional capacity of the subject, his/her ability to perform more or less codified tasks. Consciousness qua basic feeling of existence, as a background that qualifies all our waking states, seems to be exhibited by at least some living species and, as far as we know, especially by human beings. This is a feature that cannot be replicated or simulated in artifacts, which instead, in the form of software, can exhibit an intelligence superior to that of human beings. This is demonstrated by the ability of computers to defeat humans in chess and in GO, namely complex games in which human intelligence was believed to be insuperable. These examples show that appreciation for highly developed forms of intelligence also favors the illusion of seeing consciousness where there is none (as in some types of software, e.g., the one in the film *Her*, with which the protagonist falls in love) and not seeing consciousness where instead it exists (as in non-responsive individuals or, in the future, in cerebral organoids.[57]

Phenomenal consciousness amounts to the effect it causes to be a specific entity (with the variations relative to the internal and external conditions in which one finds oneself). And phenomenal consciousness seems to be the condition for the appreciation of every pleasure even at a higher level and the root of the sense of humanity. It is indeed true that an entity capable of high cognitive performance would be able to evaluate situations related to "critical interests" as preferable even if it is unable to have a direct acquaintance of "experiential interests," to use the distinction introduced by Dworkin.[58] However, it seems that the experience of pleasure and pain and the capacity for empathy based on the possibility of "reliving" the pleasure and pain of others are at the basis of the moral sense, although when this ability is not integrated with the cognitive component of judgment, it can derail from its best application.

Following this analogy, a sentient or even fully conscious computer could seem prima facie to deserve ethical protection or, in any case, not being treated as a simple artifact. The reverse ethical analogy, however, could lead us to see that only rational adult human beings have full moral status that protects them from any instrumental use. The hypothesis of expanding the moral space to new entities with specific properties could be a way to overcome this stalemate, and this is what I will try to propose in the next sections of the chapter.

---

[57] Lavazza and Massimini (2018b).

[58] "Experiential interests are, roughly, interests in having desirable felt experiences, such as enjoyment (and in avoiding undesirable experiences, such as boredom). These interests are indeed tied to the present [...]. By contrast, critical interests are not tied to the experience of their satisfaction; these are interests in having what one values or cares about become a reality, such as a parent's interest in the success and prosperity of his child or a sailor's interest in preserving his beautiful wooden boat" (Dworkin 1993, pp. 201–08).

4.4 What Is the Moral Status, If Any, of Human Cerebral Organoids?

Elsewhere, I have proposed that human cerebral organoids can acquire moral status (although likely not full moral status) once their minimum degree of consciousness is ascertained.[59] In this vein, I also stated that it does not seem ethically permissible to use human cerebral organoids as pure means if they are sentient or endowed with some rudimentary form of consciousness.[60] If it is true that current biotechnologies seem to "blur the legal distinction between human beings and other living organisms, between living human beings and dead ones, and between human tissues and cells and nonhuman ones,"[61] then sentient HCOs would "substantially" ("which is not measurable by percentages or similar specific tests but will be a judgment call") fall within the category of "human,"[62] with the resulting corollary of rights.

The point is that using an entity for an experiment, whatever the purpose of this experiment, means using this entity as a means.[63] A highly regarded ethical perspective, such as the deontological one of Kantian inspiration, establishes as a rule that human beings cannot be used as means. But when we talk about conscious HCOs, we are moving into new and unexplored ground and, probably, we should be careful with analogies. First, for the reasons I reviewed in Sect. 4.2, we cannot establish an equation without further specification between conscious human beings and HCOs (and it should still be ascertained to what extent the latter are conscious). Even the analogy between HCOs with embryos can in fact be questioned. And, in any case, many ethicists agree to allow experiments on embryos for up to 14 days, and even beyond, if the aim is to seek medical treatment that could be of great benefit to many people. And this claim is based on the difference in moral status between embryos and adult humans.

On the other hand, we know that human beings who may not have full moral status under restrictive criteria of personhood, like individuals in a vegetative state, are still not used as means. Indeed, it is generally not taken to be morally acceptable to use them as guinea pigs for experiments, although for some philosophers and bioethicists, it is legitimate to end their existence in their best interest, for them not to suffer anymore. If it is true that most if not all legislations recognize legal status—and therefore moral status—to human beings who are not able to experience

---

[59] Lavazza (2020, 2021).

[60] One can ask if I am here considering the "capability" or the "potentiality" to feel something (or to experience suffering) in order to provide grounds for the argument. My point is that in the case of HCOs the difference is small. If an entity has the capability to suffer and we do not want it to suffer, we should not do anything that can hurt it. The "potentiality" if related to our ignorance of the real inner functioning of HCOs. But if we reasonably suspect that an HCO can have some form of consciousness, then we can draw a legitimate inference that it can suffer as most living beings.

[61] Knoppers and Greely (2019).

[62] Ibidem.

[63] The insistence on the experimental use of cerebral organoids is, of course, dictated by the fact that HCOs have so far been designed and grown for the sole purpose of being models of the brain, which can be utilized as a living laboratory.

pain, including individuals in a vegetative state, this is precisely because they were conscious persons (and could become conscious again).[64]

It is also possible to consider the current use of animals in the laboratory and the ethical arguments put forward in the ethical discourse on animal welfare. Nonhuman sentient animals, perhaps endowed with rudimentary forms of consciousness, are still used for experiments, as they are not granted full moral status, although today this is generally done only when strictly necessary, while trying to minimize their suffering. The fact remains, however, that pragmatic speciesism is implicit in these practices. Nonhuman animals are used as means for research aimed at the welfare of human beings—and this research is as such justified—but the reverse never occurs, and human beings are never used as means for the purpose of animal welfare. If there is a justification for these practices, it amounts to the fact that the human species is considered to have a higher moral status than nonhuman animals, which can thus serve as means. It still happens even though, ideally, a consistency principle should induce us to use general criteria for attributing moral status that would apply to all the entities that meet the chosen criteria, without differentiating among species.

It is therefore clear that different metaethical views and different grounds for moral status are at stake here, and probably not all of them can be used together consistently. On the one hand, the criterion of sentience and interests refers to a generically utilitarian perspective, which prescribes to minimize suffering and maximize pleasure, both broadly understood. This same ethical perspective implies consequentialist criteria on the use of the entities considered. For example, although minimal sentience alone can be the basis for the attribution of a certain moral status to HCOs, this moral status will not prevent the use of brain organoids in the lab if such use is useful for relieving, even potentially, other entities with higher moral status, such as sick adult humans.

On the other hand, if we want to escape these effects of utilitarian consequentialism, we can resort to an ethical approach, as mentioned above, which can establish universal principles without exceptions, such as that of not using an entity endowed with moral value as a means. However, the main ethical theories that refer to this ethical approach, starting from the Kantian one, consider an entity worthy of full moral consideration based on specific attributes of the human being, such as the capacity for autonomy and rationality. These are characteristics that evidently not even cerebral organoids endowed with some form of consciousness can aspire to possess.

Elsewhere, I relied on Audi's interpretation of the Kantian proviso quoted above to argue that we should not inflict pain on conscious brain organoids.[65] This is a very stringent requirement that ambitiously attempts to use the idea of sentience as a basis for the attribution of moral status and then extend the basic moral protection of the deontological approach to this entity.

---

[64] Cf. Owen (2017).

[65] Lavazza (2020).

Many scientists and scholars may therefore believe that this is a too demanding position from their perspective. In the first place, for most scientists—at least according to the opinions of those who have taken a stand on the subject—brain organoids are not and will not be conscious even in the future. For most ethicists who have dealt with HCOs, the latter could be given moral status if they manifested a certain degree of consciousness, but that would not *ipso facto* mean conferring on them the rights we usually attribute to a human being, since there would be a great difference in degree among the two. Other philosophers endorsing a threshold conception of moral status, as opposed to a scalar one, would deny moral status to an entity only endowed with a rudimentary form of consciousness and not capable of expressing judgments of value.

In this sense, I think that personal moral intuitions can also play a role, since, as in the case of embryos, the human origin of cerebral organoids can deserve greater or lesser consideration; the same can be said of the fact of not being able to easily discriminate, not even at a theoretical level, what it means, what it feels like, to have a basic form of sentience or a rudimentary form of consciousness.

To move ethical reflection forward, I propose the idea that human cerebral organoids can be at least provisionally considered an entity of a new kind both from the ontological and moral viewpoint, due to their seemingly unique characteristics in the biological world (at least for now) and to the current and most common grounds for granting moral status. In this way, it will be possible to try to set the constraints to be introduced on biomedical research with HCOs.

4.5 A New Biological (and Moral) Entity?

Famously, 40 years ago, Putnam proposed the brain-in-a-vat thought experiment, but the idea that a human head could live detached from its body was not new in philosophy.[66] The plausibility of that thought experiment has paradoxically declined in the most recent decades, when biomedical knowledge and the available technology could actually make such a thing scientifically feasible in the laboratory. What we have begun to understand better is that our cognition, and presumably also our consciousness, depends heavily on interaction with the body and the external environment, and philosophical internalism, according to which meanings are formed inside the head, has suffered severe blows. From the cognitive point of view, human beings therefore seem to be in constant dialogue with their environment and, indeed, outgoing toward it, to the point of incorporating the latter into what has been called the extended mind.[67]

This premise serves to understand the skepticism toward the potential consciousness of brain organoids grown in a dish. However, science has been more imaginative than philosophy in this respect. Indeed, it is no longer a question of considering a head that loses its body and can no longer relate to its environment (or has to resort

[66] Putnam (1981).

[67] Clark and Chalmers (1998).

to environmental simulation, as Putnam hypothesized). Instead, it is a matter of a brain that grows in the laboratory in a very similar way as it would develop from the union of two gametes in human reproduction.

Can such a brain come to resemble an adult human one? Most probably not, because it will always lack a body and an external environment with which to interact. However, we have considered that a glimpse of consciousness *could* arise in it, on the grounds that spontaneous and coordinated electrical activity has already been recorded in very small organoids, and this kind of activity includes the neuronal correlate of consciousness.[68] Larger and more complex brain organoids could therefore reasonably develop sentience or an increasing degree of consciousness.

What kind of entity is an (at least partially) conscious human brain that is confined in an ad hoc apparatus inside a laboratory? It is not a human person, of course. One may wonder, however, whether it has the potential to become one as an assembloid equipped with sensory terminals, though not with behavioral effectors. Nor can it be ruled out that special brain–machine interfaces one day might enable a cerebral organoid to perform some kinds of actions in the external environment.

If it is not a person, the cerebral organoid endowed with a degree of consciousness (which may not be fully ascertainable and quantifiable, although we cannot exclude technical progress on this front) could be a *new type of entity* belonging to the human species. Such an entity would certainly be biologically unique, given both its chromosomal equipment and its specific cerebral architecture (which has individualized aspects due to the peculiar development of each organoid, even if we are striving to make them more and more homogeneous for scientific purposes), which affects the general functioning of the cerebral organoid itself.

These entities would presumably have a much poorer and more limited life than (almost) any human being, although they might experience pain or pleasure in ways that are unknown to us. The fact that they have been made to become increasingly brain-like for research purposes also puts them in a position to experience suffering and deprivation as well as to approach stages of higher consciousness, the attainment of which would, however, always be linked to human intervention (e.g., by attaching the organoid to a sensor to make it perceive the outside world).

It thus seems like we have never been ethically confronted with (only or almost exclusively) passive (i.e., incapable of proactive action) entities of this kind, which share with us humans their genetic make-up and a rudimentary basic sentience, but which are deliberately created by us for instrumental purposes. Can HCOs be compared to infants with very severe neurological disorders who retain minimal sentience and awareness? Apparently not, neither from a biological nor from a moral point of view, because, for example, brain organoids have no kinship nor they have parents developing emotional bonds toward them.

Yet, there seems to be a moral intuition that we cannot overlook: the attitude that promotes widening the moral circle to new forms of life and entities, including, for example, conscious machines. This attitude is based on tolerating and caring for

[68] Neural correlates of consciousness are probably a tiny part of the whole activity of the brain, but so far we do not have an idea of which part is the most fundamental.

different forms of sentience and embraces a moral perspective that does not privilege intelligence and efficiency but extends this welcoming outlook to the world. This does not mean, however, that every pea-sized cerebral organoid, grown in a laboratory among hundreds of others, is a fellow human being. The possibility of establishing some form of relationship, even a one-way relationship, with the entity in question is in fact one of the elements underlying the moral intuition we are talking about.

Consider also chimeras made up of a nonhuman animal and a human cerebral organoid, which develops and links up in the brain of the host. Should consciousness emerge of a higher order than that of the animal host, one might ask where such consciousness would reside, and also "whose" consciousness would that be. The idea of new entities with a special moral status is also evident in examples such as these. As Cheshire points out,[69] ethical thought has been more concerned with the humanization of nonhuman animals by inserting human neural cells in them than with the prospect of having a human consciousness, as it were, imprisoned within an animal body and brain.[70]

If brain organoids seem to escape clear-cut classification among existing entities, it could be argued that once they have acquired some form of consciousness, they would acquire a moral status of their own, not comparable to that of other entities, due to our (at least partial) ignorance of their nonmanifest characteristics and a reasonably applied recautionary principle.

4.6 Conclusion: A Tentative Proposal

An overview of the existing positions on how we should treat human cerebral organoids and an articulate discussion of the subject has led me to formulate a specific proposal, namely that we should grant HCOs, at least provisionally, the ontological and ethical status of entities of a special kind, not clearly comparable to that of other known entities. This proposal is essentially due to our current ignorance about potential development of human cerebral organoids and does not mean that HCOs have or might have greater moral status than what we are granting them now.

In this sense, given that HCOs cannot develop spontaneously, a prescription that might derive from the moral scenario described so far could be as follows. Human brain organoids should be grown for all biomedical research purposes without developing any form of sentience or consciousness. Should we have any doubt that we have unintentionally crossed this threshold, or should we wish to grow increasingly conscious organoids, we should only do so with the aim of making their existence as comfortable as possible, i.e., applying to them the bioethical principles of nonmaleficence, beneficence, and justice.

[69] Cheshire (2020).

[70] Koplin (2021).

This proposal obviously needs to be discussed and compared with other perspectives that will be put forward. And much like the other proposals I have examined above, it does not appear to be exempt from limitations.

Think of cerebral organoids in the context of personalized medicine. If a sick individual makes sure a highly developed HCO is created from his own cells (a organoid developed in a way that it has consciousness) and subsequently uses it for his treatment, might it be said that this HCO is simply a "detached" part of the individual? Or, more likely, is the brain organoid a different, autonomous entity that should not be damaged? Might this HCO be considered a kind of partial clone? Probably not, because the brain will not be the same (despite being compatible for transplantation), and in any case, clones are morally even more problematic than brain organoids. But might we compare this to a father killing his son in order to use the latter's organ? It is not a question here of assessing which entity, the sick adult or the conscious brain organoid, has a moral status that gives it greater value.

One can certainly state that the sick individual has more value. Indeed, an ethically justified order can be established between entities with full moral status in situations where a choice simply needs to be made. If we have only one respirator for two very seriously ill patients with pneumonia, all other things being equal, we will give it to the 30-year-old patient and not the 80-year-old; if a mother and her fetus are in mortal danger, we will save the mother; we will pick the doctor and not the deejay for the last available seat on a lifeboat. However, these are cases where the sacrificed would otherwise have a moral right to be saved. Might the same be said for a conscious cerebral organoid?

My answer is that we do not know, and for this reason, I propose to classify HCOs as *sui generis* entities to be granted increasing protection should some form of sentience become probable and special protection should consciousness (and cognitive capacities) be ascertained. This can be stated on the basis that HCOs are of human origin, have a human-like consciousness but have a unique biological and existential condition that is different from that of human beings.

The key point here seems to be that human brain organoids have unique characteristics as living entities. According to the current state of knowledge, we could say that the HCOs grown today do *not* have the minimum level of sentience that would enable them to acquire interests of their own. Therefore, we would have to say that they are chunks of living matter to which no moral status can be attributed. Indeed, present-day cerebral organoids are below the minimum threshold of moral status because they do not even seem to be able to develop a degree of sentience that would give them a first level of moral status (leaving aside the debate about potential capacity as an inclusive criterion for attributing moral status).

HCOs grown with new techniques or larger assembloids capable of developing complex electrical activity might acquire sentience, in which case it would seem reasonable to assign them a first degree of moral status from a utilitarian perspective, based on their ability to experience pleasure or pain. This first level of moral status involves the recognition of a relative value to HCOs, which would then be given rights *not* comparable to those of other entities, e.g., ill human adults with full moral status who might benefit from a brain organoid transplant.

However, the specific feature of brain organoids as living entities is that they can not only grow and develop individually, but they could also evolve as a *type of entity* thanks to the advancement of biomedical technologies. In the future, indeed, we may have (a) millimeter-sized HCOs with no possibility of further growth, perhaps endowed with basic sentience; (b) millimeter-sized HCOs able to grow considerably—for example, by grafting cells that allow their vascularization—which consequently might become large enough and differentiated enough to have a rudimentary form of consciousness; and (c) HCOs destined under normal laboratory conditions to become brain-like entities in a vat endowed with a more developed form of consciousness and cognitive abilities—although not comparable to those of human adults who have capacity for moral personality.[71]

Faced with this scenario, it would not be unreasonable to consider a hybrid ethical perspective, as suggested for example by Nozick, who proposed Kantianism for people and utilitarianism for nonhuman animals.[72] It would, therefore, be a matter of adopting hybrid or pluralist criteria for the attribution of moral status to different types of human brain organoids.[73] While it is true that probably HCOs will never attain the manifest conditions for having full moral status, HCOs of type (c) could be considered quasi-comparable to humans, to whom we attribute above-threshold moral status, and viewed as highly valuable because of their potential to be similar to individuals with rationality and intellectual autonomy (a real brain in a vat).

The point is not to endorse a substance view by which all human beings (and therefore properly developed HCOs that can be vaguely assimilated to human beings) possess "intrinsic value and moral status equivalent to that of an adult human being."[74] I rather support a multicriterial account such as the one proposed by Warren.[75] This type of approach highlights the growing importance of sentience, moral agency, and membership in human society. In this sense, we should not view human cerebral organoids as "persons" until they reach a certain presumptive threshold of (potential) development, above which Kantianism applies. Below this threshold, there is no need, nor is it possible, to calculate, for example, the time-relative interests of HCOs in the sense described by McMahan[76]; rather, one might seek to minimize their suffering and maximize their well-being, maintaining a consequentialist perspective that places the well-being and value of human beings with FMS (and possibly of nonhuman primates and other animal species) over that of human cerebral organoids.

However, there is still the problem of factually ascertaining the actual mental development of HCOs, as well as the fact that they are grown based on an external

---

[71] This situation is somewhat comparable to the hypothetical coexistence of different species of hominids in the same environment at the same time. In that case, it would be complicated to order the moral status of the different species according to the philosophical criteria we currently use.

[72] Nozick (1974); cf. Sebo (forthcoming).

[73] Floris (2020).

[74] Blackshaw (2019).

[75] Warren (1997).

[76] McMahan (2002).

and instrumental decision, currently for the sole purpose of research. Above the threshold of moral status in the Kantian dimension, such use could be considered morally unacceptable. Below that threshold, it would become the object of a moral evaluation weighing up its consequences.

Ultimately, what the protection suggested here looks like is both an empirical question and a matter of shared ethical sensitivity (the controversy over abortion shows how difficult it can be to reconcile disagreement, in terms of both data interpretation and, above all, at a moral level). The committees that have been set up with the participation of scientists, moral philosophers, and legal scholars,[77] together with the proposals made by the contributors to this volume, will undoubtedly contribute to further reflection on the subject. I think, however, that we are still in an uncharted territory, and it will take time for us to acquire better scientific knowledge and greater conceptual clarity to make sound moral judgements on this complicated issue.[78]

References

Baertschi B (2008) The question of the embryo's moral status. Bioethica Forum 1(2):76–80

Barnhart AJ, Dierickx K (2022) The many moral matters of organoid models: A systematic review of reasons. Med Health Care and Philos. https://doi.org/10.1007/s11019-022-10082-3

Bredenoord AL, Clevers H, Knoblich JA (2017) Human tissues in a dish: the research and ethical implications of organoid technology. Science 355:6322

Bayne T, Hohwy J, Owen AM (2016) Are there levels of consciousness? Trends Cogn Sci 20:405–413

Birch J, Browning H (2021) Neural organoids and the precautionary principle. Am J Bioeth 21(1):56–58

Blackshaw BP (2019) The ethics of killing: strengthening the substance view with time-relative interests. The New Bioethics 25(4):332–348

Boers SN, van Delden JJM, Bredenoord AL (2018) Organoids as hybrids: ethical implications for the exchange of human tissues. J Med Ethics 41:131–139

Chiaradia I, Lancaster MA (2020) Brain organoids for the study of human neurobiology at the interface of in and in vivo. Nat Neurosci 23(12):1496–1508

Chinaia AA, Lavazza A (2022) Cerebral organoids and biological hybrids as new entities in the moral landscape. AJOB Neurosci 13(2):117–119

Cheshire WP (2014) Miniature human brains: an ethical analysis. Ethics Med 30:7–12

Cheshire WP (2020) Cerebral organoids and the threshold of consciousness. Ethics Med 36(1):27–23

Clark A, Chalmers D (1998) The extended mind. Analysis 58(1):7–19

DeGrazia D (2008) Moral status as a matter of degree? South J Philos 46(2):181–198

Dehaene S (2014) Consciousness and the brain: deciphering how the brain codes our thoughts. Viking Press, New York

Douglas T (2021) Is moral status good for you? In: Clarke S, Zohny H, Savulescu J (eds) Rethinking moral status. Oxford University Press, New York

Dworkin R (1993) Life's dominion: an argument about abortion, euthanasia, and individual freedom. Knopf, New York

[77] Cf. Greely (2021).

[78] Cf. Lavazza (2016). See also Chinaia and Lavazza (2022).

Farahany NA, Greely HT, Hyman S, Koch C, Grady C, Pasca SP, Sestan N, Arlotta P, Bernat JL, Ting J, Lunshof JE, Iyer EPR, Hyun I, Capestany BH, Church GM, Huang H, Song H (2018) The ethics of experimenting with human brain tissue. Nature 556:429–432

Fessenden M (2014) We've put a Worm's mind in a Lego Robot's body. Smithsonian Magazine https://www.smithsonianmag.com/smart-news/weve-put-worms-mind-lego-robot-body-180953399/?no-ist. Accessed 21 Mar 2021

Floris G (2020) A pluralist account of the basis of moral status. Phil Stud. https://doi.org/10.1007/s11098-020-01513-z

Greely HT (2021) Human brain surrogates research: the onrushing ethical dilemma. Am J Bioeth 21(1):34–45

Hostiuc S, Rusu MC, Negoi I, Perlea P, Dorobanţu B, Drima E (2019) The moral status of cerebral organoids. Regen Ther 10:118–122

Hyun I (2013) Bioethics and the future of stem cell research. Cambridge University Press, Cambridge-New York

Hyun I, Scharf-Deering JC, Lunshof JE (2020) Ethical issues related to brain organoid research. Brain Res 1732:146653

Jaworska A (2017) Advance directives and substitute decision-making. In: Zalta EN (ed) The Stanford Encyclopedia of Philosophy. https://plato.stanford.edu/archives/sum2017/entries/advance-directives/

Jaworska A, Tannenbaum J (2018) The grounds of moral status. In: Zalta EN (ed) The Stanford Encyclopedia of Philosophy. https://plato.stanford.edu/archives/spr2018/entries/grounds-moral-status/

Kahane G, Savulescu J (2009) Brain damage and the moral significance of consciousness. J Med Philos 34:6–26

Kriegel U (2019) The value of consciousness. Analysis 79:503–520

Koplin JJ (2021) The moral relevance of humanization. Am J Bioeth 21(1):59–61

Koplin JJ, Savulescu J (2019) Moral limits of brain organoid research. J Law Med Ethics 47(4):760–767

Knoppers BM, Greely HT (2019) Biotechnologies nibbling at the legal "human". Science 366(6472):1455–1457

Lavazza A (2016) Neuroethics: a new framework—from bioethics to anthropology. In: Id. (ed) Frontiers in Neuroethics. Cambridge Scholar Publishing, Newcastle upon Tyne

Lavazza A (2019) What (or sometimes who) are organoids? And whose are they? J Med Ethics 45:144–145

Lavazza A (2020) Human cerebral organoids and consciousness: a double-edged sword. Monash Bioeth Rev 38(2):105–128

Lavazza A (2021) Potential ethical problems with human cerebral organoids: consciousness and moral status of future brains in a dish. Brain Res 1750:147146

Lavazza A, Massimini M (2018a) Cerebral organoids: ethical issues and consciousness assessment. J Med Ethics 44:606–610

Lavazza A, Massimini M (2018b) Cerebral organoids and consciousness: how far are we willing to go? J Med Ethics 44:613–614

Lavazza A, Pizzetti FG (2020) Human cerebral organoids as a new legal and ethical challenge. J Law Biosci. https://doi.org/10.1093/jlb/lsaa005

Lee AY (2018) Is consciousness intrinsically valuable? Philos Stud 20:1–17

Levy N (2014) The value of consciousness. J Conscious Stud 21(1–2):127–138

Lunshof JE (2021) Brain surrogates—empty or full makes the difference. Am J Bioeth 21(1):46–48

McMahan J (2002) The ethics of killing: problems at the margins of life. Oxford University Press, New York

Munsie M, Hyun I, Sugarman J (2017) Ethical issues in human organoid and gastruloid research. Development 144(6):942–945

Owen A (2017) Into the gray zone: a neuroscientist explores the border between life and death. Scribner, New York

Niikawa T, Hayashi Y, Shepherd J, Sawai T (2022) Human brain organoids and consciousness. Neuroethics 15, 5. https://doi.org/10.1007/s12152-022-09483-1
Nozick R (1974) Anarchy, state, and utopia. Basic Books, New York
Putnam H (1981) Brains in a vat. In: Id., reason, truth and history. Cambridge University Press, Cambridge, pp 1–21
Reardon S (2020) Can lab-grown brains become conscious? Nature 586(7831):658–661
Sawai T, Hayashi Y, Niikawa T, Shepherd J, Thomas E, Lee TL, Erler A, Watanabe M, Sakaguchi H (2021) Mapping the ethical issues of brain organoid research and application. AJOB Neurosci:1–14. https://doi.org/10.1080/21507740.2021.1896603
Sawai T, Sakaguchi H, Thomas E, Takahashi J, Fujita M (2019) The ethics of cerebral organoid research: being conscious of consciousness. Stem Cell Rep 13(3):440–447
Schneider S (2019) Artificial you: AI and the future of your mind. Princeton University Press, Princeton (NJ)
Sebo J (2017) Agency and moral status. J Moral Philos 14(1):1–22
Sebo J (forthcoming) Why should both accept and reject hybrid views. https://jeffsebodotnet.files.wordpress.com/2020/08/why-we-should-both-accept-and-reject-hybrid-views.pdf
Shepherd J (2017) The moral insignificance of self-consciousness. Eur J Philos 252:398–415
Shepherd J (2018) Ethical (and epistemological) issued regarding consciousness in cerebral organoids. J Med Ethics 44:611–612
Shepherd J, Levy N (2020) Consciousness and morality. In: Kriegel U (ed) The Oxford handbook of the philosophy of consciousness. Oxford University Press, New York
Trujillo CA, Gao R, Negraes PD, Gu J, Buchanan J, Preissl S, Wang A, Wu W, Haddad GG, Chaim IA, Domissy A, Vandenberghe M, Devor A, Yeo GW, Voytek B, Muotri AR (2019) Complex oscillatory waves emerging from cortical organoids model early human brain network development. Cell Stem Cell 25(4):558–569
Vrselja Z, Daniele SG, Silbereis J, Talpo F, Morozov YM, Sousa AM, Tanaka BS, Skarica M, Pletikos M, Kaur N, Zhuang ZW, Liu Z, Alkawadri R, Sinusas AJ, Latham SR, Waxman SGSestan N. (2019) Restoration of brain circulation and cellular functions hours post-mortem. Nature 568(7752):336–343
Walters, JM (2021) Moral Status, Encyclopediacom https://www.encyclopedia.com/science/encyclopedias-almanacs-transcripts-and-maps/moral-status#:~:text=Moral%20status%20is%20a%20concept,be%20treated%20with%20special%20regard.&text=That%20is%20because%20normal%20adults,highly%20regard%20their%20well%2Dbeing
Warren MA (1997) Moral status: obligations to persons and other living things. Clarendon Press, Oxford
Zuolo F (2016) Il problema dello status morale. APhEx 13. http://www.aphex.it/index.php?Temi=557D03012202740321040705777327
Żuradzki T (2021) Against the precautionary approach to moral status: the case of surrogates for living human brains. Am J Bioeth 21(1):53–56

What Is, or Should Be, the Legal Status of Brain Organoids? 5

Jochen Taupitz

5.1 Introduction

Special legal regulations for the production and use of (human) organoids in general and for brain organoids in particular do not exist in Germany. Their legal classification, like that of other human cells, tissues, and organs, has to take into account several areas of regulation: On the one hand, it is a question of the origin or the extraction of the source material,[1] then of the classification of the organoids themselves, and finally of the way in which they are (planned) to be used,[2] for example, transferring them to animals.[3] However, these questions are not unrelated to each other. Rather, in particular, the planned use already has a considerable influence on the requirements that are to be placed on the extraction of the source material. In the case of brain organoids cultivated from human somatic cells, for example, the question arises, which has been discussed for a long time, especially in connection with biobanks,[4] as to how specifically the donors of the starting cells must be informed about the later use and must consent to it.[5] If the brain organoids are later

[1] See Taupitz (2020a), 806 ff.

[2] See Taupitz (2020a), 810 ff.

[3] On the ethical and legal issues surrounding transplantation of human brain cells or brain organoids into animals, see Chen et al. (2019), 462 ff.; Hyun et al. (2020, p. 4); Schicktanz (2020), 203 ff.; Ethikrat (2011), in particular 110 ff.; National Academies (2021, p. 67 ff.); Taupitz and Weschka (2009).

[4] On this discussion, see, inter alia, Ethikrat (2010).

[5] Specifically on brain organoids Farahany et al. (2018) 431 f.; Boers et al. (2016), 939 f.; Cepelevicz (2020); Hyun et al. (2020), 2 f.; Jácomo (2020, p. 7); Schicktanz (2020, p. 198); Taupitz (2020a,

J. Taupitz (✉)
Institute for German, European and International Medical Law, Public Health Law and Bioethics of the Universities of Heidelberg and Mannheim, Mannheim, Germany
e-mail: taupitz@uni-mannheim.de

H.-G. Dederer, D. Hamburger (eds.), *Brain Organoids in Research and Therapy*, Advances in Neuroethics, https://doi.org/10.1007/978-3-030-97641-5_5

to be transferred to other people, the Transplantation Act (TPG) also applies to the *removal* of the cells.[6] The legal classification of the organoids produced may also be influenced both by the origin of their source material and by their intended use. Thus, organoids from human cells and those from animal cells are obviously to be treated differently. Furthermore, they are medicinal products if their subsequent use has a therapeutic purpose ("intended [...] for the curing, alleviating or preventing of human [...] diseases or disease symptoms," Sec. 2 (1)(1) Medicinal Products Act [AMG]). In particular, they can be advanced therapy medicinal products (ATMPs).[7] These cross-references can and shall be dealt with in this article only to a limited extent, since this article shall be limited to the status of brain organoids as such according to the editors' specifications.

The moral status of "mini-brains"[8] has already been the subject of a heated international debate, although not always with sufficient distinction between fictional empiricism-free thought experiments ("philosophical zombies"[9]) on the one hand and science-based ethical impact assessment on the other.[10] Since "there is hardly any field where imagination runs riot faster than in brain research,"[11] even Frankenstein fantasies are developed[12]; others speak of the "witch's kitchen of the new sorcerer's apprentices."[13]

5.2 Constitutional Framework

The German legal system is based on a presumption of freedom, which is expressed in numerous basic rights of the constitution, the German Basic Law (GG): From a (constitutional) legal point of view—in contrast to the starting point often chosen in ethics—the question is not what one may do, but what the state may prohibit.[14] This "freedom-based" starting point is so important because it allocates the burden of justification[15]: It is not freedom that must be justified, but the legal prohibition or restriction. This is especially true for the freedom of

p. 807); Taupitz (2020b, p. 212, 217).

[6] In contrast, the TPG is generally not applicable to the handling of those cells and their possible subsequent transfer to other people, see Taupitz (2020a) 810 f.

[7] Taupitz (2020a) 811 f.

[8] Viciano (2020); Parsch (2019b); see also Lavazza and Massimini (2018, p. 606): "brain in a vat"; Lavazza (2021, p. 6).

[9] Bitar (2020). On philosophical zombies, see Kirk (2009).

[10] Critical Schicktanz (2020) 194 f.—Lavazza and Massimini 2018, p. 606) on the other hand, speak openly of a thought experiment that is now becoming a laboratory experiment.

[11] Kuroczik (2018)

[12] Kurlemann (2013) and Goodall (2020).

[13] Müller-Jung (2013).

[14] Hufen (2001, p. 442).

[15] For this and the following Taupitz (2002) 23 f.

science guaranteed in Art. 5 (3)(1) GG, which is (and will be for a long time) of central importance for the development of brain organoids, which is still very much in its infancy: It is not the goals or paths of science that require argumentative legitimation, but their alleged illegitimacy; it is not science that has to justify its actions or omissions, but the legal system that has to justify why what science does or wants to do, or does not do or want to do, is illegitimate in concreto. Neither does science have to prove its benefit to society or the individual, nor does science have any argumentative obligation toward society. Rather, the constitutional legislator has placed science under the primacy of freedom for its own sake—and quite pragmatically because, viewed in the long run, society is better off with this than with any a priori restriction on the progress of scientific knowledge, no matter how noble the goals. In the end, this means that every restriction of the freedom of science requires a viable justification, whereby the goal can only be the protection of other goods with constitutional rank.[16]

In addition to scientific freedom, other fundamental rights must be mentioned in relation to the production and use of brain organoids, such as in particular the general right of personality (Art. 2 (1) in conjunction with Art. 1 (1) GG) and the right to life and physical integrity (Art. 2 (2)(1) GG). These fundamental rights are primarily rights of defense against state interference, in this case above all of the donors of the cells used for the organoids. However, in relation to (future) patients who may benefit from organoids in terms of health, it also follows that the state may not prohibit, e.g., therapeutic measures without sufficient reason. As stated in the introduction, these questions of obtaining the raw material needed for brain organoids will not be dealt with here, nor will the questions of the later use of the organoids. The same applies to animal welfare, which is anchored as a legal principle in the constitution (Art. 20a GG). It can be of importance both for the extraction of the starting material and for the transfer of brain organoids to animals.

The discussion of a special legal status of brain organoids undoubtedly centers on human dignity (Art. 1 (1) GG) and the right to life and physical integrity (Art. 2 (2)(1) GG). Against this background, the international debate is already discussing whether brain organoids may simply be destroyed: "Would destroying such an organism be murder?"[17] "Since the human cerebral organoid presents some neuronal activity, and therefore is not dead according to the brain death criterion used for legal subjects, destroying that organoid would imply to breach the dignity and the rights - first and foremost the right to life - of a legal entity."[18] Also related to the creation of brain organoids for research purposes, a violation of dignity is considered possible.[19]

---

[16] BVerfGE 141, 143 (169); 128, 1 (41); 122 89 (107).

[17] Goodall (2020).

[18] Lavazza and Pizzetti (2020, p. 10). For the argumentative reference to brain death, see infra at footnote 28.

[19] Bayne et al. (2020, p. 14).

5.3 The Legal Status of Brain Organoids De Lege Lata

5.3.1 Brain Organoids as Human Beings

The production of brain organoids would raise numerous fundamental legal questions if they had the same developmental capacity as human embryos. In particular, if they had consciousness[20] or even cognitive abilities,[21] they could legally be assigned human dignity (Art. 1 (1) GG) and protection of life (Art. 2 (2)(1) GG). This could be affirmed according to a view represented in the literature, if one assumes the beginning of "brain life" (approximately on the 57th day p.c.) as decisive for the full legal protection of the becoming human life[22] and a brain organoid would show comparable abilities at some point[23] (especially since, conversely, brain death is also widely regarded as a decisive caesura at the end of life[24]). Also with a comparative view of embryonic development, the so-called 14-day rule (i.e., no in vitro development of embryos beyond the 14th day after fertilization), which applies in other legal systems,[25] is often linked to the assumption[26] that on the 14th day of embryonic development, with the appearance of the primitive streak, the first signs of a developing nerve system and thus of pain sensation arise.[27] If the moral and also the legal status of embryos[28] is linked in this way with the beginning of

---

[20] On the controversial question of the relevance of (more or less) developed consciousness for the attribution of a special moral status, see Lavazza (2021, p. 6). Moreover, a problem results from the fact that there is so far no unanimous understanding of consciousness shared between neuroscience and philosophy/jurisprudence and even within sciences different concepts exist, see Schicktanz (2020) 200 f.; Bartfeld et al. (2020), 25 f.; Bayne et al. (2020, p. 13); Sharma et al. (2020) 49 f.; Singer (2019); Baars and Franklin (2007).

[21] Koplin and Savulescu (2019, p. 764).

[22] Sass (1989, pp. 160–191); Bartfeld et al. (2020p. 19); for further references, see Müller-Terpitz (2008, pp. 182–186); see further on different approaches Koplin and Savulescu (2019, p. 762).

[23] After all, researchers claim that they have already succeeded in keeping brain organoids alive for 10 months and measuring EEG signals. These are said to have reached a complexity similar to that of premature babies in the 28th week of gestationTrujillo et al. (2019). However, the comparison with brain activities of premature infants is rejected by other researchers as too broad, cf. Parsch (2019a); Cepelevicz (2020).

[24] On this argument, Lavazza and Pizzetti (2020) 8 ff.

[25] On the dissemination of the 14-day rule, see Matthews and Moralí (2020).

[26] However, according to recent findings, not functional neural connections or sensory system exist in the embryo at least until day 28, see Hurlbut et al (2017); Appleby and Bredenoord (2018, p. 2).

[27] Hostiuc et al. (2019, p. 119); Cepelevicz (2020); Matthews et al. (2021, p. 47, 49); McCully (2021, p.1); Lavazza (2021, p. 3) with further references. However, other authors deny that the 14-day rule was intended to be a line denoting the onset of moral status in human embryos. Rather, it is a public-policy tool designed to carve out a space for scientific inquiry and simultaneously show respect for the diverse views on human-embryo research: Huyn et al. (2021, p. 998); Hyun et al. (2016, p. 170); Cavaliere (2017, p. 3 f.); Chan (2018, p. 229); for further references, see Matthews et al. (2021, p. 48).

[28] The (German) Federal Constitutional Court, for example, also considers the caesura on the 14th day of development to be relevant, see BVerfGE 39, 1 ff. para 133: "Life in the sense of the historical existence of a human individual exists, according to established biological-physiological knowledge, at any rate from the 14th day after conception (nidation, individuation)."

brain development or with the beginning of the development of sensation,[29] it could be obvious to subject far-developed cerebral organoids or research with them to the same rules as those for dealing with embryos.[30] However, their use for research purposes would only be prohibited by the current Embryo Protection Act (ESchG) according to Sec. 2 (1) ESchG[31] if they can be subjected to the definition of an embryo in § 8 ESchG; this is more than doubtful, since the organoids did not develop via fertilization.[32] Their production would also be prohibited under certain circumstances[33] if they had the same genetic make-up as another embryo, a fetus, a human being, or a deceased person, because, then, a violation of Sec. 6 (1) ESchG (prohibition of cloning) could be affirmed. But even this is more than doubtful.[34]

Above all, however, the following is true: Since a human being (and thus also an embryo) cannot be reduced to single characteristics such as pain sensation or consciousness, and since brain organoids, unlike embryos, cannot develop into a complete organism[35] or even a human being,[36] a protection of the same kind as that provided for embryos in the ESchG can neither be derived from the applicable law nor is it constitutionally required.[37] This is true despite the fact that even the ESchG does not represent one-to-one what is constitutionally required with regard to the protection of embryos[38] and is true even if it should 1 day come to the creation of

---

[29] The ability to feel pain is generally considered to be a feature of consciousness, see Bitar (2020).

[30] Of this opinion is, in fact, Hostiuc et al. (2019, pp. 119–121); Bitar (2020); further references in Jácomo (2020, p. 7); see also Greely, cited by Cepelevicz (2020): "… the more human … [the brain organoid] gets, the more you're backing into the same sorts of ethics questions that are the reasons why you can't just use living humans". At the same time, not least in view of research with brain organoids, there is a call to extend the 14-day rule, e.g., to 28 days, see Appleby and Bredenoord (2018).

[31] Sec. 2 (1) ESchG prohibits the use of an embryo "for a purpose not serving its preservation".

[32] On the dispute as to whether entities not created by fertilization fall within the definition of "embryo," see Taupitz (2014a) paras. 48 ff.; Dederer (2020b), 55 ff.; Gassner and Opper (2020, p 260 f., 272 f.).

[33] Since the brain organoids are not created by fertilization, the provisions of the ESchG, which prohibit the use of oocytes for purposes other than the establishment of a pregnancy (§ 1 (1) (2), (2) ESchG), are not relevant.

[34] It is very controversial to what extent it depends on the way of production of the corresponding entity for the applicability of the cloning prohibition, see Günther (2014) paras. 3 ff.; Taupitz (2014a) paras. 48 ff.; Gassner and Opper (2020, p 260 f., 272 f.).

[35] Faltus (2021, p. 133).

[36] Schicktanz (2020, p. 200); Koplin and Savulescu (2019, p. 762). Since the potentiality argument, together with the species argument, the continuity argument, and the identity argument ("SCIP"), is widely believed to justify the special moral and legal status of the embryo (see the references in Müller-Terpitz (2008, pp. 49–65), the converse must also apply: entities that do not have this potential cannot enjoy comparable protection. Hostiuc et al. (2019, p. 120) consider the lack of potential for wholeness irrelevant.

[37] In this sense also Dederer (2020a, p. 43): "What is clear at this point ... is that brain organoids are not to be classified as human beings"; see also National Academies (2021, p. 96); Huyn, cited by Gogol (2018): "An organoid is not a human subject, according to federal regulations."

[38] Taupitz (2014b) B. I. para. 5 with many references; see also BVerfGE 39, 1 (45): "The legislature is in principle not obliged to take the same measures of a penal nature for the protection of unborn life as it considers expedient and necessary for the protection of born life." The legislature therefore has a wide scope of action with regard to the protection of embryos, cf. Dreier (2013) para 114; Dederer (2020b, p. 61).

human cerebral organoids "that have a larger size than the current ones, are connected both to sensory receptors and to organic (muscle) or artificial effectors, and manifest a coordinated electrical activity quite similar to that of a newborn's brain."[39] Such a strong legal protection would not be convincing because of the different way of creation in comparison to embryos by avoiding fertilization, because of the development in a completely different context and with a completely different aim than the creation of offspring,[40] because of the completely different phenotypic shape[41] and because of the not even rudimentarily existing abilities of a "normal" human brain to perform central integration, regulation and coordination services for an organism.[42] The brain organoids will not be able to develop higher brain activities or even consciousness in the foreseeable future. They lack the complexity to do so. The human brain contains a large number of core centers that perform different tasks in communicating with each other. Brain organoids in vitro can probably at best reproduce the function of single centers.[43] Their size is already limited by the fact that it has not yet been possible to integrate them into a bloodstream.[44] As a result, they do not develop further after a few months, but die from the inside because nutrients do not reach the inner cells. This is also one of the reasons why the structures hardly grow larger than a few millimeters or centimeters.[45]

Therefore, the comparative view of brain death as a decisive caesura at the end of life[46] does not lead any further. For the foreseeable future a cerebral organoid will never be able, comparable to a brain, to perform the integration necessary for a whole organism, without which this organism could not exist as a body-soul wholeness.[47] Such an integration performance includes mental as well as organismic

---

[39] Lavazza and Pizzetti (2020, p. 7); see also Koplin and Savulescu (2019, p. 764).

[40] Cf. on such aspects for the assessment of artificially created entities Taupitz (2001, p. 3440); similar later Ethikrat (2011, p. 100); Gassner and Opper (2020, p 260 f., 272 f.); further references to corresponding considerations in the Anglo-Saxon literature with regard to the moral status of early embryos in Hostiuc et al. (2019, p. 119).

[41] On the significance of likeness for the recognition of an entity as a "human being" in the sense of the human dignity guarantee Dederer (2020b, p. 66, 74).

[42] Koplin and Savulescu (2019, p. 761); Hyun et al. (2020, p. 5). But see Lavazza (2020, p. 117): "In fact, they are neither physically autonomous nor able to give rise to an adult human being.Yet the brain is the key organ of the person, the one from which one can deduce the presence of life in a person and which, if conscious, even though in a dish, should be considered a person, with increasing moral value the greater its consciousness." Slightly different however ibid. (123): "This characteristic of cerebral organoids makes them morally special, even though they cannot be considered persons in the full sense."

[43] For attempts to generate modularly assembled organoid systems, see Marton and Pasca (2019); Bagley et al. (2017).

[44] For attempts to generate vascularized organoids, see Shou et al. (2020).

[45] Chen et al. (2019) 463 f. This is one of the reasons why the transfer to animals is so interesting for researchers.

[46] See above at footnote 24.

[47] Cf. Koplin and Savulescu (2019, p. 762, 764); therefore, the brain-life criterion is not convincing as a normative starting point for the beginning of the (embryonic) protection of life, cf. Müller-Terpitz (2008, pp. 184–186).

aspects; mental processes are not conceivable without an organismic basis.[48] Brain death is regarded as the death of man precisely because no other organ can take over the function of "integrating itself with all other organs – each one of which may be necessary for the survival of the entire organism whole – into precisely that functionally interactive totality that constitutes the living organism and that is far more than a set of mutually connected individual organs. The brain is the central organ of integration, regulation and coordination. It integrates sensory and sensitive stimuli from the organism itself and from outside through the so-called afferent nerves; enables motor actions and communication via the efferent nerves (for instance, through speech, gesture, facial expressions); regulates the coordination processes within and between the other organ systems across the vegetative nervous system, including hormonal regulation; and, finally, is the basis of the mind and of subjectivity."[49] All this will not be possible in a cerebral organoid. And just as a purely mentalistic understanding of death is inappropriate,[50] a concept of status and protection based on purely mentalistic abilities is equally unconvincing.[51]

Altogether, it is thus extremely far-fetched to ascribe to cerebral organoids a status comparable to that of embryos. This is also true for very advanced brain organoids.

5.3.2 Brain Organoids as Things

From a legal point of view, brain organoids are to be treated in the same way as other human organs or cell structures.[52] They have—despite the fact that their cells are of human origin and in this sense belong to the human species[53]—no special intrinsic legal status. Substances separated from the human body are legal objects and no longer part of the legal subject.[54] According to German law, these are substances to which ownership is granted in accordance with Sec. 903 et seqq. Civil Code (BGB).[55] Although the (living) human body is not a thing in which ownership can exist,[56] with the separation from the body, the now independent

[48] Ethikrat (2015, p. 73).

[49] Ethikrat (2015, p. 73).

[50] Ethikrat (2015p. 68).

[51] In this direction also Lavazza and Pizzetti (2020 10 ff.; Koplin and Savulescu (2019, p. 762).

[52] Faltus (2021, p. 133).

[53] But see Lavazza (2020, p. 116): brain organoids "are human by definition, as they come from human cells, and this biological affiliation could grant them a moral status." Against a "species-centric" argumentation Schicktanz (2020, p. 199).

[54] Roidis-Schnorrenberg (2016, p. 56); Roth (2009, p. 65).

[55] In other jurisdictions, this is partly seen differently, cf. Boers et al. (2016, p. 938): "human tissue is neither a person nor a thing."

[56] Detailed description by Schreiber (2019) 25 ff.; see also CJEU, C-377/98, Netherlands v Parliament and Council, ECLI:EU:C:2001:523, para. 73: "Thus … an element of the human body may be part of a product which is patentable but it may not, in its natural environment, be appropriated" (emphasis by author).

body part is subject to the property law of the BGB.[57] As a rule, the researcher will have produced a new movable thing, namely the organoid, by "processing or transform[ing] [...] one or more substances [the initial cells]," so that he acquires original ownership of it according to Sec. 950 (1) BGB[58] (at this moment at the latest[59]).[60]

However, insofar as genetic material of the donor of the source material remains in the organoid, the general right of personality of the donor also extends to the organoid. This is because, according to general opinion, the general right of personality of a human being continues in the body material separated from his body.[61] There are, then, two rights to the body material with different scopes: the researcher's right of ownership enables him to exclude others from using it and to use it himself as he pleases—but only, as the law itself adds in Sec. 903 BGB concerning the "powers of the owner," "to the extent that [...] third-party rights do not conflict with this [use]," And, in fact, the original bearer of the bodily material continues to have rights to "his" bodily substances even without the ownership position, namely in the form of his general right of personality. This can be justified in two ways[62]: Either one can say that the right of personality, that accrues to the living human being, continues in the separated body parts. Or one examines whether the concrete use of the substance has "remote effects" on the person of the former bearer and thus affects the right of personality in respect of the bearer as such. Both ways lead to the same result[63]: in the German legal system it is recognized that the exercise of property rights of one person can be regarded as an encroachment on the right of personality of another person. One has to think of the "moral right" (= a form of right of personality) of an artist, on the basis of which the artist can defend himself against the current owner against a redesign of "his" work of art. The same applies to personal letters and other written documents: although the recipient regularly acquires ownership of them, the author can, on the basis of his right of personality, prohibit certain forms of use, such as publication. There can also be an intense personal relationship to human substances, which is not completely extinguished by passing them

[57] The controversial question of how the emergence of ownership at the time of separation from the body can be substantiated does not need to be addressed here, cf. Schreiber (2019) 41 ff.; for the international discussion, see for example the references in Boers et al. (2016).

[58] "A person who, by processing or transformation of one or more substances, creates a new movable thing acquires the ownership of the new thing, except where the value of the processing or the transformation is substantially less than the value of the substance. Processing also includes writing, drawing, painting, printing, engraving or a similar processing of the surface."

[59] Who acquires ownership of the bodily substances at the time of separation from the body is disputed, see Schreiber (2019) 42 ff.

[60] Cf. Schreiber (2019, p. 322); Ehrlich (2000) 57 ff.; Zech (2007) 99 ff.

[61] Schröder and Taupitz (1991) 42 ff.; for a detailed overview of the nuanced differences of opinion, see Schreiber (2019) 41 ff.

[62] See Taupitz (1991) 209 f.

[63] On the following, with references, Taupitz (1991, pp. 210–211); Schröder and Taupitz (1991, pp. 43–44).

on to third parties. Genome analysis can be used to draw a picture of the physical disposition of the former carrier. Such an inference pointing to the former carrier cannot touch his "secret" and "intimate realm" any less than an inference from written statements. And, for example, the transfer of human cells or organs to other persons is, in terms of value, on the same level as the transformation of a work of art and the resulting violation of the artist's intellectual relationship with his work. The conclusion to be drawn from this is that any use of human bodily substances must be examined to determine whether it violates the rights of personality of the former bearer. In addition to the type and aim of the use and the amount of material used, the further consequences for the original carrier and, above all, the type and extent of anonymization are decisive for the evaluation. It is also important whether the material with its (and possibly precisely because of its) individual characteristics is to be transferred to other persons and—in relation to research—whether the research project raises legal or ethical concerns.[64] Since a special—not least anthropological—significance is attached to human brain cells, the donor has the right to decide on the production and use of a brain organoid containing cells or at least genetic material from himself, in that his express consent is required for this purpose, and he can thus prohibit the production or use of the brain organoid.[65]

For the generation and use of brain organoids, it is also important that neither takes place in a legal vacuum. Research with human biological material is largely subject to the rules of medical professional law (cf. Sec. 15 of the [model] professional code of conduct for physicians working in Germany[66]), which results in particular in an obligation to obtain professional legal and ethical advice from an ethics committee before carrying out the research project. This corresponds to Art. 22 of the Council of Europe Recommendation on Research Involving Human Biological Material[67] and, according to the Animal Protection Act, also applies to research involving animals.[68] But even beyond this, many research institutions or funding institutions require the researchers to involve an ethics committee. This is justified by the responsibility of the researcher in the sense of research-accompanying self-reflection, information, and critically examining discourse[69] and enables an independent view on the project.

---

[64] Taupitz (2020a, p. 808); Taupitz (2020b, pp. 216–217) with further references.

[65] Taupitz (2020a, p. 808); Taupitz (2020b, p. 217). The possibility of a veto (which exists anyway according to general principles) is demanded by Farahany and et al. (2018) 431 f.; Schicktanz (2020, p. 198). The particularities of the use of cells by minors cannot be discussed in this article.

[66] Version of 2018, https://www.bundesaerztekammer.de/fileadmin/user_upload/downloads/pdf-Ordner/MBO/MBO-AE.pdf

[67] Recommendation CM/Rec(2016)6 of the Committee of Ministers to member States on research on biological materials of human origin, https://search.coe.int/cm/Pages/result_details.aspx?ObjectID=090000168064e8ff

[68] Taupitz (2020a, p. 813); Taupitz (2020b), 229 f., 233.

[69] Dickert (1991, p. 373).

5.3.3 Brain Organoids as a Commodity?

The question arises whether brain organoids may be classified commercial goods, i.e., whether they may be sold, in other words.

According to Sec. 17 (1)(1) Transplantation Act (TPG), no trade in human organs and tissues (which also includes individual cells, see Sec. 1a (4) TPG) is permitted; a violation is punishable under Sec. 18 TPG. However, this only applies if the organs or tissues are "intended to be used for the treatment of another person." The prohibition of trafficking therefore does not apply to cells used for the production of brain organoids, provided that these serve purely research purposes.[70] This will be the case for a long time to come.

Moreover, the prohibition of trafficking does not apply to autologous transplants[71] and to brain organoids even in cases of allogeneic transplantation, since these, as entities produced in vitro, are not organs within the meaning of the TPG.[72] Rather, they are medicinal products manufactured from or using organs or tissues[73] and to which the prohibition of trade does not apply (see Sec. 17 (1)(2)(2) TPG).[74] Overall, therefore, no prohibition of trade in brain organoids can be derived from the Transplantation Act. And even within the scope of application of the prohibition of trade, the law does not prohibit "the granting or acceptance of reasonable remuneration for the measures required to achieve the goal of curative treatment, in particular for the removal, preservation, further processing including measures to protect against infection, storage and transport of the organs or tissues" (Sec. 17 (1)(2)(1) TPG).[75]

It is questionable, however, whether brain organoids are not to be classified as objects with which trade is not permitted beyond the current Transplantation Act. After all, some international rules expressly disapprove of making the human body and its parts as such the object of financial gain. These include, in particular, Art. 3(2) of the EU Charter of Fundamental Rights,[76] Art. 21 of the Council of Europe's Convention on Human Rights in Biomedicine,[77] and Art. 6 of the Council of

[70] Cf. Häberle (2020) para. 2.

[71] Baumann and Kügele (2020) para. 1

[72] More detailed Gerke (2020, p. 295); Taupitz (2020a) 810 f.

[73] Taupitz (2020a) 811 f.

[74] More detailed Scholz and Middel (2018) paras. 13 et seq.; on the inapplicability of the prohibition of trafficking to substantially engineered tissues such as induced pluripotent stem cells and products derived therefrom also Harder (2020, p. 170).

[75] The same applies to the prohibition of financial gain in Art. 21 of the Human Rights Convention on Biomedicine of the Council of Europe, which will be presented below, cf. Harder (2020, p. 164).

[76] "In the fields of medicine and biology, the following must be respected in particular: [...] - the prohibition on making the human body and its parts as such a source of financial gain, ..."

[77] "Prohibition of financial gain: The human body and its parts shall not, as such, give rise to financial gain." Art. 21 of the Convention for the Protection of Human Rights and Dignity of the Human Being with regard to the Application of Biology and Medicine: Convention on Human Rights and Biomedicine, http://www.coe.int/de/web/conventions/search-on-treaties/-/conventions/rms/090000168007cf98

Europe's Recommendation on Research Involving Human Biological Material.[78] Admittedly, it is not clear from these provisions whether they refer only to the living human body as a whole (in this case, therefore, to the removal of cells from the human body), or whether substances already separated from the body in the subsequent chain of use are also to be covered by the prohibition of making profit. As, according to widespread opinion, human dignity forms the background to the prohibitions on the realization of profits,[79] but as human body substances that are already separated from the human body have no human dignity and their use does, in general, not constitute a violation of human dignity,[80] according to a convincing opinion, the prohibitions do not refer to biological material separated from the body.[81] Even if one sees this differently,[82] consonant with the prevailing opinion, body substances that are separated from the body, at least below a "relevance threshold,"[83] i.e., in case of body substances of minor normative relevance, cannot be covered by the prohibitions. In addition to shed hair, fingernails, or toenails,[84] this also includes small amounts of residual blood[85] or a few cells that were removed in the course of a diagnostic or therapeutic intervention. This applies all the more to cells taken from a deceased person, because, in this case, at most the much weaker postmortem right of personality can come into play.[86] All this also applies to cells that can realistically be used for the production of brain organoids, if these were generated at all from brain cells and not from induced pluripotent stem cells. As far

[78] "Prohibition of financial gain: Biological materials of human origin should not, as such, give rise to financial gain." Art. 6 of the Recommendation CM/Rec(2016)6 of the Committee of Ministers to Member States on research on biological materials of human origin, https://search.coe.int/cm/Pages/result_details.aspx?ObjectID=090000168064e8ff

[79] See for example CJEU, C-377/98, Netherlands v Parliament and Council, ECLI:EU:C:2001:523, para. 77; Deutscher Bundestag (1996, p. 29); further references at Roidis-Schnorrenberg (2016) 80 ff., and Fröhlich (2012) 139 ff.

[80] See Zech (2007, p. 118); Schnorrenberg (2010, p. 236); Roidis-Schnorrenberg (2016) 113 f.; Taupitz (2000, p. 157); Fröhlich (2012, p. 161 and 208), with the indication that a violation of human dignity can be considered if the body substance is used with humiliating intent. However, this is not specific to the use of human body substances, but can also occur with other material (image, sound recording, etc.).

[81] Roidis-Schnorrenberg (2016) 113 f.; see also Halasz (2004) 122 f.

[82] See for example Borowski (2019) para. 46; Heselhaus (2017) para. 24; Breithaupt (2012) 40 ff., 59.

[83] Borowski (2019) para. 46; see also Kranz (2008) 183, according to which the prohibitions on commercialization are to be interpreted restrictively.

[84] On these bodily substances, see explicitly the Explanatory Report to the Convention for the protection of Human Rights and Dignity of the Human Being with regard to the Application of Biology and Medicine, May 1997, para. 133, https://rm.coe.int/16800ccde5; concerning hair see also the Commentary of the Charter of fundamental rights of the European Union, June 2006, 40, https://sites.uclouvain.be/cridho/documents/Download.Rep/NetworkCommentaryFinal.pdf; see also Breithaupt (2012) 43 ff., 57, 59; Heselhaus (2017) para. 24.

[85] Taupitz (2017p. 357). For an extension of the exceptions listed in the official explanations (see previous footnote) to other bodily substances, see for instance Schwarzburg (2012) 184 f.

[86] Taupitz (1996), 7 f.

as a new thing (here the brain organoid, see Sect. 5.3.2]) has been generated from the removed cells in a complex manufacturing process, it is even more remote to regard this artificially produced thing as a subject of a prohibition of profit making. This is especially true in view of the fact that the prohibition on the realization of profit is primarily justified by the lack of suitable donor organs and the resulting temptation to exploit the health problems of fatally or critically ill patients in a particularly reprehensible manner for economic reasons. In addition, financial incentives to potential living donors to impair their health and physical integrity for the sake of economic advantages are to be prevented.[87] None of this applies to artificially created brain organoids.

From a legal point of view, it is also important to note that the EU Charter of Fundamental Rights is directly applicable only to the implementation of Union law[88] and that its Art. 3 (2) in particular has no direct effect on private law relationships.[89] The Convention on Human Rights in Biomedicine[90] is not binding on the Federal Republic of Germany in any case because Germany has not signed and ratified this treaty, and the Council of Europe's Recommendation has not been made binding by its authors.

Overall, a prohibition of trafficking does not exist in relation to brain organoids. Even if a prohibition of profit-making were to be applied to them, this would not change the legal situation described above, namely that ownership can exist in them and that such ownership can also be transferred. The acquisition and transfer of ownership rights are not covered by a prohibition on the realization of profits.[91]

5.3.4 Brain Organoids as Data in the Sense of Data Protection Law?

In part, human body materials as such are (also[92]) classified as data because of the genetic material they contain, which includes information about the former carrier.[93] Data has a different status than things, which is already shown by the fact that there is no "ownership" in them according to current law.[94] As far as bodily materials are (also) classified as data, any handling of them is relevant under data protection law, which means, e.g., that even the physical acquisition of the samples is to be regarded

[87] Tag (2017) para. 3.

[88] Art. 51 (1) of the Charter; see also Jarass (2021a) para. 9.

[89] Jarass (2021b) para. 3; Augsberg (2015) para. 7; Roidis-Schnorrenberg (2016) 120 ff.; Fröhlich (2012) 176 with further references; different view: Heselhaus (2017) para. 20.

[90] About the convention in detail Taupitz (2002).

[91] Borowski (2019) para. 46; Heselhaus (2017) para. 24; Jarass (2021b) para. 16.

[92] Those who classify human bodily substances as data do not thereby at the same time deny their character as corporeal things in which ownership may exist.

[93] Büchner (2010) 123 f.

[94] Hoeren (2019); Determann (2018).

as data collection.[95] However, this contradicts the recognizable intention of the German legislator, who explicitly distinguishes between genetic samples and genetic data in Sec. 3 (10), (11) of the Genetic Diagnostics Act.[96] The same applies to the European legislator, who differentiates accordingly in the definition of genetic data in Art. 4 (13) of the General Data Protection Regulation (GDPR).[97] Nor can it be cited as a counter-argument that data carriers containing sensitive data fall within the scope of the GDPR.[98] This is because human-created data carriers are not involved in the case of human-created material. Data protection law is intended to cover mentally mediated or mediatable facts, as it serves to protect informational self-determination and thus the relationship of the individual with his or her data.[99] If data protection law were to be extended accordingly to natural data carriers, data protection law would have to apply "to the storage of almost any object."[100] Therefore, bodily materials as such do not qualify as data.[101]

However, it is undoubtedly a matter of data collection if (new) data are generated by the examination of the material[102] or even the donors of the source material are re-identified.[103] Accordingly, for this conduct and for the subsequent use of the data, the data protection law is relevant, provided that it does not concern anonymous material.[104] However, this concerns the use of the body material, in this case the organoid, but not the status of the material (organoid) itself.

5.4 The Legal Status of Brain Organoids De Lege Ferenda

Finally, it must be examined whether a change in the law is necessary with regard to brain organoids. After all, animals are (also) attributed a special legal status, since they are not things and are protected by special regulations (the animal welfare laws) (Sec. 90a (1) BGB), even though they are widely treated as things in law (Sec. 90a (2) BGB). And due to the increasing intelligence and autonomy of

[95] Vossenkuhl (2013, p. 6), provided that the biological cell material is obtained from the outset for the purpose of obtaining information.

[96] See also Deutscher Bundestag (2008) 22 f.

[97] Accordingly, genetic data are personal data relating to the inherited or acquired genetic characteristics of a natural person which provide unique information about the physiology or health of that natural person and which have been obtained, in particular, from the analysis of a biological sample from the natural person concerned.

[98] Ziebarth (2018) para 8.

[99] Deutscher Bundestag (2008) 22 f.; Schreiber (2019) 102 f.

[100] Breyer (2004, p. 660).

[101] Schreiber (2019) 104 f.; Breithaupt (2012) 240 f.; Fink (2005, p. 60); Halasz (2004) 263 f.; Koch (2013, p. 117).

[102] Schreiber (2019) 106 f.

[103] For example, in order to be able to contact genetic relatives, e.g. in the case of dementia research with the source cells used, cf. Ooi et al. (2020, p. 450).

[104] According to some opinions, however, human genetic material is never anonymous, see (rejecting) Taupitz (2020c) 608 f.

artificial intelligence (AI) systems, calls for a change in the law to recognize the legal capacity of (certain) autonomous systems are growing louder.[105] The proposals range from a narrowly defined partial legal capacity to full legal capacity ("e-person").[106]

However, both areas must be clearly separated from each other, as the rationale is different in each case: The idea of introducing an "e-person" is driven by the notion that the decision-making processes of AI systems are increasingly approaching those of humans and that the systems are achieving a certain degree of autonomy. In this context, the term "autonomy" expresses the fact that humans have relinquished a certain degree of control with regard to the intransparency of AI decisions.[107] On the other hand, brain organoids are mainly considered in terms of their ability to feel and, in particular, to feel pain (as animals do). From this, it is widely concluded that they have "interests" which are to be considered morally.[108] Moreover, the intended consequences are different in each case: In the case of the e-person, it is primarily a matter of their liability for causing harm to third parties[109] (so that the e-person would have to be endowed with his or her own assets), whereas in relation to brain organoids, it is primarily a matter of preventing actions that cause harm (e.g., pain) to the organoid itself.

Regardless of these differences, in both areas arises the cardinal question, which extent of "autonomy" or "sensibility" or "consciousness" is required to grant a special "status." After all, "autonomy" of AI-systems and "sensibility" or "consciousness" in brain organoids can be developed more or less gradually,[110] so that the thresholds would have to be determined decisionistically by the legislator. It would also have to be decided how to demarcate the AI system in question when it is networked with other systems, while in relation to brain organoids, conversely, it would have to be asked which additional attributes such as pain receptors or sensory organs that convey information content about the external world[111] are required in order to grant the special status. As regards brain organoids, the additional problem arises how the corresponding ability—beyond the measurement of e.g. pure electrical currents—can be determined, since a brain organoid as such cannot communicate and

[105] The European Parliament's resolution of February 16, 2017, with recommendations to the Commission on Civil Law Rules on Robotics (2015/2103(INL)) called on the Commission to study the implications of "creating a specific legal status for robots" and "applying electronic personality to cases where robots make autonomous decisions or otherwise interact with third parties independently", https://eur-lex.europa.eu/legal-content/EN/TXT/HTML/?uri=CELEX:52017IP0051&from=DE

[106] Teubner (2018) 160 ff.; Zech (2020) A 95.

[107] Mühlböck and Taupitz (2021, p. 183).

[108] Koplin and Savulescu (2019) 763 f.; Lavazza (2021, p. 3, 6); Lavazza and Massimini (2018, p. 609).

[109] Mühlböck and Taupitz (2021, p. 214).

[110] Lavazza (2020, p. 107); regarding AI systems Mühlböck and Taupitz (2021, p. 215).

[111] See Faltus (2021, p. 133).

"pain sensation" or "consciousness" can at best be "measured" via (for their part disputed[112]) correlates.[113]

On the basis of German (and European[114]) law, it is not at all about the (intrinsic) rights of brain organoids (or AI systems), certainly not comparable to those of embryos or even born humans (see Sect. 5.3.1), so that also the occasionally expressed consideration to install a guardian for brain organoids[115] is absurd. Comparable to animals, it is at most a question of possible (protective) duties of humans toward brain organoids[116] or of objective-legal limits of research for supra-individual reasons—just as, by the way, the granting of a special "moral status" is not binary (all-or-nothing[117]) but can have quite different forms and, above all, consequences,[118] i.e., by no means per se the granting of rights or the "status" of a born human being.[119] Especially with regard to brain organoids, in the international discussion, it is predominantly demanded to grant only a "certain degree of moral status" to highly developed cerebral organoids, if they would really develop characteristics like consciousness, pain perception, or self-knowledge.[120] At the same time, however, it is emphasized even in this respect that one is still very far away from this and even more from "brain in the dish"[121] and that the scientifically inaccurate designation of organoids as "mini-brains"[122] has caused completely exaggerated concerns.[123]

With regard to German law, it is further doubtful whether (additional) legal rules (namely restrictions for research) such as those being considered in the ethical debate abroad (such as a ban on cultivating highly developed brain organoids beyond a certain period of time[124] (such as the 14-day time limit widely applied to

---

[112] Doerig et al. (2019); Lavazza (2021) 5 f.; Birch and Browning (2021, p. 5).

[113] Bayne et al. (2020) 11 ff; Lavazza (2020, p. 112, 114); Koplin and Savulescu (2019, p. 762); Singer (2019); Farahany and et al. (2018, p. 431); Lavazza and Massimini (2018, p. 608).

[114] Lavazza and Pizzetti (2020, p. 22): "... based on Italian law and European law as a superordinate system ..., it must in fact be concluded that HCOs [human cerebral organoids] have no right to any special legal protection, as they do not fall into any category other than that of biological material ...".

[115] But see Kaulen (2018).

[116] Indeed, the international debate often calls for protection of brain organoids similar to that of animals, cf. Birch and Browning (2021, p. 6); Bitar (2020); Koplin and Savulescu (2019, p. 763; Sawai et al. (2019, p. 440) with further references.

[117] On this view Iltis et al. (2019, p. 11)

[118] DeGrazia (2008) 181 ff.; Lavazza (2020, p. 116); Lavazza (2021, p. 6); Koplin and Savulescu (2019, p. 764). Shepherd (2018, p. 15) describes the attribution of a moral status as "a kind of placeholder for attribution of reasons to regard and treat an entity in certain ways". Similar Iltis et al. (2019, p. 9 with further references).

[119] Lavazza (2020, p. 123): "This characteristic of cerebral organoids makes them morally special, even though they cannot be considered persons in the full sense." Cf. also Schicktanz (2020, p. 198).

[120] Koplin and Savulescu (2019, p. 760); Lavazza and Pizzetti (2020, p. 14, 22)S.

[121] Weiler (2020), Koplin and Savulescu (2019, p. 761, 765), Cepelevicz (2020) and Lange (2019).

[122] See above Footnote 8.

[123] Lunshof and Greely, cited by Cepelevicz (2020).

[124] Lavazza and Pizzetti (2020, p. 20).

embryos[125]) and/or beyond the formation of certain key areas for the emergence of consciousness[126] or a ban on conducting certain experiments with the brain organoid[127]) would even be constitutional. This is because the freedom of science under Article 5 (3) of the Constitution may only be interfered with to protect other goods with constitutional rank.[128] However, since brain organoids have no special constitutional status (see Sect. 5.3.1) and alleged protected goods such as the "image of man in our culture" or "cultural self-image," which are referred to in connection with the protection of embryos and which could also be brought into play in relation to brain organoids, just like the "precautionary principle" as such[129] do not legitimize any restrictions on fundamental rights,[130] only the general right of personality of the donors of the source material (Art. 2 (1) in conjunction with Art. 1 (1) GG) or animal welfare in the context of animal experiments (Art. 20a GG) could lead to a justification of restrictions on brain organoid research. However, the general right of personality of the donors of the source material is already sufficiently protected by the requirement of their specific consent to the production of brain organoids and by the right of veto to which the donors are thus entitled (see Sect. 5.3.2), and animal welfare is also sufficiently taken into account by the current Animal Protection Act. That means, that there is no need or justification for further restrictions of research.[131]

The constitutional concerns about restrictions on brain organoid research are strengthened by the view of possible therapeutic options for patients.[132] This is because the state may not prohibit against the background of patients' rights to life, physical integrity, and health (Art. 2 GG), the development and use of potential medical therapies without sufficient reason. Also the possible avoidance of research with embryos (insofar as it is permitted abroad) and fetuses[133] or primates through the use of brain organoid[134] has to be taken into account. In the international debate, there is a justified warning that the great potential of research with brain organoids should not be prematurely dried up by restrictive rules.[135] This does not contradict

---

[125] Jácomo (2020, p. 7); Bitar (2020).

[126] Bitar (2020); Lavazza (2021, p. 7); National Academies (2021, p. 97).

[127] Reardon (2020, p. 661) and Goodall (2020).

[128] See above 5.2.

[129] The application of the precautionary principle is required with respect to brain organoids by Birch and Browning (2021), 2 ff.

[130] Taupitz (2014b), B. III. para. 23.

[131] But differently Birch and Browning (2021, p. 3): "regulation is urgently needed".

[132] Lavazza and Pizzetti (2020, p. 20); Koplin and Savulescu (2019, p. 763); National Academies (2021, p. 97).

[133] Lavazza and Massimini (2018, p. 607).

[134] Sidhaye and Knoblich (2021, p. 53); differentiating Schicktanz (2020) 196 f.

[135] Koplin and Savulescu (2019, p. 761); Farahany et al. (2018, p. 432). However, this warning would also have to cover guidelines of research self-control/self-restraint, as they are often demanded (as an alternative to legal regulations): Reardon (2020) 661; Koplin and Savulescu (2019, p. 761); Cepelevicz (2020); Chen et al. (2019, p. 463).

the provision of advice[136] to researchers by ethics committees,[137] as is good practice in Germany (see Sect. 5.3.2). If necessary, the ethics committees should be expanded to include appropriate expertise.[138]

5.5 Summary

5.5.1. German law does not contain any special legal rules concerning brain organoids.

5.5.2. Brain organoids have the same status as other human organs. They are things in which ownership can exist. As a rule, ownership belongs to the person who produced the brain organoid. In addition, the donor of the cells used for their production has, on the basis of his general right of personality, the right to determine the production and use of the brain organoid by requiring his specific consent for this purpose.

5.5.3. Even highly developed brain organoids do not have a legal status comparable to that of embryos or even born human beings. It is also not appropriate to grant them such a status.

5.5.4. Research with brain organoids generated from human cells is subject to evaluation by ethics committees (which may have to be expanded to include appropriate expertise); this applies also in the case of the inclusion of animals.

5.5.5. There is no need to amend or supplement German law specifically with regard to brain organoids in the foreseeable future, if only in view of the fact that the process of producing complex organoids is only just beginning.

5.5.6. Restrictions on research with brain organoids that go beyond the current law would also hardly be justifiable with regard to the freedom of science under Article 5 (3) of the Constitution.

[136] If the researcher only has to obtain the advice of an ethics committee, he is not bound by the advice as a matter of law. This is different if—e.g. according to the Medicinal Products Act (AMG)—he must obtain approval before carrying out the research project.

[137] Munsie et al. (2017, p. 944); National Academies (2021, p. 99 ff.); on the constitutional requirements for the need to involve an ethics committee see Hufen (2017, p.1266 ff.).

[138] Skepticism as to whether existing local research ethics committees are competent enough to assist with Farahany et al. (2018, p. 432).

References

Appleby JB, Bredenoord A L (2018) Should the 14-day rule for embryo research become the 28-day rule? EMBO Mol Med. 10(9):e9437
Augsberg S (2015) Kommentierung zu GRC Art. 3. In: von der Groeben H, Schwarze J, Hatje A(Hrsg.) Europäisches Unionsrecht, 7. Auflage. Nomos, Baden-Baden
Baars BJ, Franklin S (2007) An architectural model of conscious and unconscious brain functions: global workspace theory and IDA. Neural Netw 20:955–961
Bagley JA et al (2017) Fused cerebral organoids model interactions between brain regions. Nat Methods 14(7):743–751
Bartfeld S et al (2020) Kernaussagen und Handlungsempfehlungen zu Organoiden. In: Bartfeld S et al (eds) Organoide, 2020. Nomos, Baden-Baden, pp 13–28
Baumann G, Kügele S (2020) Kommentierung zu § 17 TPG. Nomos Bundesrecht, Erläuterungen, In
Bayne T et al (2020) Are there islands of awareness? Trends Neurosci 43:6–16
Birch J, Browning H (2021) Neural organoids and the precautionary principle. Am J Bioeth 21:56–58. https://doi.org/10.1080/15265161.2020.1845858
Bitar M (2020) Are brain organoids equivalent ot Philospphical zombies? Preprints 2020. https://doi.org/10.20944/preprints202004.0028.v1
Boers S et al (2016) Organoid biobanking: identifying the ethics. EMBO Rep 17:938–941
Borowski M (2019) Kommentierung zu GRC Art. 3. In: Meyer J, Hölscheidt S (Hrsg.) Charta der Grundrechte der Europäischen Union, 5. Auflage. Nomos, Baden-Baden
Breithaupt J (2012) Rechte an Körpersubstanzen und deren Auswirkung auf die Forschung mit abgetrennten Körpersubstanzen. Nomos, Baden-Baden
Breyer P (2004) Der datenschutzrechtliche Schutz von Körpersubstanzen, die Patienten zu Analysezwecken entnommen wurden. MedR 22:660–667
Büchner B (2010) Körpersubstanzen als Forschungsmaterialien. Dr Kovac, Hamburg
Cavaliere G (2017) A 14-day limit for bioethics: the debate over human embryo research. BMC Medical Ethics 18(1):38. https://doi.org/10.1186/s12910-017-0198-5
Cepelevicz J (2020) An ethical future for brain organoids takes shape, Quantamagazine. https://www.quantamagazine.org/an-ethical-future-for-brain-organoids-takes-shape-20200123/
Chan S (2018) How and why to replace the 14-day rule. Curr Stem Cell Rep 4(3):228–234
Chen H et al (2019) Transplantation of human brain organoids: revisiting the science and ethics of brain chimeras. Cell Stem Cell 25(4):462–472
Dederer H-G (2020a) Über die Rechte von Gehirn-Organoiden, Laborjournal. https://www.laborjournal.de/rubric/special/special/2020_05_03.php
Dederer H-G (2020b) Der manipulierte Embryo: Konsequenzen für das Recht. In: Jahrbuch für Recht und Ethik. Duncker & Humblot, Berlin 8:53–81.
DeGrazia D (2008) Moral status as a matter of degree? South J Philos 46:181–198
Determann L (2018) Kein Eigentum an Daten. MMR 21:277–278
Deutscher Bundestag (1996) Entwurf eines Gesetzes über die Spende, Entnahme und Übertragung von Organen (Transplantationsgesetz - TPG), BT-Drucksache 13/4355
Deutscher Bundestag (2008) Entwurf eines Gesetzes über genetische Untersuchungen bei Menschen (Gendiagnostikgesetz – GenDG), BT-Drucksache 16/10532
Dickert T (1991) Naturwissenschaften und Forschungsfreiheit, 1991. Duncker & Humblot, Berlin
Doerig A et al (2019) The unfolding argument: why IIT and other causal structure theories cannot explain consciousness. Conscious Cogn 72:49–59
Dreier H (2013) Kommentierung zu GG Art. 1 Abs. 1. In: Dreier H (Hrsg.) Grundgesetz-Kommentar, 3. Auflage
Ehrlich S (2000) Gewinnabschöpfung des Patienten bei kommerzieller Nutzung von Körpersubstanzen durch den Arzt? Peter Lang, Frankfurt am Main
Ethikrat D (2010) Humanbiobanken für die Forschung. Eigenverlag, Berlin
Ethikrat D (2011) Mensch-Tier-Mischwesen in der Forschung. Eigenverlag, Berlin
Ethikrat D (2015) Hirntod und Entscheidung zur Organspende. Eigenverlag, Berlin

Faltus T (2021) Organoide, Assembloide, Gastruloide, SHEEFs & Organ-on-Chip. MedR 39:125–134
Farahany N et al (2018) The ethics of experimenting with human brain tissue. Nature 556:429–432
Fink S (2005) Einwilligung und vertragliche Regelungen zur Entnahme von Körpersubstanzen, deren Aufbewahrung und Verwendung in Biobanken. Jur. Diss, Mannheim
Fröhlich A (2012) Die Kommerzialisierung von menschlichem Gewebe. Lit, Münster
Gassner U M, Opper J (2020) Zur Zulässigkeit therapeutischen Klonens mittels Zellkerntransfer, in: Opper J et al., Chancen und Risiken der Stammzellforschung. Berliner Wissenschaft-Verlag, Berlin. p 255–278
Gerke S (2020) Die klinische Translation von hiPS-Zellen in Deutschland. In: Gerke S et al (eds) Die klinische Anwendung von humanen induzierten pluripotenten Stammzellen. Springer, Berlin, pp 243–327
Gogol A (2018), A human brain model in a petri dish? EurekAlert! 16.10.2018 https://www.eurekalert.org/pub_releases/2018-10/cwru-ahb101618.php
Goodall R (2020), A lab-grown brain: a mind of its own? https://theboar.org/2020/11/a-lab-grown-brain-a-mind-of-its-own/
Günther H-L (2014) Kommentierung zu § 6. In: Günther H-L, Taupitz J, Kaiser P (Hrsg.) Embryonenschutzgesetz, 2. Auflage. Kohlhammer, Stuttgart
Häberle P (2020), Kommentierung zu § 17 TPG. In: Erbs G, Kohlhaas M (Hrsg.) Strafrechtliche Nebengesetze, Werkstand: 233. EL, Beck, München
Halasz C (2004) Das Recht auf biomaterielle Selbstbestimmung. Springer, Berlin
Harder M (2020) Herausforderungen innovativer Gewebemedizin aus unternehmerischer Sicht. In: Gerke S et al (eds) Die klinische Anwendung von humanen induzierten pluripotenten Stammzellen. Springer, Berlin, pp 157–178
Heselhaus S (2017) Kommentierung zu GRC Art. 3. In: Pechstein M, Nowak C, Häde U (eds) Frankfurter Kommentar. Mohr Siebeck, Tübingen
Hoeren T (2019) Datenbesitz statt Dateneigentum. MMR 22:5–8
Hostiuc S et al (2019) The moral status of cerebral organoids. Regenerative Therapy 10:118–120
Hufen F (2001) Präimplantationsdiagnostik aus verfassungsrechtlicher Sicht. MedR 19:440–451
Hufen F (2017) Wissenschaft zwischen Freiheit und Kontrolle NVwZ 36:1265–1268
Hurlbut J B et al. (2017) Revisiting the Warnock rule. Nature Biotechnology 35(11) 1029–1042 https://doi.org/10.1038/nbt.4015
Hyun I et al (2016) Revisit the 14-day rule. Nature 533:169–171
Hyun I et al (2020) Ethical issues related to brain organoid research. Brain Res 1732:146653
Hyun I et al (2021) Human embryo research beyond the primitive streak. Science 371(6533): 998–1000
Iltis A S et al (2019) Human Embryo Research beyond Day 14: Ethical Questions and Considerations, Rice University's Baker Institute for Public Policy
Jácomo A (2020) The promises of brain organoids in the context of bioethical redefinition. Embryo status, consent and animal research. EIDON 54(Dec):3–10
Jarass H (2021a) Einleitung. In: Jarass H (Hrsg.) Charta der Grunderechte der EU, 4. Auflage. Beck, München
Jarass H (2021b) Kommentierung zu GRC Art. 3. In: Jarass H (Hrsg.) Charta der Grunderechte der EU, 4. Auflage. Beck, München
Kaulen H (2018) Braucht dieses Hirn einen Vormund? FAZ 25(5):2018. https://www.faz.net/-gx3-9ado5
Kirk R (2009) Zombies. In: Zalta EN (ed) Stanford Encyclopedia of Philosophy. Metphysics Research Lab, Stanford
Koch N (2013) Das Biobankgeheimnis, Schutz der Persönlichkeitsrechte in der biomedizinischen Forschung https://doi.org/10.25358/openscience-4218
Koplin J, Savulescu J (2019) Moral limits of brain organoid research. J Law Med Ethics 47:760–767
Kranz S (2008) Biomedizinrecht in der EU. Dr. Kovac, Hamburg
Kurlemann R (2013) Neues Mini-Gehirn weckt Frankenstein-Fantasien. WirtschaftsWoche vom 29(8):2013. https://www.wiwo.de/technologie/forschung/stammzellenforschung-neues-mini-gehirn-weckt-frankenstein-fantasien/8712192.html

Kuroczik J (2018) Herr, wirf Hirn herunter, Frankfurter Allgemeine Zeitung vom 8.5.2018, https://www.faz.net/aktuell/wissen/medizin-ernaehrung/wie-weit-darf-die-forschung-mit-labor-gehirnen-gehen-15575367.html
Lange M (2019) „Mini-Gehirne" aus Stammzellen: Was denkt ein Organoid in der Petrischale?, Deutschlandfunk 1.1.2019. https://www.deutschlandfunk.de/mini-gehirne-aus-stammzellen-was-denkt-ein-organoid-in-der.740.de.html?dram:article_id=436772
Lavazza A (2020) Human cerebral organoids and consciousness: a double-edged sword. Monash Bioeth Rev 38:105–128
Lavazza A (2021) Potential ethical problems with human cerebral organoids: consciousness and moral status of future brains in a dish. Brain Res. https://doi.org/10.1016/j.brainres.2020.147146
Lavazza A, Massimini M (2018) Cerebral organoids: ethical issues and consciousness assessment. J Med Ethics 44:606–610
Lavazza A, Pizzetti F (2020) Human cerebral organoids as a new legal and ethical challenge. J Law Biosci. https://doi.org/10.1093/jlb/lsaa005
Marton R, Pasca S (2019) Organoid and Assembloid Technologies for Investigating Cellular Crosstalk in human brain development and disease. Trends Cell Biol. https://doi.org/10.1016/j.tcb.2019.11.004.(2019)
Matthews KRW, Moralí D (2020) National human embryo and embryoid research policies: a survey of 22 top research-intensive countries. Regen Med 15(7):1905–1917
Matthews KRW et al (2021) Rethinking human embryo research policies. Hastings Cent Rep 51(1):47–51. https://doi.org/10.1002/hast.1215
McCully S (2021) The time has come to extend the 14-day limit. Journal of Medical Ethics 47(12) e66–e66. https://doi.org/10.1136/medethics-2020-106406
Mühlböck L, Taupitz J (2021) Haftung für Schäden durch KI in der Medizin. AcP 221:179–218
Müller-Jung J (2013) "Zerebrale Organoide" Was macht man mit so wenig Hirn? Frankfurter Allgemeine Zeitung 28(8):2013. https://www.faz.net/-gx5-7h03o
Müller-Terpitz R (2008) Der Schutz des pränatalen Lebens. Mohr Siebeck, Tübingen
Munsie M et al (2017) Ethical issues in human organoid and gastruloid research. Development 144:942–945
National Academies of Sciences, Engineering, Medicine (2021) The emerging field of human neural organoids, transplants, and chimeras: science, ethics, and governance. The National Academies Press, Washington, DC
Ooi L et al (2020) If human brain organoids are the answer to understanding dementia, what are the questions? Neuroscientist 26(5–6):438–454
Parsch S (2019a) Hirn aus dem Labor. Süddeutsche Zeitung 29.8.2019a. https://www.sueddeutsche.de/wissen/gehirn-labor-mini-brain-organoid-1.4580139
Parsch S (2019b) Forscher züchten Mini-Gehirne im Labor, Der Spiegel 30.8.2019b, https://www.spiegel.de/wissenschaft/mensch/forscher-zuechten-mini-gehirne-im-labor-a-1284395.html
Reardon S (2020) Can lab-grown brains become conscious? Nature 586:658–661
Roidis-Schnorrenberg H-E (2016) Das Verbot der Kommerzialisierung des menschlichen Körpers und seiner Teile. Verlag Universität Mannheim Service und Marketing, Mannheim
Roth C (2009) Eigentum an Körperteilen. Springer, Berlin
Sass H-M (1989) Hirntod und Hirnleben. In: Sass H-M (ed) Medizin und Ethik. Reclam, Stuttgart, pp 160–191
Sawai T et al (2019) The ethics of cerebral organoid research: being conscious of consciousness. Stem Cell Reports 13:440–447
Schicktanz S (2020) Sind menschliche zerebrale Organoide moralisch schützenswert? In: Bartfeld S et al (eds) Organoide. Nomos, Baden-Baden, pp 190–211
Schnorrenberg H-E (2010) Zur Kommerzialisierung menschlicher Körpersubstan-zen: Verstößt die Vereinbarung der Zahlung eines Entgelt an den Substanzspender gegen die Menschenwürde? In: Potthast T et al (eds) Wem gehört der menschliche Körper? Mentis, Paderborn, pp 223–242
Scholz K, Middel C-D (2018) Kommentierung zu § 17 TPG. In: Spickhoff A (Hrsg.) Medizinrecht, 3. Auflage. Beck, München

Schreiber M (2019) Die medizinische Forschung mit abgetrennten Körpersubstanzen Minderjähriger. Lit, Münster

Schröder M, Taupitz J (1991) Menschliches Blut: verwendbar nach Belieben des Arztes? Enke, Stuttgart

Schwarzburg K (2012) Die Menschenwürde im Recht der Europäischen Union. Nomos, Baden-Baden

Sharma A et al (2020) Scientific and ethical uncertainties in brain organoid ResearchThe. Am J Bioeth 21(1):48–51

Shepherd J (2018) Counsciousness and moral status. Routledge, London

Shou Y et al (2020) The application of brain organoids: from neuronal development to neurological diseases, front. Cell. Dev Biol 22. https://doi.org/10.3389/fcell.2020.579659

Sidhaye J, Knoblich JA (2021) Brain organoids: an ensemble of bioassays to investigate human neurodevelopment and disease. Cell Death & Differentiation 28:52–67

Singer W (2019) A naturalistic approach to the hard problem of consciousness. Front Syst Neurosci 25. https://doi.org/10.3389/fnsys.2019.00058

Tag B (2017), Kommentierung zu § 18 TPG. In: Münchener Kommentar zum StGB, 3. Auflage. Beck, München

Taupitz J (1991) Wem gebührt der Schatz im menschlichen Körper? - Zur Beteiligung des Patienten an der kommerziellen Nutzung seiner Körpersubstanzen. AcP 191:201–246

Taupitz J (1996) Das Recht im Tod: Freie Verfügbarkeit der Leiche? - Rechtliche und ethische Probleme der Nutzung des Körpers Verstorbener. Humanitas, Dortmund

Taupitz J (2000) Forschung mit menschlichen Zellen in Österreich. JBl 122:152–161

Taupitz J (2001) Der rechtliche Rahmen des Klonens zu therapeutischen Zwecken. Neue Juristische Wochenschrift 54:3433–3440

Taupitz J (2002) Biomedizinische Forschung zwischen Freiheit und Verantwortung. Springer, Berlin

Taupitz J (2014a) Kommentierung zu § 8 ESchG. In: Günther H-L, Taupitz J, Kaiser P(Hrsg.) Embryonenschutzgesetz, 2. Auflage. Kohlhammer, Stuttgart

Taupitz J (2014b) Juristische Grundlagen. In: Günther H-L, Taupitz J, Kaiser P(Hrsg.) Embryonenschutzgesetz, 2. Auflage. Kohlhammer, Stuttgart. BI-III

Taupitz J (2017) Verkauf von Restblut an die Medizinprodukteindustrie: nur mit Einwilligung des Patienten? MedR 35:353–361

Taupitz J (2020a) Organoide. MedR 38:805–813

Taupitz J (2020b) Organoide: Die deutsche Rechtslage, in: Bartfeld et al. (Hrsg.), Organoide, Nomos, Baden-Baden, p 212-236

Taupitz J (2020c) Biobanken: Forschungsprivilegien im Datenschutzrecht. In: Katzenmeier C (ed) Festschrift für Dieter Hart. Springer, Berlin, pp 603–625

Taupitz J, Weschka M (eds) (2009) CHIMBRIDS - chimeras and hybrids in comparative European and international research. Springer, Berlin

Teubner G (2018) Digitale Rechtssubjekte? AcP 218:155–295

Trujillo C et al (2019) Complex oscillatory waves emerging from cortical organoids model early human brain network development. Stem Cell 25(4):558–569

Viciano A (2020) Das Mini-Gehirn wird durchleuchtet. SZ vom 20(10):2020. https://www.sueddeutsche.de/wissen/hirnforschung-genetik-genschere-crispr-cas9-organoide-1.5098193

Vossenkuhl C (2013) Der Schutz genetischer Daten. Springer, Berlin

Weiler N (2020) Not 'brains in the dish': cerebral organoids flunk comparison to developing nervous system. Research. https://www.ucsf.edu/news/2020/01/416526/not-brains-dish-cerebral-organoids-flunk-comparison-developing-nervous-system

Zech E (2007) Gewebebanken für Therapie und Forschung: Rechtliche Grundlagen und Grenzen. Cuvillier, Göttingen

Zech H (2020) Entscheidungen digitaler autonomer Systeme: Empfehlen sich Regelungen zu Verantwortung und Haftung?, Gutachten A zum 73. Dt. Juristentag, Beck, München

Ziebarth W (2018) Kommentierung zu Art. 4 DSGVO. In: Sydow G (Hrsg.) Europäische Datenschutzgrundverordnung, 2. Auflage. Nomos, Baden-Baden

Part III

The Informed Consent Challenge

Ethics Considerations Regarding Donors' and Patients' Consent

6

Jeremy Sugarman

6.1 Introduction

Informed consent is a crucial factor in determining whether particular uses of brain organoids for research and clinical translation are ethically acceptable. However, while appropriate consent is a necessary condition for determining ethical acceptability, it is not alone sufficient to do so. Scientifically exciting and interesting potential research uses of brain organoids include experiments designed to enhance understanding of human brain development, elucidating the pathogenesis of diseases and conditions, identify potential drug candidates to pursue for possible clinical development (e.g., infectious diseases, dementias) and examining the foundations of consciousness. Promising pathways for the potential clinical translation of brain organoids include personalized medicine (e.g., selecting drugs likely to be safe and effective in particular patients with cancers, psychiatric diseases, and dementias) and transplantation (e.g., degenerative neurologic diseases, stroke, and trauma). In the context of basic research, consent of donors whose tissues are used to derive brain organoids is of primary concern, whereas in clinical translation the consent of both allogeneic donors and patients may be relevant.

In this chapter, I examine key ethical issues related to informed consent for brain organoid research and clinical translation. In order to do so, I first describe both a standard conceptual approach to informed consent that aims at meeting the ethical goal of respecting the autonomy of persons and some of the other ethically relevant functions of informed consent. This conceptual work provides a foundation for mapping some of the ethical issues related to informed consent in regard to the decision-making capacity and voluntariness of those being asked to consent, disclosure requirements associated with the use of brain organoids in general and

J. Sugarman (✉)
Berman Institute of Bioethics, Johns Hopkins University, Baltimore, MD, USA
e-mail: jsugarman@jhu.edu

H.-G. Dederer, D. Hamburger (eds.), *Brain Organoids in Research and Therapy*, Advances in Neuroethics, https://doi.org/10.1007/978-3-030-97641-5_6

for particular proposed uses in particular, threats to understanding that must be overcome, and considerations for authorization. Finally, I conclude by offering some suggestions for grappling with such informed consent challenges related to brain organoids.

6.2 Conceptual Considerations

From an ethics perspective, informed consent is a means of respecting the autonomy of persons. While there are a variety of conceptual models and practices regarding informed consent, to be ethically sound it arguably should generally include a set of necessary elements. Following the work of Tom Beauchamp and James Childress, these elements can be categorized as: Threshold, Information, and Consent. The Threshold, or precondition category, includes Competency and Voluntariness; the Information category includes Disclosure and Understanding, as well as offering a Recommendation (in clinical, but not research contexts); and the Consent category includes Decision and Authorization.[1] Although it is beyond the scope of this chapter to comprehensively explore the justifications and scope of each element, nor exceptions to the requirement to obtain informed consent, they provide a helpful framework for capturing ethically essential aspects of the informed consent process. In addition, in practice, these categories can generally be understood as steps in the process of obtaining informed consent. At the risk of oversimplification, each of these steps will be briefly described, recognizing that there is substantial scholarship regarding all of them.

Competence, or decision-making capacity, is a precondition for an informed consent process. Whereas competence is a legal status in many jurisdictions, decision-making capacity captures the ethically salient criteria for informed consent. In general, adequate decision-making capacity includes the ability to understand current circumstances, appreciate the implications of particular choices, make a rational choice, and express that choice. Voluntariness requires not being under the control of others, which precludes the use of coercion and undue influence.

The disclosure element involves providing necessary information about a proposed clinical intervention or research use. Although there are specific jurisdictional legal requirements in particular clinical circumstances and research settings, disclosure generally includes providing information about the nature of an intervention or use, its risks and burdens as well as its potential benefits and alternatives. The understanding element demands that this information be comprehended by the person being asked to consent. As mentioned earlier, in clinical settings, consistent with clinicians' fiduciary obligations towards patients, it can be appropriate to provide a recommendation about a proposed approach.

[1] Beauchamp and Childress (2019).

Once these other elements have been satisfied, a decision can be taken about whether or not to proceed. Finally, the decision is authorized, which may be oral or written depending upon the context. When written authorization is obtained, the consent document typically includes key information that was disclosed during the consent process.

Even though the standard or primary ethical justification of consent is based on the ethical principle of respect for autonomy, recent scholarship regarding consent in the research setting makes evident that consent can serve additional "participant-centered ethical functions: (1) providing transparency; (2) allowing control and authorization; (3) promoting concordance with participants' values; and (4) protecting participants' welfare interests. In addition, … [there are] three systemic or procedural functions that are more policy focused: (5) promoting trust; (6) satisfying regulatory requirements; and (7) promoting the integrity of research and researchers."[2] Recognizing these other ethical functions can help identify ethically relevant considerations for informed consent and underscore the necessity of obtaining consent for uses of brain organoids as well as facilitate helping to meet these goals in practice.

6.3 Threats to Decision-Making Capacity and Voluntariness

Brain diseases and conditions obviously can, but do not necessarily, undermine decision-making capacity. Consequently, ensuring decision-making capacity warrants special consideration in brain organoid research and clinical translation. While formal assessments of capacity are unlikely to be necessary when obtaining tissue from persons unaffected with brain diseases and conditions, there should be a rebuttable presumption for doing so with affected patients. While trained and experienced clinicians are generally able to make determinations of decision-making capacity, sometimes the special expertise of psychiatrists or neurologists may be necessary.

In situations where an affected person lacks decision-making capacity, where permissible by law, proxy consent for such a use must be obtained. Similarly, if the proposed use involves children, parental permission, ideally with the assent of the child for nontherapeutic research uses, substitute for individual informed consent.

In addition, given the devastating nature of many brain diseases and conditions as well as the lack of viable curative options, patients and their family members may face challenges related to voluntariness. Accordingly, those seeking to obtain consent must be sensitive to this concern and take measures to address this. This could include emphasizing that proposed research is optional, that research options are unproven, and that standard care will still be provided regardless of a decision regarding a proposed use.

[2] Dickert et al. (2017).

6.4 General Disclosure Requirements

As mentioned earlier (under Sect. 6.2), disclosure involves providing information about the nature of the proposed use(s), its associated risks and burdens as well as its potential benefits and alternatives. For brain organoid research and clinical translation, an essential starting point involves providing information about the nature of organoids.

6.4.1 The Nature of Organoids

Although research with a wide range of organoids is burgeoning, the vast majority of those asked to contribute tissues to make organoids and participate in their clinical translation are currently unlikely to have an accurate understanding of them. Consequently, the disclosure process must include an explanation of the nature of organoids. However, emerging empirical research regarding patients' perspectives on organoids suggests it will be challenging to do so in a manner that will be truly understandable. Of note, interview studies in both the Netherlands and the USA have found that patients tend to imagine both positive and negative attributes associated with organoids, ranging from their being markedly beneficial in ways that exceed current capabilities to frightening scientific fictions.[3] Furthermore, interviewees generally view brain organoids as ontologically and morally distinct from other types of organoids.[4] These findings reinforce the need for careful explanation.

In describing organoids, information must be provided about how organoids are made, including the types of cells used to produce them (e.g., resident "adult" stem cells in tissues, induced pluripotent stem cells, and human embryonic stem cells). When induced pluripotent or human embryonic stem cells are used to make organoids, consistent with differing legal requirements and professional guidelines,[5] information specific to them must be provided.[6] While using induced pluripotent stem cells to create brain organoids does not raise concerns related to the destruction of human embryos inherent to deriving human embryonic stem cells, they nevertheless can be morally salient to patients.[7]

6.4.2 Other General Disclosure Requirements

Consistent with general expectations of disclosure for related life sciences research and clinical translation, information must be provided about immortalization, genetic

---

[3] Haselager et al. (2021); Bollinger et al. (2021).

[4] Bollinger et al. (2021).

[5] ISSCR (2021).

[6] Lowenthal et al. (2012).

[7] Dasgupta et al. (2014).

modification, sharing of materials, and measures to protect privacy and their limitations.

In addition, any commercial uses of brain organoids and financial conflicts of interest should be disclosed. While disclosure and consent may not resolve all of the ethical tensions when there are financial interests at stake, they are minimum requirements in any management plan regarding them. Of note, the need for this disclosure in the context of commercial uses of organoids is reinforced by the fact that such uses can raise concerns among those asked to participate in brain organoid research and clinical translation.[8] Moreover, early data suggest that commercial use is of relevance to patients who have been involved with organoid research and see the informed consent process as one safeguard for it.[9]

6.5 Specific Disclosure Requirements Based on Proposed Use

In addition to general disclosure requirements, providing information about the proposed use is a core part of the informed consent process. Of course, the information to be disclosed is contingent upon the type of use (i.e., basic science, biobanking, personalized medicine, and transplantation) and then tailored to it.

6.5.1 Basic Science

Although some in vitro basic science research involving brain organoids is unlikely to raise significant ethical concerns, the ethical implications of other basic science research efforts are currently unsettled as is the appropriate type of oversight of them.[10] Hyun and colleagues'[11] recent observations are sobering:

> Ethical concerns also arise when research teams generate brain organoids using iPS cell lines derived from anonymized or de-identified tissues samples procured from tissue banks. At this time, it is not a standard practice that the informed consent for tissue collection used by most tissue banks actually discloses to tissue donors the possibility that their biological specimens could be used for iPS cell derivation and use in general, and much less to generate brain organoids. It is currently unknown whether tissue bank donors approve of the use of their biospecimens for brain organoid creation and their subsequent use for nearly limitless future applications, as this is a very recent application and data on donor preferences and objections are lacking. The main ethical concern here is that, while donors' tissue samples can be anonymized or de-identified by a tissue storage facility, it cannot be assumed that tissue donors have given their consent for their participation specifically in brain organoid research.

[8] Boers et al. (2016).

[9] Boers et al. (2018).

[10] Chen et al. (2019) and Chapman (2019).

[11] Hyun et al. (2020), p. 3.

As such, ensuring that proposed basic science uses are at least consistent with the provenance and consent of the biomaterials being used is a minimal requirement. However, absent data on previous tissue donors' attitudes and potential concerns about brain organoid research with their tissues, given emerging data regarding patients' perspectives on brain organoids (see Sect. 6.4), using materials that have been obtained with prospective consent that satisfied the general disclosure requirements delineated above is ethically preferable to relying on broad consent that could not have anticipated the full range of uses that some people find to be morally troublesome. Relatedly, the types of brain organoid research that can raise moral concerns should be disclosed during the consent process. These include research involving chimeras, complex organoids, and assembloids and work directed towards understanding consciousness.

While research involving chimeras is commonplace, it can raise moral concerns, especially when organoids "humanize" a resulting chimera.[12] As summarized elsewhere, there are some important settled and unsettled considerations in determining the ethical appropriateness of specific neurologic experiments involving chimeras.[13] Regardless, a necessary, but clearly not sufficient, criterion for conducting such research is consent for this proposed use. While much of the scholarship related to these issues has been in the setting of stem cell and brain tissue research, research with human brain organoids should at least *prima facie* be held to the same standards at least in regard to consent.

Nonetheless, the conceptual literature regarding brain organoids includes substantial debates about the moral status of brain organoids as they become more mature and complex due in large part to concerns about consciousness and sentience.[14] There are related normative issues regarding assembloids. Specifically, given uncertainties regarding the moral status of complex organoids and assembloids,[15] explicit consent for these types of experiments is indicated.

While the valence of most ethics discussions raises concerns about the development of consciousness or sentience in brain organoids, paradoxically brain organoids may be the most preferable scientific means of understanding the nature of human consciousness. Of course, such research would raise complex ethical considerations that would need to be addressed, yet given these uncertainties and moral concerns about creating consciousness, explicit consent for this work would also be needed.

6.5.2 Biobanking

In addition to standard biobanking of biological materials used to make brain organoids, living biobanking of brain organoids holds great promise for basic

---

[12] Munsie et al. (2017).

[13] Greely et al. (2007).

[14] Munsie et al. (2017).

[15] Hyun et al. (2020).

research and clinical translation.[16] As in biobanking in general, specific issues to be disclosed during the consent process will be predicated in large part on the structure and function of the biobank and permissible uses of banked of organoids or the human biological materials used to generate them. Accordingly, of great relevance will be the scope of consent and permissible uses, the governance model for determining use and distribution of banked organoids, including the permissibility of commercial uses. In addition, since brain organoids will have a genetic relationship to tissue donors, this must be disclosed along with precautions taken to protect privacy of any associated clinical or demographic information. Furthermore, any provisions for providing results that may be of clinical significance to donors as well as benefit sharing should be transparently described.

Building upon earlier approaches employed in other research settings,[17] Boers and Bredenoord[18] have argued for obtaining "consent for governance" for organoid biobanking. This deviates from most conventional approaches to consent that tend to encapsulate potential uses at the time of consent by obtaining consent to particular approaches to future decision-making through an articulated governance mechanism. As such, consent for governance includes an initial consent procedure incorporating the information delineated above with emphasis placed on describing privacy measures (given the actual inability to anonymize biomaterials), participant engagement, benefit sharing, and ethical oversight.

6.5.3 Personalized Medicine

Organoids can be used to help select medications that are likely to be effective in particular patients. A paradigmatic example derives from the use of gastrointestinal organoids to select medications in patients with cystic fibrosis.[19] There is hope that such an approach might also be useful for selecting medications for a variety of conditions effecting the brain (e.g., schizophrenia) as well as brain cancers (e.g., glioblastoma), where efficacy of particular treatments across populations of patients is variable, yet treatment toxicity is high. However, challenges to such use will likely require the generation of patient-specific brain organoids with known correlates of efficacy or toxicity. Because obtaining brain tissue requires an invasive procedure with some risk, where scientifically appropriate, the likely approach to making brain organoids for many diseases and conditions will probably employ skin biopsies and the derivation of induced pluripotent stem cells from them. In this setting, disclosure should include information about the current uncertainties associated with the possibility of producing a suitable organoid, the time needed for organoid maturation and testing, the lack of data on predictability in selecting

[16] Li et al. (2020).

[17] Lavori et al. (2002).

[18] Boers and Bredenoord (2018).

[19] Dekkers et al. (2013) and Berkers et al. (2019).

medications, and the alternative of simply trying another medication without using brain organoids.

6.5.4 Transplantation

Brain organoid transplantation might eventually provide viable treatment options for certain neurologic diseases and conditions. For example, autologous organoid transplants might prove useful for certain types of cerebrovascular accidents and Parkinson disease. Experience garnered in similar settings, such as the use of fetal substantia nigra transplants and pluripotent stem cell derivatives for the treatment of Parkinson disease, helps to identify information that should be disclosed for such research. These include the inherent risks related to interventions into the human brain, including collateral damage due to transplantation and uncontrolled cell growth, which can have profound effects in the brain that may not be reversible.

6.6 Ensuring Understanding

As described earlier (see Sect. 6.2), it is essential that those being asked to provide informed consent understand the information that has been disclosed. This can be especially challenging in situations where the science is novel and in clinical settings where standard treatment options are not ideal.

Emerging empirical data on patients' perception of organoids in general and brain organoids in particular (see Sect. 6.4) hint at some of the challenges that will be encountered in this setting. Specifically, the proclivity for patients to use science fiction when conceptualizing and describing organoids needs to be countered by current realities. To make matters worse, the hype associated with the use of brain organoids must also be overcome so that informed consent can be obtained.

Given this state of affairs it may be prudent to develop and use balanced standard materials describing brain organoids during the consent process. Such materials could help trigger discussion about organoids as well as the specific proposed use, which promises to be helpful since extended discussions during the consent process are associated with enhanced understanding.[20] In situations where the risks are particularly high, consideration should be given to formally assessing understanding prior to seeking consent.

6.7 Authorization

The documents used when obtaining written authorization should include key aspects of what has been disclosed during the consent process. The International Society for Stem Cell Research has offered sample consent documents that have

[20] Nishimura et al. (2013).

some relevance to brain organoid research. Similarly, at Johns Hopkins Medicine, the Institutional Review Board has drafted templates to be used in specific research settings, which can then be tailored based upon the proposed research and certain regulatory requirements. For example, the informed consent template research involving pluripotent stem cells suggests the following text in regard to basic science research organoids: "We may use the cells taken from your [specify source of cells, e.g. skin] to create what is sometimes called an 'organoid.' An organoid is an organized cluster of cells, grown in the lab, which are designed to mimic organ structure and function. Organoids can be used to help understand diseases and treatments for them."[21]

6.8 Concluding Comments

While the ethical requirement to obtain explicit informed consent for brain organoid research and clinical translation seems clear, doing so in practice may be challenging due to rapid scientific progress, changing policies and practices, baseline understanding of brain organoids, and local contexts. Properly addressing these challenges will be facilitated by empirical data and sharing experiences regarding effective (and ineffective) approaches. For example, gathering additional data regarding patients' knowledge, attitudes, and beliefs about brain organoid research in different settings is needed. In addition, the materials used to disclose information about brain organoid research and clinical translation could be developed using formative research methods to help ensure understanding.[22] Similarly, novel approaches to consent (e.g., consent for governance) should arguably be tested rather than simply implemented since even well-considered interventions aimed at improving consent can fail in practice.[23] While such efforts will require time and resources, they should help to meet the ethical justification for informed consent as well as some of its ethically important goals.

References

Beauchamp TL, Childress JF (2019) Principles of biomedical ethics, 8th edn. Oxford U Press, New York

Berkers G, van Mourik P, Vonk AM et al (2019) Rectal organoids enable personalized treatment of cystic fibrosis. Cell Rep 26(7):1701–1708

Boers SN, Bredenoord AL (2018) Consent for governance in the ethical use of organoids. Nat Cell Biol 20(6):642–645. https://doi.org/10.1038/s41556-018-0112-5

Boers SN, van Delden JJ, Clevers H, Bredenoord AL (2016) Organoid biobanking: identifying the ethics: organoids revive old and raise new ethical challenges for basic research and therapeutic use. EMBO Rep 17(7):938–941. https://doi.org/10.15252/embr.201642613

[21] Johns Hopkins Medicine IRB (2016).

[22] Taylor et al. (2007).

[23] Lavori et al. (2007).

Boers SN, de Winter-de Groot KM, Noordhoek J et al (2018) Mini-guts in a dish: perspectives of adult cystic fibrosis (CF) patients and parents of young CF patients on organoid technology. J Cyst Fibros 17(3):407–415
Bollinger J, May E, Mathews D et al (2021) Patients' perspectives on the derivation and use of organoids. Stem Cell Reports 16(8):1874–1883. https://doi.org/10.1016/j.stemcr.2021.07.004
Chapman AR (2019) Brain models in a dish: ethical issues in developing brain organoids. AJOB Neurosci 10(3):113–115. https://doi.org/10.1080/21507740.2019.1632963
Chen HI, Wolf JA, Blue R, Song MM, Moreno JD, Ming GL, Song H (2019) Transplantation of human brain organoids: revisiting the science and ethics of brain chimeras. Cell Stem Cell 25(4):462–472
Dasgupta I, Bollinger J, Mathews DJ et al (2014) Patients' attitudes toward the donation of biological materials for the derivation of induced pluripotent stem cells. Cell Stem Cell 14(1):9–12. https://doi.org/10.1016/j.stem.2013.12.006
Dekkers JF, van der Ent CK, Beekman JM (2013) Novel opportunities for CFTR-targeting drug development using organoids. Rare Dis 11(1):e27112. https://doi.org/10.4161/rdis.27112
Dickert NW, Eyal N, Goldkind SF et al (2017) Reframing consent for clinical research: a function-based approach. Am J Bioeth 17(12):3–11. https://doi.org/10.1080/15265161.2017.1388448
Greely HT, Cho MK, Hogle LF et al (2007) Thinking about the human neuron mouse. AJOB 7(5):27–40. https://doi.org/10.1080/15265160701290371
Haselager DR, Boers SN, Jongsma KR et al (2021) Breeding brains? Patients' and laymen's perspectives on cerebral organoids. Regen Med. https://doi.org/10.2217/rme-2020-0108
Hyun I, Scharf-Deering JC, Lunshof JE (2020) Ethical issues related to brain organoid research. Brain Res 1732:146653. https://doi.org/10.1016/j.brainres.2020.146653. Epub 2020 Feb 1
International Society of Stem Cell Research (2021) ISSCR Guidelines for stem cell research and clinical translation. Retrieved May 8, 2022, from https://www.isscr.org/docs/default-source/all-isscr-guidelines/2021-guidelines/isscr-guidelines-for-stem-cell-research-and-clinical-translation-2021.pdf?sfvrsn=ced254b1_4
Johns Hopkins Medicine Institutional Review Board. (2016). Human pluripotent stem cell (hPSC research informed consent/authorization template. Retrieved February 2, 2021, from https://www.hopkinsmedicine.org/institutional_review_board/forms/
Lavori P, Wilt T, Sugarman J (2007) Quality assurance questionnaire for professionals fails to improve the quality of informed consent. Clin Trials 4:638–649
Lavori PW, Krause-Steinrauf H, Brophy M et al (2002) Principles, organization, and operation of a DNA bank for clinical trials: a Department of Veterans Affairs cooperative study. Control Clin Trials 23:222–239
Li S, Wang M, Zhou J (2020) Brain organoids: a promising living biobank resource for neuroscience research. Biopreserv Biobank 18(2):136–143. https://doi.org/10.1089/bio.2019.0111
Lowenthal J, Lipnick S, Rao M, Hull SC (2012) Specimen collection for induced pluripotent stem cell research: harmonizing the approach to informed consent. Stem Cells Transl Med 1(5):409–421. https://doi.org/10.5966/sctm.2012-0029
Munsie M, Hyun I, Sugarman J (2017) Ethical issues in human organoid and gastruloid research. Development 144:942–945. https://doi.org/10.1242/dev.140111
Nishimura A, Carey J, Erwin PJ et al (2013) Improving understanding in the research informed consent process: a systematic review of 54 interventions tested in randomized control trials. BMC Med Ethics 14:28. https://doi.org/10.1186/1472-6939-14-28
Taylor HA, Sugarman J, Pisetsky DS, Bathon J (2007) Formative research in clinical trial development: attitudes of patients with arthritis in enhancing prevention trials. Ann Rheum Dis 66:542–544

The Legal Requirements for—and Limits to—the Donor's and the Patient's Consent

7

Silvia Deuring

7.1 Basic Issues of Informed Consent

7.1.1 "The" Consent in the Context of Generating Brain Organoids

The cells required for the generation of brain organoids can stem from different sources. They can be adult, multipotent stem cells obtained from donor material, which can only differentiate into certain cell types; pluripotent stem cells; stem cells artificially produced from differentiated soma cells, the so-called induced pluripotent stem cells (hiPSCs); and, finally, embryonic stem cells.[1]

If one considers the process of creating brain organoids, it quickly becomes clear that it is hardly possible to speak of "the" consent of the cell donor. In particular when brain organoids are generated from adult stem cells or hiPSCs, a distinction must be drawn between the consent for the collection of the corresponding cells on the one hand and the consent for specific-purpose use after the collection on the other hand. This need to differentiate already results from the fact that different legal interests of the cell donor are affected; moreover, the person collecting the cells and the person using them do not have to be identical.[2] These consents do not always coincide; for instance, the cell collection may originally have been carried

[1] Lancaster et al. (2013), p. 374; Bartfeld and Clevers (2018), p. 91; Taupitz (2020a), p. 805.

[2] On the rights regarding severed bodily substances, see below, Sect. 7.2.2.2. See also Halàsz (2004), p. 216; Central Ethics Committee (2003), pp. 5-6, both also arguing that the removal interferes with bodily integrity, whereas the further use of seperated body substances can violate (only) the right of personality; and Parliamentary Document 16/5374 (2007), p. 72, which also differentiates between removal and further use.

S. Deuring (✉)
Faculty of Law, Ludwig Maximilian University, Munich, Germany
e-mail: silvia.deuring@jura.uni-muenchen.de

H.-G. Dederer, D. Hamburger (eds.), *Brain Organoids in Research and Therapy*, Advances in Neuroethics, https://doi.org/10.1007/978-3-030-97641-5_7

out for completely different purposes, e.g., for diagnostic or therapeutic reasons, with the decision to use the cells for research purposes arising later. This problem does not occur, however, in case of embryonic stem cell use, at least in Germany, where only imported and thus already extracted embryonic stem cells may be used. Incidentally, this study will not touch upon the use of embryonic cells.

In the following, we will proceed as follows: First, we will address general principles of informed consent in German and international law. The concerned person's right to physical integrity as well as of personality will play an important role in this exposition (Sect. 7.1.2). We will subsequently examine the effectiveness of informed consent to the removal of cells specifically for the purpose of generating brain organoids (Sect. 7.2.1). In so doing, we must distinguish between different research objectives and assess how specific the consent must be. This is followed by the question of whether bodily substances that were removed for completely different purposes can also be used without consent for the generation of brain organoids. The answer will depend on the scope of the right of personality regarding separated bodily substances (Sect. 7.2.2). We will then deal with questions of informed consent in autologous and allogeneic transplantation of brain organoids; in the case of allogeneic transplantation, we must differentiate between the donor and the recipient (Sect. 7.3). The study concludes with a brief look at issues of consent in data protection law (Sect. 7.4).

The requirement of informed consent as well as the criteria which the consent must meet differ according to the concerned group of persons. For that reason, the following sections will distinguish between adults, minors who are capable of advanced reflection, and incapacitated persons.

7.1.2 Informed Consent in International and German Law

7.1.2.1 International Standards

At the international level, there are numerous sets of rules predicated on the principle of informed consent, be it in the medical field in general or in the field of scientific research in particular.

For example, the Convention on Human Rights and Biomedicine of the Council of Europe ("Oviedo Convention") from 1997 provides in its Art. 5 that an intervention in the health field may only be carried out after the person concerned has given free and informed consent to it. Although this convention is not binding for Germany,[3] which has not signed nor ratified it, it is binding on a large number of other European states.

[3] In the opinion of the German legislature, it sets too low requirements in some areas and provides too little protection for certain groups of persons. For that reason, some argue that the Convention should be taken into account to the extent it sets higher, not lower, requirements than the German legal system. See, e.g., Breithaupt (2012), p. 243.

The "Oviedo Convention" as well as additional protocols[4] and recommendations[5] to this convention also contain specific regulations on consent to research projects: Pursuant to Art. 16 para. 5 of the Convention, research may only be conducted after informed consent has been given. This requirement is reiterated in Art. 14 of the Additional Protocol to the Convention of Human Rights and Biomedicine Concerning Biomedical Research, and Art. 13 of the Additional Protocol sets out the informed-consent requirements in more detail. This Additional Protocol applies not only to research performed directly on human beings but also to research on body materials taken for the purpose of carrying out that specific research project.[6]

For research on other bodily materials of human origin, i.e., those removed for initial storage and later use in a (still undetermined) research project, or those removed for any other reason other than to carry out the planned research project, the Committee of Ministers of the Council of Europe has drawn up Recommendation CM/Rec(2016)6 on research on biological materials of human origin. Art. 10 and Art. 11 para. 1 of this Recommendation regulate the requirements for informed consent for storage-related collection. Art. 11 para. 2 then deals with consent to storage if the bodily substances were removed for purposes other than storage for research; para. 3, furthermore, addresses the storage of bodily materials that can no longer be identified. Finally, Art. 21 regulates the use of the stored materials for research purposes and establishes the principle, in para. 1, that research on these substances may only be carried out following appropriate consent. Para. 2 deals with exceptions to this principle.[7] In this context, the World Medical Association's (WMA) Declaration of Taipei on Ethical Considerations Regarding Health Databases and Biobanks (2016), which deals with informed consent to the collection, storage, and use of data and biological material from individuals, also deserves mention.

The Declaration of Helsinki of the WMA, as amended by the 64th General Assembly in Fortaleza in 2013, refers specifically to scientific research. It applies to research on humans, including research on identifiable human materials (para. 2 of the preamble) and thus ultimately to the removal and further use of bodily substances, provided they remain identifiable. In particular, para. 26 of the Declaration deals with the scope of informed consent and the obtaining of consent, preferably in writing. In Germany, this declaration must be observed (only) by physicians, pursuant to sec. 15 MBO-Ä (Model Professional Code of Conduct of the German Medical Association) or, to be more precise, the corresponding

---

[4] Additional Protocol to the Convention on Human Rights and Biomedicine Concerning Biomedical Research (2005).

[5] Recommendation Rec(2016)6 of the Committee of Ministers to member states on research on biological materials of human origin.

[6] This can be deduced from Art. 2 no. 2 of the Recommendation CM/Rec(2016)6: "This recommendation does not apply to (...) the use in a specific research project of biological materials of human origin removed for the sole purpose of that project. This is within the scope of the Additional Protocol concerning Biomedical Research (CETS No. 195)."

[7] See Sect. 7.2.2.3.

references in the professional regulations of the federated state medical associations.[8]

The various documents also deal with medical interventions on persons incapable of consent, be they minors or adults. The Oviedo Convention, for example, leaves it to national law to determine when a minor or adult is incapable of giving consent (Art. 6 of the Convention), but provides that in cases of incapacity the consent of the representative must be obtained before any medical intervention (Art. 6 para. 2 and 3 of the Convention). The participation of persons incapable of consenting to research interventions is possible but subject to restrictions (Art. 17 of the Convention): Research on persons incapable of consenting is subsidiary and may only be carried out, furthermore, if the research has the potential to produce real or direct benefits for the health of the person concerned or for other persons of the same age afflicted with the same disease or disorder or having the same condition. Moreover, the research may entail only minimal risk and minimal burden for the individual concerned, and the representative must have given their authorization explicitly and in writing. Furthermore, the person concerned may object to the intervention.

The Declaration of Helsinki also considers permissible, to a limited extent, research interventions on persons incapable of consent; in addition to proxy consent, it requires either an individual benefit, or, like the Oviedo Convention, a benefit for the health of the group represented by the subject, provided the research entails only minimal risks and minimal burdens (para 28). The incapacitated person must be involved in the decision, and his or her objection must be respected (para. 29). Finally, Art. 12 and Art. 21 of Recommendation CM/Rec(2016)6 deal with the removal and storage of bodily substances from persons not able to consent for research purposes. Art. 12 para. 1 and 2 and Art. 21 para. 5 also assume the subsidiarity of such research and require a benefit for the person unable to consent or, failing that, a group benefit. In addition, the removal of the materials may only be accompanied by a minimal risk and a minimal burden.

Moreover, there are specific regulations on the conduct of clinical trials, particularly in the area of medicinal products law. The Regulation (EU) No 536/2014 is particularly relevant in this regard. We will consider these provisions in greater detail below.[9]

7.1.2.2 German Law

Any interference with a person's physical integrity requires informed consent. In German law, this results in particular from the Basic Law ("*Grundgesetz*," GG), which includes a fundamental right to physical integrity (Art. 2 para. 2 sent. 1 GG) and a general right of personality (Art. 2 para. 1 in conjunction with Art. 1 para. 1

[8] Which is why, within the German legal system, the declaration is not only a recommendation. See Kern (2019a), § 4 III.6. para. 36.

[9] See in particular Sect. 7.3.1.3.

GG).[10] Of course, the fact that individuals *must* consent to physical interventions also means that they *may* consent to such measures. Thus, individuals may dispose of their bodies as they see fit.[11] Thus, cell donors may, within the context analyzed in this chapter, consent to the removal of cells not only for therapeutic and diagnostic purposes but also for research purposes, for transplantation purposes, for storage in tissue banks, or for any other kind of processing.[12]

The requirement of informed consent specifically for medical treatments follows from sections 630d and 630e of the Civil Code ("*Bürgerliches Gesetzbuch,*" BGB). The Model Professional Code of Conduct—which is not binding in itself but becomes binding through the adoption of the corresponding clauses in the professional regulations of the federated states' medical associations—also establishes, in its section 8, the requirement of informed consent for medical treatment. Should the person be incapable of consenting, the legal representative or the person authorized for this purpose must be informed and give his or her consent (sec. 630d para. 1 sent. 2, para. 2, sec. 630e para. 4 BGB); failure to do so results in claims for damages under sec. 280 para. 1 BGB and sec. 823 para. 1 BGB.

The Civil Code does not define when a person is capable or incapable of giving consent. According to the explanatory memorandum to the law, patients must be capable of understanding the information so that they may make a self-determined decision and be able to assess the benefits and risks of the specific intervention.[13] Some special laws, however, provide specific definitions: For example, sec. 40b para. 3 sent. 1 of the German Medicinal Products Act (version as of 27 January 2022) defines the capacity of minors to give consent as the ability to grasp the nature, significance and scope of the clinical trial and to act accordingly. The same definition, of course with regard to transplantation, is found in sec. 8 para. 1 sent. 1 no. 5 and sec. 8c para. 2 sent. 1 of the Transplantation Act.

As is already clear from the cited provision of the German Medicinal Products Act, minors are not per se incapable of consent under German law. Instead, their ability to consent depends on their mental maturity.

It is uncontested that parents cannot force a medical intervention against the will of their child if it is capable of giving consent.[14]

It is unclear, however, whether the consent of mature minors is sufficient, or whether their parents must always consent as well.[15] This question arises in equal

---

[10] Spickhoff (2008), p. 385; Taupitz (2000), p. A12. On the problem of which specific article provides the basis for the right to self-determination, see Müller (2013), p. 175 et seq. Müller argues that the right to consent to physical interventions follows from Art. 2 para. 2 p. 1 GG, and that the right to be informed follows from Art. 2 para. 1 in conjunction with Art. 1 para. 1 GG.

[11] Schroth (2009), p. 722; Halàsz (2004), p. 19.

[12] On possible uses of collected somatic cells, Dettmeyer and Madea (2004), p. 86.

[13] Wagner (2020a), sec. 630d BGB para. 21; for more details, see Taupitz (2012), p. 585; Taupitz (2000), pp. A58 et seq.

[14] Schreiber (2019), p. 223.

[15] On this issue, Katzenmeier (2020), sec. 630d BGB para. 13 et seq; Fink (2005), pp. 78–79; Schreiber (2019), p. 154 et seq.

measure with respect to all personal rights, be it physical integrity or the general right of personality.[16] The starting point is secs. 1626 et seq. BGB. According to these provisions, parents are entitled to parental care for their children; at the same time, however, they must consider the child's increasing ability and need to act independently and responsibly (sec. 1626 para. 2 BGB). In addition, the child has a constitutionally guaranteed right to self-determination. In any event, we can observe a clear tendency: Where there is a risk of grave danger or interventions are irreversible, the requirements for the minor's capacity to consent are either toughened to the point where it can no longer be assumed,[17] or co-consent is deemed necessary.[18]

Some scholars go so far as to argue that both the minor and their legal representative must always give their consent; this, they claim, considers the minor's right to self-determination, by actively involving him or her, and at the same time does not dilute the parents' right of care. Only where risks for the minor are so low that the parents' "interference" is no longer justified should the minor be regarded as autonomous.[19] Yet there is no reason to correct decision of minors if they can exercise their right to self-determination.[20] In principle, therefore, as is also the prevailing doctrine in criminal law,[21] we should allocate an exclusive right to consent, at least in cases in which the minor is not in serious danger. To insist on the principle of co-consent would yield the contradictory outcome that the minor is considered to be, but not treated as, fully capable of consenting.[22] Incidentally, the more a decision implicates the privacy of minors, the more likely it is that their right of personality will override the parents' right of custody, even in the case of irreversible interventions.[23]

For adult persons under custodianship, the custodian must give consent if the person under custodianship is incapable of doing so (secs. 1896 et seq. BGB).[24]

It is also important to note that consent can be revoked at any time informally and without giving reasons (sec. 630d para. 3 BGB). We can also find this provision in specific regulations such as sec. 40b para. 1 of the German Medicinal Products Act (version as of 27 January 2022), Art. 29 para. 2 lit (a) (ii) of the Regulation (EU) No 536/2014.

---

[16] Fink (2005), p. 79; on the consent with regard to the right of personality, Schreiber (2019), p. 223 et seq.

[17] Spickhoff (2018a), sec. 107 BGB para. 15; Wagner (2020a), sec. 630d para. 43.

[18] Katzenmeier (2020), sec. 630d BGB para. 14; Lipp (2021c), XIII. D. para. 38; limited to cases of risk of death or considerable damage to health: Taupitz (2000), p. A63 et seq.; on the minor's right of veto, Federal Court of Justice (2007), p. 218. The right to veto differs from co-consent in that the minor does not have to actively exercise his or her right of self-determination. Schreiber (2019), p. 166.

[19] Fink (2005), pp. 79–80; Schreiber (2019), pp. 212, 223–224, likewise argues in favor of a right of co-decision.

[20] Kern (1994), p. 755.

[21] Spickhoff (2018b), sec. 630d para. 8.

[22] Spickhoff (2018a), sec. 107 para. 15; Spickhoff (2008), pp. 389–390; see also Wagner (2020a), sec. 630d para. 43–44.

[23] Thus, in the case of abortion, Higher Regional Court Hamm (2020), p. 1374. Generally, Spickhoff (2018a), sec. 107 para. 15.

[24] Katzenmeier (2020), sec. 630d para. 18 et seq.; Taupitz (2000), pp. A67–A68.

7.2 Generating Brain Organoids for Research

Brain organoids can be generated from cells that have been harvested either for that specific purpose (Sect. 7.2.1) or for another reason, e.g., for therapeutic reasons, with the wish to conduct research with these cells arising at a later point (Sect. 7.2.2). The following sections will address the requirement and scope of informed consent in both situations.

7.2.1 Removal of Body Material Specifically for the Generation of Brain Organoids

Brain organoids can be used for different research purposes. We will therefore differentiate between consent to basic research *in vitro* (Sect. 7.2.1.1), to transplanting brain organoids to animals (Sect. 7.2.1.2) and to drug research as well as personalized medicine (Sect. 7.2.1.3). Finally, we will address the possible scope of the consent (Sect. 7.2.1.4).

7.2.1.1 Removal for the Generation of Brain Organoids for Basic Research In Vitro

7.2.1.1.1 Consent to Research of Adults Capable of Giving Consent

In Germany, only certain fields of research on humans are regulated by specific laws. Where specific regulations do not apply, the legal requirements for human research thus follow from the general provisions of private, criminal, and public law.[25] Sec. 15 para. 3 of the Model Professional Code of Conduct for Physicians, or more precisely the corresponding clauses in the professional regulations of the federated states' medical associations, moreover, establishes the binding force of the Declaration of Helsinki.

The following section addresses basic in vitro research, that is, investigations into the development of certain brain diseases or the development of the human brain in general. It is important to bear in mind, as mentioned in Sect. 7.1.1, that we must distinguish the consent to the physical intervention from the consent to specific further uses of the removed materials.

First of all, the removal of cells to create brain organoids for whatever reason is only possible following informed consent. The concept of voluntary consent is an expression of the right of self-determination and is recognized across the world.[26] The information given before consent has, in the words of the Declaration of Helsinki, to be "adequate" (para. 26): It should enable the subject to weigh the pros and cons of participating in the research project and to make up his or her mind freely. For this reason, the physician must set out the scientific justification of the research project; the planned interventions and the risks involved and how the substances and the concerned person's data will be used, including whether the data

[25] Lipp (2021c), XIII. B. para. 13.

[26] Lipp (2021c), XIII. E. I. para. 50; see above, Sect. 7.1.2.

will be disclosed to third parties. In short, all circumstances that are relevant for the person concerned must be explained to her.[27]

To ensure that the consent remains voluntary, the person concerned can revoke it at any time (see, for example, para. 26 of the Declaration of Helsinki).[28] The physician must also inform the individual about what will happen to the samples, data, and research results should he or she indeed revoke his or her consent.[29] However, the consent cannot be revoked if the substances have been anonymized; the physician must inform the individual of this possibility.[30]

Despite this right to self-determination, it follows from both national and international rules that in vitro brain research is not without limits. The preamble of the Additional Protocol to the Convention on Human Rights and Biomedicine Concerning Biomedical Research (2005) explicitly states that biomedical research that is contrary to human dignity and human rights should never be carried out, and that the paramount concern is the protection of the human being participating in research. According to its Art. 1, the overall aim of the Convention of Oviedo is to protect the dignity and identity of all human beings with regard to the application of biology and medicine. Para. 9 of the Declaration of Helsinki, moreover, likewise emphasizes the protection of dignity.

In the following, I will lay out the criteria that, according to German law, make inadmissible the consent to a physical intervention or to a research project in general.

First of all, the basis for the removal and further use of body cells is a research contract. Its validity may depend on ethical standards, which are translated into law by sec. 134 (violation of a statutory prohibition) and sec. 138 BGB ("immorality").[31] However, we must distinguish the consent regarding the physical intervention and the further use of the substances from the contract. Thus, consent is possible up to the point of "immorality." It is unclear, however, how to ascertain immorality.

Under criminal law, for instance, there is sec. 228 Criminal Code ("*Strafgesetzbuch,*" StGB), according to which a person cannot consent to immoral bodily harm. A physical intervention is considered immoral—at least according to prevailing doctrine and case law[32]—if it is accompanied by a serious danger to health with the risk of death. The objective of the intervention is, in principle, irrelevant; in other words, an immoral objective does not vitiate the consent to a minor

[27] Lipp (2021c), XIII. E. I. para. 50; see also, National Ethics Council (2004), pp. 16–17, 64–65.

[28] Lipp (2021c), XIII. E. I. para. 51.

[29] National Ethics Council (2004), p. 65; see also Baston-Vogt (1997), p. 274 et seq.

[30] See, e.g., para. 12 of the WMA Declaration of Taipei on Ethical considerations regarding Health Databases and Biobanks (2016); for revocation and its consequences, see National Ethics Council (2004), pp. 69–70.

[31] Kern and Rehborn (2019), sec. 42 para. 79.

[32] For this interpretation, see Stock (2009), p. 155; Suhr (2016), p. 172; Sternberg-Lieben (2019), sec. 228 StGB para. 17–18; Federal Court of Justice (2004), p. 2459.

intervention.[33] However, the objective does play a role when the intervention in question is severe: A severe intervention may be justified following consent if there is a medical reason for it. Consequently, the consent is vitiated if no such reason exists. Yet the immorality of the consent follows from the absence of such reason, not from an "immoral" objective.[34]

In private law, on the other hand, we are left with sec. 138 BGB and sec. 134 BGB, which decree the nullity of immoral contracts and of contracts that violate a statutory prohibition. Because consent is not a contract, however,[35] the question arises as to how we should gauge the immorality of consent, e.g., in the context of sec. 823 BGB, which grants claims for damages for physical interventions carried out without consent. Often scholars will simply re-state that consent must not be immoral; some fail to relate immorality to a specific rule,[36] while others refer to both sec. 228 StGB and sec. 138 BGB.[37] The Federal Court of Justice has also occasionally referred to both provisions;[38] in other instances, it has suggested an analogy to sec. 138 BGB,[39] while in others still it has applied that provision directly.[40]

We will here follow the prevailing approach and apply the principles developed in sec. 228 StGB in private-law contexts as well. This indeed guarantees a uniform standard in matters of the right freely to dispose of one's own body.[41] Furthermore, in light of the right to self-determination, this liberty should only be restricted within narrow limits, namely only when the very objective of protecting the body so demands. The threshold is only reached where the life of the person is at stake, as German law prohibits killing upon request (sec. 216 StGB).[42]

---

[33] On the irrelevance, in principle, of ulterior objectives, Sternberg-Lieben (2019), sec. 228 StGB para. 19; Federal Court of Justice (2004), p. 2459; Förster (2020), sec. 823 BGB para. 34; and, as regards the result, Ohly (2002), p. 421.

[34] Hardtung (2017), sec. 228 StGB para. 47; see also Ohly (2002), pp. 421 et seq.

[35] Förster (2020), sec. 823 BGB para. 35; but see Ohly (2002), p. 408.

[36] Spindler (2020), sec. 823 BGB para. 84; Förster (2020), sec. 823 para. 34; Deutsch and Spickhoff (2014), para. 419.

[37] Halàsz (2004), p. 216; Wenzel (2019), chapter 4 para. 158; Prütting and Merrem (2019), sec. 630d BGB para. 11.; diese Uneinheitlichkeit auch feststellend: Ohly (2002), pp. 397–398.

[38] Federal Court of Justice (2017), p. 2686.

[39] Federal Court of Justice (1976), p. 1790.

[40] Federal Court of Justice (1953), p. 701.

[41] Sec. 138 BGB has a much broader scope of application than sec. 228 StGB. Cf. Stock (2009), p. 158. See the rulings of the Federal Court of Justice (1976), p. 1790; Federal Court of Justice (1953), p. 701), which measured the immorality of consent in private law against sec. 138 BGB (applied analogously) and did not focus on a danger to life. Instead, they asked whether the interventions in the body generally violate "what moral conduct is required of the individual within the social community according to the prevailing views of our legal and cultural society" (see Federal Court of Justice (1976), p. 1790, on the case of sterilization). On the application of criminal law restrictions to private law, Ohly (2002), pp. 400 et seq. According to Ohly (2002), we should ask whether the restriction seeks to protect the person consenting, in which case the invalidity of the consent extends to private law (pp. 405–407).

[42] Stock (2009), p. 156, Suhr (2016), p. 172, and Schroth (2009), p. 726, also argue that societal aspects should be left out of the equation when applying sec. 228 StGB, as they would render moot

It follows from this that research subjects may not be exposed to the concrete danger of death or serious bodily harm. Within these limits, medical research on humans is, however, permissible.[43] Accordingly, the removal of cells for the generation of brain organoids is not subject to any reservations, since the removal does not involve any such risks for the cell donor—at least not if the cells are, e.g., blood or skin cells. Whether the person consents for immoral purposes is, as shown above, irrelevant.

Whether the consent to the specific further use of the substances can be considered immoral presents a distinct question. In this context, sec. 228 StGB does not offer any help: first, the use of substances that have already been removed no longer constitutes an encroachment on bodily integrity; second, the provision may not be extended to include other legal interests. As severed bodily substances are covered not by the right of bodily integrity but "only" by the right of personality (and by the right to property),[44] it is questionable whether disposing over those substances can ever be considered immoral: After all, the core of the right of personality consists in the right to define freely what constitutes one's personality, which includes the relationship one desires with one's bodily substances. There is, in particular, no provision in criminal law that sanctions "destroying" another person's right of personality even when that person has granted his or her consent (as is the case, pursuant to sec. 216 StGB, with the right to life).

For that reason, only one other aspect requires consideration: Since the production of brain organoids can at least be considered controversial from an ethical perspective, we must examine whether Art. 1 para. 1 GG, which protects the dignity of the human being, can stand in the way of giving consent to such research. Can Art. 1 para. 1 GG, in other words, restricts the right freely to develop one's own personality and to live accordingly? More, does the consent to the creation of brain organoids from one's own cells even conflict with human dignity?

There is certainly no universal understanding of what human dignity entails. Accordingly, I will briefly attempt to outline a concept of dignity and to apply it to the issue before us.

The German Constitution entrenches the protection of human dignity in its very first provision. The basic premise is that every human being is entitled to recognition of his or her unique value.[45] The special value of human beings follows from their capacity to reason as well as their autonomy. This approach goes back to Immanuel Kant in particular, but can be traced back even further.[46] From a theological perspective (not only that of Christianity[47]), human beings hold a

---

the right to self-determination.

[43] Lipp (2021b), VII. E. I. para. 41; Ohly (2002), p. 426.

[44] See Sect. 7.2.2.2.

[45] Dederer (2009), p. 109.

[46] Dederer (2009), p. 107 with reference to John Locke, Giovanni Pico de la Mirandolla and Marcus Aurelius; Lackermair (2017), p. 293. In the field of ethics, the moral status of human beings is often justified with reference to *Kant* or to the autonomy of human beings as well: Chen et al. (2019), p. 466; Karpowicz et al. (2004), p. 334.

[47] Dederer (2009), p. 108.

special position among all creatures (*imago-dei doctrine*), regardless of their respective abilities.[48] From a historical perspective, moreover, Art. 1 GG was a reaction to the Nazi regime; following this period of absolute disregard for the human value of individuals, it seeks to make the respect for the freedom and self-determination of the individual—and indeed of every individual, irrespective of his or her mental or physical characteristics, religion, or other features—the matrix of the state order.[49]

For its part, the Federal Constitutional Court has not explicitly predicated its jurisprudence on specific philosophical or theological ideas. On the basis of the same considerations,[50] however, it has emphasized that dignity is a "value which belongs to man by virtue of his being a person"[51] (although it is best to speak of "human-ness," not of persons[52]). It has also held that Art. 1 GG "is based on the concept of a spiritual and moral being which is designed to determine and develop itself in freedom."[53]

Consequently, we can speak of a violation of dignity when persons are subjected to treatment that calls into question, in a fundamental manner, their right to be respected as a self-determined individual entitled to all humans rights, or when the treatment of human beings bespeaks a wilful disregard for their dignity. The treatment, in other words, must express contempt for the value that an individual has by virtue of his or her personhood.[54] This must be established on a case-by-case basis.[55] To conclude, human dignity implicates the respect for the freedom, autonomy, and uniqueness of each human being as well as respect for the equality of all individuals.[56]

In light of these principles, one could argue that the creation of artificial brains from the cell material of a donor calls into question that person's uniqueness as this creation "duplicates," as it were, his or her brain and thus, ultimately, his or her personality. I do not consider this objection persuasive, however, at least as long as the brain organoids remain as rudimentary as they are today: As things currently stand, there is no evidence that brain organoids have any degree of consciousness or could be able to generate more complex information of any kind.[57] Because of inadequate nutrient, gas, and waste exchange, they are only the size of a few millimeters.[58]

---

[48] Dederer (2009), pp. 107–108; Lackermair (2017), p. 290.

[49] Lackermair (2017), pp. 293–294.

[50] Thus also, Dederer (2009), p. 108.

[51] Federal Constitutional Court (1970), p. 26.

[52] Dederer (2009), p. 108; Lackermair (2017), pp. 296–270.

[53] Federal Constitutional Court (1977), p. 227; cf. Dederer (2009), p. 108.

[54] Federal Constitutional Court (1970), p. 26. For criticism of the "wilfulness" personhood criteria, see Lackermair (2017), pp. 269–270.

[55] For more details: Dederer (2009), p. 118 et seq.

[56] Lackermair (2017), p. 297.

[57] Chen et al. (2019), p. 463.

[58] Chen et al. (2019), p. 463.

Furthermore, brain organoids are not even necessarily a miniature version of the entire donor's brain, since they can be limited to region-specific parts.[59] The image of the so-called brain in a vat—that is, "a disembodied organ capable of perception and thought imprisoned in a dehumanizing existence"—is currently far from realistic, given the "lack of sophisticated sensory inputs into developing brain organoids" necessary for "the iterative learning and conditioning that cultivate cognitive processes."[60]

But even once artificial brains become more sophisticated, we should not prematurely assume a violation of the donor's dignity. The following points deserve to be considered:

Since the direct effect on the cell donor is ultimately limited to the removal of the cells, it already seems questionable whether the manner in which the body substances are used can call into question the intrinsic value of the donor.[61]

As brain organoids, in principle, share the individual's nuclear genetic set, they may come close to a cloned version of the donor's brain.[62] Even "real" cloning, however, does not automatically violate the dignity of the copied original—at least not simply because the original now shares his or her genetic setup with another living being (and thus also with a brain of the same genetic origin): the case of identical twins demonstrates that the same genome does not entail the same identity and personality.[63] Thus, one's identity and personality flow above all from one's environment and one's own history, and not just from one's genetic material, whose effects are also considerably influenced by external factors.[64] What is true for cloning probably applies *a fortiori* for brain organoids. If only the brain is "reproduced"—be it a fully functional human brain—far more characteristics that determine the identity and individuality of a human being are lacking, such as his or her appearance, behavioral patterns, etc. The identity of the new brain is new, not an imitation of that of the donor. Arguing otherwise overestimates the importance of the origin of the brain cells for the formation of a personality.[65] More, the mere uncertainty of the research outcome—how far developed the brain organoid may be—does not in itself render a given consent

59 Qian et al. (2016), pp. 1238 et seq.; Qian et al. (2018), pp. 565 et seq.

60 Chen et al. (2019), p. 465. See also Farahany et al. (2018), p. 430, who note that it is unclear whether brain organoids will attain consciousness in the future, and Lavazza and Massimini (2018), p. 608, who compare the challenge of detecting brain activity in cerebral organoids with the efforts to assess consciousness in brain-injured non-communicating patients. Incidentally, researchers have already managed to produce neural activity on a region where cells of the retina had formed together with cells of the brain. See Farahany et al. (2018), p. 430; Quadrato et al. (2017), pp. 48–53.

61 Thus, in the context of human–animal hybrids, Lackermair (2017), p. 299.

62 Cf. Lavazza and Pizzetti (2020), p. 11, who emphasize that brain organoids are not human beings who are genetically identical to the cloned "original."

63 Lackermair (2017), p. 301.

64 Kersten (2004), p. 491; Lackermair (2017), p. 301.

65 Lackermair (2017), p. 299.

invalid, provided it is ensured that the person giving consent is aware of the uncertainty.[66]

The question, finally, is whether the cultivation of artificial brains constitutes a form of arbitrary manipulation of human beings in general.[67] However, we should be wary of arguments of such general nature. Humans are arbitrarily manipulable in many respects, especially in medical ones: both medical treatments and research measures influence the human body in a more or less artifical way.[68] Furthermore, it is also questionable to assume a dignity of humanity as a whole that can prevail over the rights of individuals.[69] Were we to consider the research measures described here contrary to a "dignity of humanity," we would simply give in to a more or less vague feeling of unease. In so doing, we would disregard, in a paternalistic manner, the will of the donor—who, in the end, consents to an ultimately harmless intervention (such as blood sampling) in order to further important research purposes (e.g.,[70] studying human brain development or modeling central nervous system disorders such as microcephaly,[71] autism spectrum disorders,[72] and Zika virus infections[73]).

Generally speaking, then, it is doubtful whether human dignity, which is supposed to ensure autonomy, can be used to frustrate the will of donors who have autonomously determined what their cells are to be used for. If there is a violation of dignity at all, then only of the "personality" that is "trapped" in the organoid; the only reason the corresponding research would have to cease would be to protect this personality, as opposed to the voluntary donor.[74]

7.2.1.1.2 Consent to Research Involving Persons Incapable of Giving Consent

I will now address questions of consent to the removal of cells from persons incapable of giving consent. Again, consent bears upon two distinct rights: the right to physical integrity, which is implicated by the removal, and the general right of personality, which bears upon the further uses of the collected material. The problem

[66] As regards human cloning Frankenberg (2000), p. 330; as regards human–animal hybrids: Lackermair (2017), p. 288.

[67] See Lackermair (2017), p. 301, who asks that question with regard to hybrids and chimeras.

[68] Lackermair (2017), pp. 301–302.

[69] For arguments in favor, see German Ethics Council (2011), pp. 61–62. For objections, see Lackermair (2017), p. 350 et seq.

[70] Listed by Daviaud et al. (2018), p. 2.

[71] Lancaster et al. (2013), p. 373 et seq.; Li et al. (2017), p. 823 et seq.

[72] Forsberg et al. (2018), p. 1 et seq.

[73] Qian et al. (2016), p. 1238 et seq.; Watanabe et al. (2017), p. 517 et seq.

[74] Kersten (2004), p. 509 et seq.; Buchanan et al. (2012), p. 199; Dreier (2013), Art. 1 sec. 1 para. 109; Lackermair (2017), p. 302 et seq.; Spranger (2001), p. 242; and Schroth (2009), p. 722, all reject the idea of human dignity as a constraint on that person's own rights. For a more cautious approach, see Ohly (2002), p. 414. On Ohly's view, whether one has the right to dispose of one's own rights depends on whether this disposition would cause irrevocable loss of liberty, personal self-determination, or the essential factual prerequisites of a life lived autonomously. There is no such risk in our case, however.

here is that the provisions on the legal representation of minors obligate the parents to consider "the best interest" of the child (sec. 1627 BGB). As regards incapacitated adults, moreover, the law provides that the custodian must attend to the affairs of the person under custodianship in a manner that is conducive to his "welfare" (sec. 1901 para. 2 sent. 1 BGB); to do so, the custodian must consider the wishes of that person (sec. 1901 para. 2 sent. 1 and para. 3 BGB).

For that reason, many scholars doubt that research on persons incapable of giving consent is admissible. The research, they argue, is not in the concerned persons' best interest. Instead, it instrumentalizes them in violation of Art. 1 GG, at least if the research is not expected to be of direct benefit to them (as would be the case with therapeutic research). The inhumane experiments conducted during the Nazi regime suggest proceeding cautiously.[75]

Certainly, historical experience teaches us that the problem of research on persons incapable of giving consent should be handled with sensitivity, and that research with persons incapable of giving consent should be subject to strict conditions. It would go too far, however, to ban it altogether whenever it does not promise any benefit to the person concerned: If researching certain diseases or conditions necessarily involves persons incapable of giving consent and the results of these projects could allow to cure or at least alleviate the suffering of people with the same condition, research on this group of persons should be possible, provided it necessitates only minor physical interventions.[76] To ban research on persons incapable of consent altogether would neglect the right to life and health of other individuals affected by the same diseases and conditions.[77]

Of course, the end cannot justify the means, but we should refrain from labeling a minor intervention carried out for important reasons a violation of dignity. Whether there is a violation of dignity always depends on the circumstances, which means we must consider the intensity and the effects of the intervention as well as its objective.[78] Specific legislation, moreover, already permits some research for the benefit of others. According to the Medicinal Products Act, clinical trials on minors for the benefit of other minors are possible if the research is absolutely necessary, relates to a clinical condition from which the minor suffers, and is only associated with a minimal risk and a minimal burden (sec. 40b para. 4 sent. 1 lit. (a) of the Act in its version as of 27 January 2022). Thus, if we do not wish to consider this Act unconstitutional, we should not deem group-beneficial research implicating persons incapable of giving consent an automatic violation of their dignity.

Admittedly, the Medicinal Products Act deliberately does not allow the research for the benefit of others on *adults* who are incapable of giving consent (sec. 40b para. 4 sent. 3 of the Act in its version as of 27 January 2022). Yet we should not draw any conclusions from this omission; as a specific law, the Medicinal Products

---

[75] Spranger (2001), pp. 242–243 (see p. 242 for the permissibility of therapeutic research); see also Taupitz (2012), pp. 585–586, and the references cited therein.

[76] Taupitz (2012), p. 586. On the permissibility of research on minors that does not benefit them directly but involves only minor physical interventions, Lipp (2021c), XIII E. I. para. 107.

[77] Taupitz (2012), p. 586.

[78] Taupitz (2012), p. 586.

Act has no effect on research in other areas. Since the welfare of the person under custodianship also includes taking into account his or her preferences, and since these preferences are determinative if higher-ranking legal interests do not stand in the way, participation in research projects by persons under custodianship must be permissible, therefore, if the person concerned wishes to participate and if the research measure entails few, if any, risks.[79] However, even absent such a wish, I believe the well-being of the person is not endangered if he or she does not object,[80] the research involves only minor risks and burdens, and his or her legal representative consents.

It bears emphasizing that these findings are in line with international standards, which do not ban research on persons incapable of giving consent (see Art. 17 of the Oviedo Convention, paras. 28–30 of the Declaration of Helsinki, Art. 12 and 21 para. 5 of Recommendation CM/Rec(2016)6).[81]

From this we can draw the following conclusions for the collection of cells for the generation of brain organoids for research purposes: To begin with, the collection on persons incapable of consent is only permissible if it is associated with minimal risks and burdens, which is likely to be the case with a blood sample, or at least if blood has to be taken anyway.[82] In addition, however, the research objective must require involving persons incapable of giving consent, for instance, because it aims to investigate a disease or condition present in the individual concerned. So this prerequisite will, for example, not be met if the genetic disease to be investigated can simply be "programmed," through genetic modifications, into the cells taken from a person capable of giving consent. For that reason, genetic engineering of this sort should have precedence.

Of course, we must also consider whether the *further use* of bodily substances threatens the well-being of the person incapable of giving consent, thereby preempting his or her representative's consent to such projects. But I do not think that such use impairs the person's right to informational self-determination or of personality in general. Regarding the right to informational self-determination, the data protection regulations already provide sufficient protection against the illegal or improper use (including re-use) of the data.[83] Anonymization, insofar as it is compatible with the research objective, may serve to increase this protection.

Nor do I see a violation of the "right not to know" about one's own genetic makeup, a right that partakes in the protection of informational self-determination.

---

[79] On the relevance of the incapacitated person's wish to participate in the research, Lipp (2021c), XIII. E. I. para. 104. But see Spranger (2001), p. 243, who, to protect the person under custodianship from self-harm, argues that his or her wish is irrelevant if it involves any form of research.

[80] For the right to object, see below.

[81] See Sect. 7.1.2.1.

[82] On this controversial point, cf. Spranger (2001), p. 243.

[83] As regards minors, Schreiber (2019), p. 278 et seq, in particular p. 284-285. See Sect. 7.4, on the data protection requirements, pursuant to which the further use of data without the data subject's consent is only permissible if there is an appropriate legal basis and particular protective requirements are met; on the view I advocate, the further use of genetic data is unlawful if the data subject does not consent. Violations may incur onerous financial sanctions; see sect. 41 of the Federal Data Protection Law and Art. 83 para. 5 lit. (a) Reg. (EU) No 2016/679.

In principle, unsolicited feedback on suspicious findings following the genetic analysis of the samples might affect this right. But there is no consensus as to when this right is violated. Thus, it is disputed whether the violation of the "right not to know" requires the explicit prior statement, by the person concerned or his or her representative, that they "do not wish to know."[84] It is also doubtful whether the right is infringed if the information in question—as will often be the case—is given to the legal representative and not to the person to whom it refers.[85] (Although it remains possible, of course, that the person concerned will gain knowledge of the information at a later point, thereby imperiling his or her psychological integrity.)

Moreover, violations of the "right not to know" are somewhat hypothetical because the researchers will anticipate the issue of transmitting random findings and will include them in the consent procedure. If the representative agrees to the feedback, the person incapable of giving consent will then fall within the protection of the Gene Diagnostics Act ("*Gendiagnostikgesetz,*" GenDG)[86]: As this Act provides that genetic examinations of persons incapable of giving consent may only be carried out for diseases that can be treated or prevented, sec. 14 para. 1 GenDG, it follows that only such findings may be communicated.[87]

Finally, a non-consensual further use of the body substance itself for purposes other than those originally planned does not pose any significant danger to incapacitated persons either—concrete to their general right of personality—because further use without consent is only permissible under extremely narrow conditions.[88]

Since research on incapacitated persons is not per se unlawful, then, a few remarks concerning the consent procedure are in order: It is important to note that the researchers must inform not only the legal representatives prior to consent but also the research participants themselves, should the participants have the requisite mental capacity. The persons concerned have information rights as well.[89] Their veto, moreover, must also be taken into account. This applies not only to adults[90] but also to minors, whose wish not to participate in the research measure should be determinative if it is sufficiently clear, serious, and continuous—at least if the research will not produce any direct benefit for the minor.[91] If the minor would not benefit from the research at all, there is no justification, not even pursuant to the right of parental care, to "break" this will.[92]

---

[84] Taupitz (1998), p. 597 et seq.

[85] See Federal Court of Justice (2014), p. 2192, which in this case denied a violation.

[86] On the applicability of the GenDG—whose sec. 2 para. 2 no. 1 states the Act does not apply to "research"—Schreiber (2019), p. 94.

[87] Schreiber (2019), p. 286.

[88] See Sect. 7.2.2.3.

[89] See Schreiber (2019), p. 293, and the references cited therein; Taupitz (2000), p. A79.

[90] For adults, cf. Lipp (2021c), XIII. E. I. para. 104; for the general context, National Ethics Council (2004), p. 21; Taupitz (2000), p. A75 et seq.

[91] Schreiber (2019), p. 154; Spickhoff (2018b), sec. 630d para. 7; Taupitz (2000), p. A75 et seq.

[92] Schreiber (2019), p. 154.

The Oviedo Convention, for its part, also assumes in Art. 17 para. 1 v that research on persons incapable of consent is only possible if the person concerned does not object. Para. 29 of the Declaration of Helsinki provides, "[w]hen a potential research subject who is deemed incapable of giving informed consent is able to give assent to decisions about participation in research, the physician must seek that assent in addition to the consent of the legally authorized representative. The potential subject's dissent should be respected." Specific legislation, finally, likewise provides for the consideration of the opposing will of research participants who are incapable of giving consent.[93] Besides, these groups of persons must also be expressly included in the information procedures when participating in clinical trials of medicinal products.[94]

7.2.1.1.3 Consent to Research Involving Minors Capable of Giving Consent

If minors are capable of giving consent, we must ask ourselves whether they alone can consent to research interventions, or whether the consent of their legal representatives is always necessary as well. Following our findings in Sect. 7.1.2.2, minors with the appropriate mental capacity are authorized to make decisions on their own, unless high risiks—e.g., to the minor's life—make parental involvement appear justified. Specific legislation, such as the rules of the Medicinal Products Act related to clinical trials, which has already been mentioned many times, provides, however, that both minors and their legal representatives must consent (sec. 40b para. 3 sent. 1of the Medicinal Products Act in its version as of 27 January 2022). Some scholars extend this principle to all research interventions; in all matters of research, then, the minor's right to decide would be limited to a right of *co*-decision.[95] I do not find this conclusion persuasive, however. Even in the case of research interventions, the minor's right to exclusive consent should depend on the concrete potential risks the research bears for him or her.

Let us take the example of a blood sample. While the physical intervention itself poses little risk, thereby implying that the consent of minors may suffice, the concrete further use may still imperil the right of personality of minors as well as their property rights concerning the cells collected from their body.[96] As we will see in the following, these property rights make it impossible for minors to give consent without their legal representatives.

Mature minor can consent to interferences with their general right of personality. Crucially, the ethical issues surrounding a research project do nothing to change that: If minors are capable of understanding the significance and scope of the research, including the uncertainty of the research result as well as its ethical significance, then they must have the right to make an autonomous decision. We can draw this conclusion from the idea of the religious maturity of minors, which sec. 5 of the

---

[93] Such as in Art. 31 para. 1 lit. (c) and Art. 32 para. 1 lit. (c) of the Reg. (EU) No 536/2014.

[94] See Art. 31 para. 1 lit. (b), para. 3 and Art. 32 para. 1 lit. (b), para. 2 of the Reg. (EU) No 536/2014.

[95] Lipp (2021c), XIII. E. I. para. 108.

[96] See Sect. 7.2.2.2.

Act on the Religious Education of Children sets at the age of 14. From this age, the parents' influence in matters of religion, and therefore also in ethical questions, is less legitimate.[97] For that reason, the collection and use and re-use of genetic—and therefore particularly sensitive—data in the context of the generation and examination of brain organoids does not justify depriving minors of their sole decision-making authority, since there is no serious danger to their right to informational self-determination.[98]

What we should not underestimate, conversely, are the consequences of an unwanted feedback regarding genetic aberrations. The physical and psychological integrity of minors will suffer if they fail to come to terms with this information. Moreover, prior discussions between the researcher and the minor regarding possible feedback information will force the minor to decide in advance whether he or she wishes to exercise his or her right not to know. This, too, may prove challenging. For these reasons, we should not prematurely affirm a minor's capacity to give consent.

In the end, however, what frustrates the minor's exclusive right to consent are the property rights implicated by the research we are concerned with here:[99] Research measures impair the minor's property rights if they destroy or process the collected cells,[100] thereby granting the researcher original ownership of the newly created thing (sec. 950 of the Civil Code); this will likely be the case with brain organoids.[101]

This raises the problem that only persons with the capacity to contract, so individuals of age (over 18), may consent to the infringement of their property rights: Secs. 104 et seq. of the Civil Code, which require the parents' involvement for legally disadvantageous transactions, would be undermined if minors could consent to property-altering research without involving their parents. Since the loss of property by means of processing is legally disadvantageous, the consent of the legal representative is necessary,[102] therefore, also for generating brain organoids from cells; as is always the case, the economic value of the bodily substance is irrelevant for the question of whether a legal transaction is legally advantageous or not.

It follows from this that minors are prevented from consenting to research on separated body substances without the consent of their parents, at least if we assume that minors have property rights to substances separated from their body.

7.2.1.2 Removal of Cells for the Generation of Brain Organoids Transferred to Animals

The following section addresses an individual's consent to the transplantation of brain organoids derived from his or her own body cells to animals. The transfer of human brain organoids to animals is not pure science fiction: Research groups have

[97] Fink (2005), p. 80.

[98] As just discussed in the case of research on incapacitated persons.

[99] See Sect. 7.2.2.2, on rights over severed bodily substances.

[100] For the legal consequences of altering a thing by processing it, see secs. 947 BGB et seq.

[101] Cf. Faltus (2021), p. 131.

[102] See generally Klumpp (2017), preliminary remarks to §§ 104 ff., para. 100, and the references cited therein; Fink (2005), p. 77; c.f. Schreiber (2019), p. 319 et seq.

already carried out transplants into rodents. While brain organoids cultivated purely in vitro have a limited lifespan and developmental capacity, as they lack a supply of oxygen and nutrients (vascularization), an appropriate environment in vivo can overcome these limitations.[103] As the experimental transplantations provided the positive finding "that human brain organoids can integrate and form functional circuits in mouse brains," it seems possible that organoids may "provide an alternative to pure populations of a particular cell type, espescially for the treatment of complex brain disorders or injuries,"[104] as "conventional cell-based transplant methods face the hurdles of poor graft survival and inadequate neural differentiation."[105]

As regards the consent requirement for the removal and use of the cells, the principles established above remain applicable[106]: Since it interferes with the concerned person's physical integrity, the removal of cells requires consent; this consent extends to the agreed further use, if only implicitly. As the removal itself cannot be considered immoral within the meaning of sec. 228 StGB, the consent is valid. Again, however, we should take a brief look at the requirements that flow from Art. 1 para. 1 GG. After all, experiments that transplant brain organoids into animals raise the question of where the boundary between humans and animals runs, and to what extent we may cross this boundary.[107]

To begin with, the transplantation of cells to an animal does not entail the sort of "animalization" of the donor that could be incompatible with his dignity[108]—provided one considers "animalization" violative of human dignity in the first place. As we saw above, the simple use of a human being's cells does not call into question his or her quality as a human being.[109] This assessment does not change just because a transplantation creates the so-called neurochimeras. The transplantation does not represent a disregard for the uniqueness of that human being. Thus, the kind of transplantions that merely insert organoids into the brain of a host—e.g., a mouse—almost certainly does not lead to the formation of a complete human personality. In experiments in which mouse embryos were injected with human stem cells, for example, some of these stem cells (0.1%) developed into neuronal cells but did not have any effect on the cognitive abilities of the mice.[110]

Admittedly, the transplantation of organoids differs from that of simple cells: the transfer of brain organoids, and thus of small brain parts, is in principle far more likely to influence the cognitive abilities of the recipient than the transfer of single

103 Mansour et al. (2018), p. 432 et seq.; Daviaud et al. (2018), p. 1 et seq.

104 Mansour et al. (2018), p. 440; Daviaud et al. (2018), pp. 2 and 3, also report positive findings.

105 Daviaud et al. (2018), pp. 1 and 2.

106 The German Transplantation Act does not apply to the transfer of human tissue or organs to animals, which means that no specific requirements for the informed consent of the cell donor arise from more specific legislation.

107 See also Farahany et al. (2018), p. 431.

108 Generally on the creation of chimeras, Lackermair (2017), p. 299.

109 Lackermair (2017), p. 299.

110 Muotri et al. (2005), p. 18644 et seq.; Lackermair (2017), p. 68 et seq., 299–300, refers to this experiment to argue against "humanizing" animals through human neuronal cells.

cells, especially because the organoids can generate region-specific neuronal organoids.[111] At the same time, however, the effect is also likely to be limited to a "single discrete function" (if at all) because of the small size of the organoid compared with the host's brain: The achievable effect depends on several variables, including "the percentage of the animal brain that is of human origin, the specific site of brain integration, and host factors such as species and age."[112] Furthermore, current studies suggest that brain organoid transplantations, by causing a surgical cavity, "are more likely to worsen brain function than to improve it."[113] Accordingly, cerebral enhancement is a purely theoretical issue at the moment.[114]

But even if the animal containing a human brain organoid would show rudimentary human behavior (such as the chickens that were transplanted with parts of quail brains in the embryonic stage and then made typical quail sounds[115]), it is doubtful whether this outcome would truly imitate, let alone duplicate, the human personality. A individual's personality is much more complex than generalizable and, above all, rudimentary human behavior. Even if the entire brain of, say, a mouse consisted purely of human cells, moreover, a complete humanization of animals is considered (rather) improbable.[116]

This may be different as regards primates.[117] Even then, however, I submit one should not speak of a disregard for the uniqueness of the donor: As mentioned above, the assumption of such a direct causal connection between the formation of personality and the genetic content of neuronal cells overestimates the influence of the genome.[118] Taking into consideration that personality also, or rather primarily, develops through one's own history and environment, one can at best speak of the emergence of a *new* personality.[119]

Moreover, we should bear in mind once again that the experiments at issue here require the donors' consent, and that it is hardly justifiable to deny them the right to consent on the ground that it would imperil their dignity. If anything requires protection under the right to human dignity, it is the being that results from the research, as the latter may yield a being whose status as human or animal is unclear. We will need to clarify, therefore, where to draw the line between acceptable and

[111] On the possibility of generating not only whole-brain organoids but also region-specific brain organoids, Chen et al. (2019), pp. 463–464, and Daviaud et al. (2018), p. 17 (and the references cited therein). Karpowicz et al. (2004), p. 334, presume there might be a transfer of functional behavior when entire brain regions are transplanted between closely related, functionally and morphologically similar beings, such as chimpanzees and humans.

[112] Chen et al. (2019), pp. 465, 467.

[113] Chen et al. (2019), p. 466.

[114] Chen et al. (2019), p. 467.

[115] Balaban et al. (1988), p. 1339 et seq.

[116] Greely et al. (2007), p. 35; Chen et al. (2019), p. 468; cf. Lackermair (2017), p. 70.

[117] Chen et al. (2019), p. 469, do not answer that question.

[118] Lackermair (2017), p. 299.

[119] Lackermair (2017), p. 300.

non-acceptable forms of cerebral enhancement of animals. This, however, is a question for another day.[120]

In conclusion, then, it is not per se impossible to consent to research involving the transplantation of human brain organoids into animals, especially given the current state of research and the exclusive use of rodents. As regards persons incapable of giving consent and minors capable of giving consent, finally, the principles set out under Sects. 7.2.1.2 and 7.2.1.3 apply here as well.

7.2.1.3 Removal for the Generation of Brain Organoids in the Field of Drug Research and Personalized Medicine

Brain organoids can also be used for testing drugs. In the field of toxicological screening, for instance, they can be used to assess the toxicity of substances. To do so, tissue samples are taken from defined patient groups and propagated.[121] Furthermore, patient-specific organoids can also be used in the field of personalized medicine, where they help find the ideal treatment for a specific patient. Examples include rare diseases for which there are no clinical trials, given the high cost and low benefit, such as rare gene mutations leading to cystic fibrosis.[122] Diseases of the brain may profit from the use of organoids as well.

The question we will now address is whether specific regulations establish special requirements for the consent and information of the cell donor in this context.

With regard to toxicological screenings, first of all, we need to inquire whether the corresponding tests constitute clinical studies.

The version of the German Medicinal Products Act ("*Arzneimittelgesetz*," AMG) prior to 27 January 2022, which was based on Directive 2001/20/EC (now replaced by Reg. (EU) No 536/2014), provided in sec. 40 para. 1 sent. 3 no. 3 AMG (now replaced by Art. 29 Reg. (EU) No 536/2014 and the amended version of the Medicinal Products Act in its version as of 27 January 2022) that a clinical trial in humans may only be conducted "if and as long as the person concerned (a) has come of age and is capable of recognizing the nature, significance and scope of the clinical trial and of acting accordingly, (b) has been informed in accordance with paragraph 2 sentence 1 and has given his or her written consent [...] and (c) has been informed in accordance with paragraph 2a sentences 1 and 2 and has given his or her written or electronic consent; the consent must also expressly refer to the processing of health data." Special provisions applied to minors and to adults who are incapable of giving consent (sec. 41 para. 3 AMG[123] (now replaced by Art. 31 Reg. (EU) No 536/2014 and the amended version of the Medicinal Products Act in its version as of 27 January 2022) and sec. 40 para. 4 AMG[124] (now replaced by Art. 32

[120] See Chen et al. (2019), p. 469. Farahany et al. (2018), p. 431, suggest a case-by-case evaluation. For Greely et al. (2007), p. 38, a mouse with human language capacities and self-consciousness would at least be "troubling."

[121] Bartfeld and Clevers (2018), p. 93.

[122] Bartfeld and Clevers (2018), p. 93.

[123] Based on Art. 5 Directive 2001/20/EU.

[124] Based on Art. 4 Directive 2001/20/EU.

Reg. (EU) No 536/2014 and the amended version of the Medicinal Prodcuts Act in its version as of 27 January 2022)).

Sec. 4 para. 23 sent. 1 AMG in its version prior to 27 January 2022 defined a clinical trial. To do so, it drew on Directive 2001/20/EC,[125] whose Art. 2 defined clinical trials as "any investigation *in human subjects* intended to discover or verify the clinical, pharmacological and/or other pharmacodynamic effects of one or more investigational medicinal product(s), and/or to identify any adverse reactions to one or more investigational medicinal product(s) and/or to study absorption, distribution, metabolism and excretion of one or more investigational medicinal product(s) with the object of ascertaining its (their) safety and/or efficacy." Since the substances we are concerned with are not tested "in human subjects," but only on organoids produced from human cells, I argue that the procedure could not qualify as a clinical trial within the meaning of that provision.

Now, Art. 2 para. 2 of Reg. (EU) No 536/2014 differentiates between a "clinical study" and a "clinical trial." Both, however, relate to investigations "in *relation to humans* (...)."[126] I see two arguments why this revised definition does also not cover the substances relevant here. First, the revision likely did not seek to modify the meaning or scope of the Regulation; the German version, for instance, still uses the words "in humans." Second, the term "in relation to humans" continues to suggest a close relationship to the human body, which means that the use of medicines on substances that have been separated from the body no longer constitutes use "in relation to humans."[127]

I submit that a purposive interpretation yields the same answer. The Regulation seeks to ensure the safety of the test persons,[128] which is not affected if the medicinal product does not enter the body itself. Para. 11 of the recitals states that the risk to subject safety in a clinical trial mainly stems from two sources, the investigational medicinal product and the intervention, but it does not explicate what it means by "*the* intervention"; I contend that it refers to the investigational measures that are carried out on test persons within the framework of a clinical trial to test the effect of the medicinal product; these measures do not apply, of course, when substances are tested on brain organoids only.

Another argument against the classification of drug tests on separated body materials as clinical trials is that some articles—e.g., in Reg. (EU) No 536/2014—are premised on a drug application directly in the human body.[129] Finally, the con-

[125] Wachenhausen (2016a), sec. 4 AMG para. 184.

[126] The French version of the Directive referred to "chez l'homme," while the Regulation refers to "en rapport avec l'homme."

[127] There does not seem to be any scholarship as yet regarding the changed wording. Neither this term nor the term "in humans" used in the Directive appear anywhere else in the Regulation itself, nor is there any reference to this amendment.

[128] Lipp (2021c), XIII.E.IV.1. para. 71; Wachenhausen (2016b), sec. 40 AMG para. 7.

[129] See, e.g., Art. 31 and 32 (clinical trials on incapacitated subjects and minors), as such a trial can only be carried out if either the subject has a *direct benefit* or there is at least a benefit for the population represented by the subject, provided it imposes only a minimal burden on the subject *in comparison with the standard treatment*. However, neither can the procedure examined here yield a direct benefit nor can the prerequisite be fulfilled that only a minimal burden may exist in com-

ception of the pharmaceutical regulations as a whole suggests that toxicology studies and clinical trials are distinct operations. Thus, sec. 22 AMG requires the submission of documents from both the clinical trial (sec. 22 para. 2 no. 3) and the toxicological studies (sec. 22 para. 2 no. 2) for the authorization of medicinal products. In fact, toxicological tests have always been carried out, as the so-called preclinical studies, on animals or in vitro.[130]

The targeted testing of efficacy for a specific patient in the context of personalized medicine is not a clinical trial either, which means that secs. 40 et seq. AMG as well as Reg. (EU) 536/2014 do not apply. For one, the researchers do not use the medicinal product "in humans." Second, it is already doubtful whether such tests constitute an experiment within the meaning of medicinal products law. It is possible to combine research and therapeutical purposes, thereby conducting a so-called therapeutic experiment. However, an intervention cannot be considered research if it is aimed only at curing an individual person, regardless of whether the measure yields, as a side effect, new insights.[131] Individual healing attempts are therefore not subject to the regulations on clinical trials.[132] It can be difficult to distinguish individual healing attempts from research, of course. A multitude of individual healing attempts does not necessarily constitute research, the threshold is indeed crossed once the new method involves a pilot study or a planned and organized series of healing attempts.[133]

Thus, there are no specific statutory requirements for informed consent in the field of drug research on brain organoids and personalized medicine. Again, then, minors and persons who are incapable of giving consent are subject to the general principles (Sects. 7.2.1.1.2 and 7.2.1.1.3).

7.2.1.4 The Scope of Consent

Yet another question is how specific the person's consent must be. It is unclear, for instance, whether one can consent ex ante to research projects that, at the time of consent, are still unknown.

On the one hand, blanket consents are considered problematic because they lack the specificity that inheres in the concept of informed consent and is based on Art. 2 para. 1 in conjunction with Art. 1 para. 1 GG.[134] Instances of "broad"

parison with the standard treatment. The latter means that no further burdensome interventions may be carried out compared to the standard therapy, such as the collection of samples to test the drug's mode of operation. See Schreiber (2019), p. 248.

[130] Winnands (2016), sec. 22 para. 54; Rehmann (2020), sec. 22 para. 20; Franken (2020), sec. 12 A. para 2 and A.IV. para. 9.

[131] A so-called individual healing attempt. See Lipp (2021c), XIII.E.III. paras. 59 et seq.; Kern (2019b), § 131 I.3. para. 20.

[132] Kern (2019b), § 131 I.3. para. 20; Bender (2005), p. 512.

[133] Lipp (2021c), XIII.E.III. para. 59–61. On the healing attempt as a therapeutic study, if the results are evaluated systematically, Schreiber (2019), p. 8. Bender (2005), p. 515, considers a healing attempt a therapeutic study and therefore research if it comprises at least 10 persons, as this number suggests a certain degree of standardization.

[134] Halàsz (2004), p. 231.

consent—which may refer to a specific research objective or "medical research" in general—are viewed more favorably, on the other hand.[135]

I concur with this approach. It would go too far to void any consent that does not relate to specific research projects, for the right to self-determination must also include the right to accept uncertainty.[136] We can still speak of "informed consent" if the persons concerned are informed about the scope of their consent and know, therefore, what they are getting into.[137] In addition, the persons concerned can always exclude certain areas of research or revoke their consent.[138]

Some scholars suggest that informed consent extends only to research projects that the donor of the body materials can expect and that do not violate legal prohibitions.[139] The question is, however, whether anything is gained by this restriction. After all, research that violates legal prohibitions is always inadmissible. And it is hard to define what research the donor of the substance can expect without accepting precisely the sort of uncertainty the scholars tried to avoid.

Nor does Art. 10 sec. 1 of Recommendation CM/Rec(2016)6 prohibit broad consent. It merely states that "the person concerned should be provided with comprehensible information that *is as precise as possible* with regard to the nature of any envisaged research use and the possible choices that he or she could exercise (...)." As precise as possible includes information that can still be vague because the research project itself is still unknown.

Furthermore, para. 12 of the WMA Declaration of Taipei on Ethical Considerations Regarding Health Databases and Biobanks (2016) likewise permits a "multiple and indefinite use" of body materials donated for research; this, then, encompasses the possibility of broad consent.[140] The Declaration, incidentally, specifies the points about which the researcher must inform the patient. They include in particular "the risks and burdens associated with collection, storage and use of data and material;" "the nature of the data or material to be collected; the procedures for return of results including incidental findings;" "and when applicable, commercial use and benefit sharing, intellectual property issues and the transfer of data or material to other institutions or third countries."

It follows that researchers may generate brain organoids from donated body material, provided the person concerned donated the material for research *in general*. However, donors should be made aware that their substances could become the object of ethically controversial research. For that reason, the information given before consent should comprise some examples, thereby allowing the donors to get a general idea and, if necessary, to restrict the scope of their consent.

---

[135] See Schreiber (2019), pp. 294–295; Central Ethics Committee (2003), p. 9, which allows for consent that covers "all possible studies"; National Ethics Council (2004), pp. 14, 58–59.

[136] Taupitz and Schreiber (2016), p. 307.

[137] Schreiber (2019), pp. 294–295; Taupitz and Schreiber (2016), p. 307.

[138] Taupitz and Schreiber (2016), p. 307; Schreiber (2019), p. 295.

[139] Halàsz (2004), p. 232.

[140] Schreiber (2019), p. 295, refers to this Declaration in the context of broad consent.

The consent given by representatives in the case of persons incapable of giving consent is limited by default, as research on these groups of persons is per se subsidiary. Moreover, if the research does not bring any benefit to the individual, it must at least produce benefits for members of the same group of persons, as is also explicitly established in Art. 12 para. 1 and Art. 21 para. 5 Recommendation CM/Rec(2016)6 for the collection and use of separated body substances. This restriction, consequently, limits the scope of any instance of "broad" consent. Again, however, the information must only be *as precise as possible* (Art. 12 para. 2 lit. b) ii)). There is some leeway, then, for "broad" consent after all. Particularly in the case of research on brain organoids, however, it remains doubtful whether it is even necessary to use body material from persons incapable of giving consent, since the researchers may be able to artificially produce the relevant diseases if they genetically modify other donated cells.

7.2.1.5 Interim Conclusion

Adults capable of giving consent can consent to the removal of bodily substances for the generation of brain organoids for research purposes. "Immorality" is the only restriction on their right to consent to interventions in their physical integrity; the threshold of immorality is only crossed if the removal procedure threatens the concerned person's life. The right to consent to the use of the substances thus extracted is not subject to this restriction, however, as this would violate the donor's right of personality.

If the substances are to be taken from persons incapable of giving consent, the legal representative must consent to this procedure as well as to the further use of the substances. However, research on persons incapable of giving consent is always subsidiary. It is doubtful, therefore, given the possibility of modifying genes through genetic engineering, whether cells need to be taken from incapacitated persons at all.

If minors are capable of giving consent, they can, generally speaking, decide for themselves whether to participate in research projects. Research projects involving brain organoids constitute an exception, however, as such research impairs the property rights the minor has over the separated substances; to this impairment they cannot consent. Here, then, the consent of both the parents and the child is necessary.

In all cases, the participants, even if they are incapacitated, must be informed of all important aspects of the research project.

7.2.2 Removal of Body Cells for Other Purposes

7.2.2.1 Introduction

If the objective of generating brain organoids from the harvested materials did not yet exist when the harvesting took place, the consent initially granted only justifies the removal of the material and its use within the scope of the purpose of removal.[141]

[141] Lippert (2001), p. 407.

Additional consent may be required for other uses, including the production of brain organoids. We will take a closer look at this requirement in the following.

7.2.2.2 The Right Over Severed Bodily Substances Under German Law

Whether the further use of severed bodily substances requires consent depends on the rights the person concerned continues to hold over severed substances.

First of all, the living human body is not subject to property rights: The body is not a thing.[142] The right to dispose of one's own body follows from the right of personality (Art. 1 para. 1 GG, Art. 1 para. 1 in conjunction with Art. 2 para. 1 GG and Art. 2 para. 1 GG)[143] and the right to physical integrity (Art. 2 para. 2 GG). Most scholars argue that the general right of personality continues to apply to the bodily parts and substances separated from the body.[144] There are, however, good arguments in favor of the substances simultaneously becoming things, thereby falling within the domain of property rights. This means that the donor of the substances can dispose of them in a legal transaction. Thus, he can transfer ownership of the substances to research centers[145] while simultaneously retaining the right of personality regarding said substances.[146] The new owner must take into account the right of personality when exercising his ownership rights: According to sec. 903 sent. 1 BGB, owners can use an object at their discretion, provided there are no conflicting rights of third parties—such as, of course, the right of personality.[147]

In its so-called "sperm decision," the Federal Court of Justice took a somewhat different path. It held that substances which are removed from the body in order to be transferred back at a later point remain part of the body during separation, as they continue to form a "functional unit" with it. Consequently, the destruction of the substances without or against the will of the person from whom they originate constitutes an infringement of physical integrity under sec. 823 para. 1 BGB.[148] However, this decision remains contested (and rightly so). First, the court's

[142] On this now outdated opinion, see Halàsz (2004), p. 15 et seq.

[143] Halàsz (2004), p. 19. According to the so-called superposition thesis, the living body is also a thing, but the property over it is subsumed by the right of personality until a part of the body is separated. See Schünemann (1985), p. 86 et seq.

[144] At least if they contain genetic material, Taupitz (1991), p. 210.

[145] However, it is unclear how separated substances become a thing. For the different approaches, see Halàsz (2004), p. 31 et seq.; Lippert (2001), p. 407.

[146] The opinions on how the right of personality continues to cover the substances after their removal differ. Taupitz (1991), p. 209 et seq., proposes two different solutions: either the right of personality continues to exist, by analogy to sec. 953 BGB, *in the substance* or the use of the substance affects *the donor's* right of personality; Halàsz (2004), p. 36 et seq., emphasizes the connection that remains between the substance and the donor.

[147] On this approach, which combines both rights, Schröder and Taupitz (1991), p. 40 et seq.; Taupitz (1991), p. 209 et seq.; Halàsz (2004), p. 26 et seq.; Lippert (2001), p. 407; Baston-Vogt (1997), p. 285 et seq. On the approach that emphasizes the right of personality, Schröder and Taupitz (1991), p. 38 et seq.; Halàsz (2004), p. 20 et seq. On the approach that emphasizes property rights, Schröder and Taupitz (1991), p. 35 et seq.; Halàsz (2004), p. 22 et seq.

[148] Federal Court of Justice (1994), pp. 127–128.

interpretation goes far beyond the (narrow) wording of sec. 823 para. 1 BGB.[149] Second, the right of personality also protected by sec. 823 para. 1 BGB renders such an extensive reading unnecessary.[150]

Accordingly, every use of bodily substances must be examined for compliance with the former substance bearer's right of personality.[151] Both under private law, i.e., sec. 823 para. 1 BGB, and in the context of constitutional law, a violation of the right of personality is determined by balancing, in a comprehensive manner, the interests concerned.[152] These interests include the researcher's fundamental right to free research under Art. 5 para. 3 GG. In doing so, one should keep in mind that every research activity which involves human material without consent encroaches on the rights of another person. It is doubtful, therefore, whether the right to free research even extends to actions that constitute such an encroachment.[153] Why should a balancing of interests be necessary at all? Would it not make more sense to say that research may not occur absent consent? After all, if research is carried out on a living human being, there is no question that research interests can never justify infringements to the right to self-determination regarding one's body; why this should suddenly be the case once some substances have been separated from the body is certainly a good question.[154]

One possible explanation is that the downgrading of the body part to a thing and the separation from the body weakens the right of personality by changing the relationship with the different kind of substances.[155] But even then we must keep in mind that a weighing of interests only becomes necessary because the holder of the right of personality witnesses an "invasion" of his or her sphere of interests. More, the right to free research is certainly not more valuable, a priori, than the right of personality.[156]

The following observations will disregard property rights regarding bodily substances. Instead, we will assume that patients, by leaving the substances with the physician, either transferred their ownership to said physician (sec. 929 BGB), if need be implicitly, or that they abandoned it (sec. 959 BGB).[157]

---

[149] Laufs and Reiling (1994), p. 775.

[150] See Laufs and Reiling (1994), p. 775, who fail to appreciate, however, that this aspect of the right of personality is not about "family planning" as an activity but about the bond that connects the person concerned with his or her body part even though it is separated from him or her.

[151] Taupitz (1991), p. 210.

[152] Schröder and Taupitz (1991), pp. 44, 54; Taupitz (1991), p. 210-211; Taupitz and Schreiber (2016), p. 305; Fink (2005), p. 56.

[153] See, e.g., Halàsz (2004), p. 195, who argues that the self-determination of the rights holder and the physician's right to free research must limit each other. Why should that be so?

[154] See also von Freier (2005), pp. 325–326.

[155] Baston-Vogt (1997), p. 289 et seq.

[156] Schröder and Taupitz (1991), p. 67.

[157] On dereliction and the transfer of ownership, Halàsz (2004), pp. 258–259; Breithaupt (2012), p. 215 et seq.; Schreiber (2019), pp. 320–321.

7.2.2.3 Is Further Use for the Generation of Brain Organoids Compatible With the Donor's Right of Personality?

7.2.2.3.1 Introduction

We should now ask whether every further use of bodily substances without explicit consent infringes the concerned person's right of personality. Were this to be the case, every further use would, in principle, have to be covered by informed consent.[158]

It has to be emphasized that German law does not prejudge the outcome of the weighing process. Sec. 8b of the Transplantation Act, which deals with the donation of organs and tissue removed for other purposes than for transplantation, constitutes an exception, as it explicitly requires consent if the physician decides, at a later point, to transplant the removed material. From this exception in this specific area we cannot draw the overall conclusion, however, that the law mandates informed consent with regard to every conceivable application of body substances.

Finally, it has to be clarified that leaving substances with the physician who analyzed them for therapeutical reasons does not imply consent to other uses than destruction, which is the only further use the patient presumably expects.[159] We should bear in mind that other areas of the law likewise distinguish between destruction and other types of use and that a consent to destroying a substance is distinct from the consent to other uses.[160]

7.2.2.3.2 The Use of Identifiable and Anonymized Material

Most scholars argue that the use of identifiable material absent consent violates the concerned person's right of personality. I agree with this position,[161] in particular if the research is accompanied by genetic analysis, especially of coding regions.[162] In this case, the right of personality—in its manifestation as the right to informational self-determination—is particularly affected, as this type of research may even explore the core of what the right of personality seeks to protect. No interference with this core is permissible without consent.[163] Since even partial genome analysis

[158] Thus Lippert (2001), p. 407, according to whom a separate informed consent is required "in normal cases," but who later limits this statement to research projects in which genetic dispositions are examined (p. 409). See also Dettmeyer and Madea (2004), pp. 85–86. I set to the side other ways—other than giving consent—to allow the use of separated body substances. On this matter, Fink (2005), p. 154 et seq.; Halàsz (2004), p. 233 et seq.

[159] For persuasive arguments against assuming implied consent, Schröder and Taupitz (1991), p. 62; Taupitz (1991), pp. 218–219; Breithaupt (2012), pp. 254–255; von Freier (2005), p. 326.

[160] Thus, e.g., in copyright law, or when a person uses a thing that its owner threw away in the expectation that it would be destroyed. See Taupitz (1991), p. 219.

[161] Based on the principle that research always requires prior consent. See Art. 21 para. 1, para. 2 lit. (a) Recommendation CM/Rec(2016)6, and Art. 22 Oviedo Convention.

[162] Contrary to non-coding material, coding material, which codes for the synthesis of certain proteins, allows drawing conclusions about personal characteristics. See Halàsz (2004), pp. 201–202.

[163] See Halàsz (2004), p. 202, who assigns this data to the core area protected by the right of personality. For a less extensive view, see Fink (2005), p. 66. Generally on research with genetic data, Schröder and Taupitz (1991), p. 64. Schreiber (2019), p. 124 et seq., suggests differentiating, especially with regard to the purpose of the data collection.

can collect personal data, the patient should have the right to decide on the use and disclosure of the data.

Genome analysis poses further risks for the patient, namely when the physician or researcher confronts the dilemma, absent prior clarification, of whether to disclose abnormalities that have an impact on the patient's health. It is possible that the patient does not want to be burdened with such information.[164]

Nevertheless, there are cases in which the researcher should have the right to use identifiable substances without the concerned person's consent. The Declaration of Helsinki (para. 32) and the Recommendation CM/Rec(2016)6 (Art. 21 para. 2 lit. b)), for instance, posit such a right when obtaining consent involves unreasonable effort and when there are clearly overriding research interests.[165] It is essential, however, to bear in mind that both criteria have to be met. Researcher will have to accept administrative efforts and the corresponding lack of time, therefore, provided they do not jeopardize their research objective.[166] Nor should in fact the impossibility to obtain consent—e.g., because there is no way to reach the person concerned—be sufficient in my opinion. Otherwise, the right to self-determination would effectively be negated. The violation of the right of personality does not depend on the possibility to obtain consent but on how incisive the research activity is and which interest the researcher can assert.[167]

The Central Ethics Committee of the German Medical Association and numerous scholars have proposed considering the following criteria in addition to those mentioned in Art. 21 para. 2 lit. b) CM/Rec(2016)6[168]—i.e., a significant research objective and the unreasonableness or impossibility of obtaining consent: the emotional and symbolic significance of the body substance used; a possible further benefit of the substance for the donor; the question of whether the research involves use of the substances that is ethically and legally controversial; the question of whether the substance will be transferred to another human being or whether another use is planned that interferes in a particularly intense way with interests of the donor, such as the collection of personal data, in particular genetic data, or a duplication of the substance due to its special properties; as well as, finally, the possibility of conducting the research on other available substances for which consent has been obtained.[169]

After having dealt with the use of identifiable substances without consent, we should also ask how the anonymization of bodily substances affects the concerned

---

[164] Schröder and Taupitz (1991), p. 64.

[165] Halàsz (2004), p. 197–199; Schreiber (2019), p. 309; National Ethics Council (2004), pp. 13, 57–58

[166] See Halàsz (2004), p. 196.

[167] See also von Freier (2005), p. 326.

[168] "(...) evidence is provided that reasonable efforts have been made to contact the person concerned (i.); the research addresses an important scientific interest and is in accordance with the principle of proportionality (ii); the aims of the research could not reasonably be achieved using biological materials for which consent or authorisation can be obtained (iii); and there is no evidence that the person concerned has expressly opposed such research use (iv)."

[169] Central Ethics Committee (2003), p. 6; Taupitz (2020a), p. 808; Schröder and Taupitz (1991), pp. 82–83; Schreiber (2019), pp. 309–310.

person's right of personality. Anonymization might facilitate the use of the substances without the donor's consent, as the donor's interest in consenting may be less forthright. Some argue indeed that there are no restrictions in law on research conducted on anonymized substances;[170] others make the more modest claim that the requirements for waiving the consent requirement are more lenient than in the case of identifiable materials.[171] Their background assumption is that the only purpose of extending the right of personality to the further use of bodily subtances is to prevent their individualization, and that anonymization preempts this risk.[172]

This assumption, however, is not entirely correct. Since every cell of a body contains the same genetic code, a complete anonymization may be impossible if corresponding comparison material is available; future medical and electronic developments may exacerbate this problem.[173] It is doubtful, for that reason, whether the current de facto anonymization sufficiently takes into account the donor's privacy interests, especially if the research includes the genetic examination of coding material, thereby allowing the scientist to research the personality traits of identified (or at least identifiable) persons.[174]

The prevailing opinion is also wrong to suggest, moreover, that right of personality over severed bodily materials merely protects the persons concerned against the identification or unauthorized dissemination of their data. I believe the bond that connects the former substance bearers with their now separated substance goes beyond that. As regards the living body, the right of personality transcends a mere right to "data privacy." They also encompass the right to determine who may perform what actions on the body. I do not see any reason why this should change after the separation of materials from the body. Instead, it makes more sense to argue that the right to physical self-determination continues to be effective.[175]

Consequently, any use of bodily substances, identifiable or not, for research purposes requires, in priniciple, the consent of the person concerned.[176] The protection against the identification and attribution of certain characteristics and the right to

---

[170] This is also the case in Art. 21 para. 4 Recommendation CM/Rec(2016)6. See Breithaupt (2012), pp. 209, 262; Dettmeyer and Madea (2004), pp. 92–93; National Ethics Council (2004), pp. 12–13, 52, 56–57. Halàsz (2004), p. 203, for whom the right of personaltiy protects only "genetically relevant" substances (pp. 56–57), argues that this holds true at least for non-coding bodily materials. Nitz and Dierks (2002), pp. 402-403, also seem to argue that consent is not "normally" required for research on anonymized material. See also Taupitz and Schreiber (2016), p. 306, who argue, however, that the waivability of consent also depends on the type of use. von Freier (2005), p. 323, finally, refers to the statement of the National Ethics Council (2004).

[171] See, e.g., Central Ethics Committee (2003), p. 6, which emphasizes the criterion of anonymization. See also Taupitz (2020a), p. 808; Schreiber (2019), pp. 307–308 and 311.

[172] Cf. von Freier (2005), p. 323

[173] Halàsz (2004), p. 200; Fink (2005), p. 62.

[174] Halàsz (2004), p. 203.

[175] von Freier (2005), p. 324 et seq.; also Halàsz (2004), p. 87 et seq.; Fink (2005), p. 56.

[176] Thus also Freund and Weiss (2004), p. 317; Schreiber (2019), pp. 310 et seq.; von Freier (2005), p. 327; Fink (2005), p. 75, also does not consider consent per se indispensable. Taupitz (2020a), p. 808, focuses on the degree of anonymization but also emphasizes additional aspects.

determine the use of one's own bodily substances are two equally valid rights. Each requires examining whether research without consent violates its precepts. Anonymization can at best overcome the lack of consent as regards the right not to be identified; it does nothing to prevent a violation of the right to decide for oneself how one's bodily substances should be used. Thus, research without consent is only permissible if, on the basis of the criteria established for identifiable materials, the researcher's right to free research proves more important than the donor's right of personality. The exact use of the substances will prove especially important in the weighing process. Ethically or legally controversial uses will tilt the balance in favor of the right of personality.[177]

But does the passage of time or geographic separation maybe weaken—or void—the right of personality?[178] Scholars who support this proposition argue that an individual's legitimate interest in his or her bodily substances—identifiable or not—wanes over time.[179] Yet the question of interest (or disinterest) should not be our point of departure, since the substance carrier is simply unaware of the researchers' intention to use the substances for research purposes.[180]

What if persons incapable of giving consent are concerned? Research on substances already removed for other purposes is of course permissible if the legal representative has consented to that use. Nevertheless, the researchers should prioritize bodily materials removed from individuals who can give consent. This rule, which we can also find in Art. 21 para. 5 (read together with para 2) of Recommendation CM/Rec(2016)6, reflects the principle that only as a last resort should persons incapable of giving consent be exposed to the risks of research interventions. This holds even though research on donated materials poses few risks to the patient—namely, identifiability, even in the case of (de facto) anonymization, data use in violation of data protection regulations, and unwanted feedback regarding genetic aberrations. In my opinion, however (subsidiary) research on these substances should be permissible even without the consent of the legal representative if it meets the criteria that apply to persons capable of giving consent. After all, the criteria limit the options for research and the risks for the concerned individual are negligible.[181]

7.2.2.3.3 The Case of Generating Brain Organoids Without Consent

To conclude, I submit that the use of separated bodily substances to generate brain organoids requires consent, regardless of whether the substances are anonymized or not.

Although brain organoids are not yet very developed, the very principle of generating artificial brain organoids is likely to be ethically controversial, especially since it is not clear at what point scientists will be capable of generating a more developed

177 Explicitly Fink (2005), p. 75.

178 For this stance, see Breithaupt (2012), pp. 208–209, and Nitz and Dierks (2002), p. 402.

179 Baston-Vogt (1997), pp. 291–292.

180 See also von Freier (2005), p. 326.

181 For minors, see also Schreiber (2019), p. 312.

brain. Moreover, the question of which status to attribute to the organoids remains unresolved not only from an ethical but also from a legal perspective.[182] Furthermore, although the original material (which may consist of individual cells) may not hold any particular symbolic value, the manufactured product does. For these reasons, we should not assume that everyone will remain indifferent to such research.

The further the development of these organoids progresses, the closer one comes to the problem of cloning. Crucially, the absence of consent in cases of cloning raises the specter of a violation of human dignity: Cloning without consent disregards the genetic uniqueness of the donor by demonstrating to a person that he or she can be duplicated.[183] Of course, the generation of brain organoids does not constitute cloning in the strict sense of term: The brain (organoid) alone does not constitute an entire human being.[184] Still, the concerns that attach to unconsented cloning also apply if researchers generate a large number of (functional?) brains from the cells of one person.

In addition, research on and with brain organoids may also involve genetic analyses. That is the case, for example, of measures to test whether medicines are effective or whether genetic corrections are feasible. If the substances used are identifiable, the lawfulness of the research also clearly turns on the patient's right to informational self-determination.

Research involving brain organoids is particularly controversial from an ethical and legal perspective, of course, when the organoids are transferred to animals.[185] But the unsolicited transfer to other humans also fundamentally affects personality interests,[186] especially if brain areas are affected. It is up to the patient alone to decide whether and in which person his bodily substances should continue to exist.[187] Therefore, a transplantation of brain organoids (or parts thereof) requires consent as well. This requirement, incidentally, already follows from the Transplantation Act (sec. 8b).

Using the donated body materials or the brain organoids produced from them for economic purposes raises questions as well. Since they are things, body substances can (subject to the prohibition of tissue trade according to sec. 17 of the Transplantation Act[188]) be disposed of in legal transactions.[189]

If we stipulate a "right to exploit, for economic gain, one's right of personality," the rights holder may want to transfer the removed substances only in return for

---

[182] See Lavazza and Pizzetti (2020), p. 13 et seq.; Farahany et al. (2018), p. 432.

[183] Dreier (2013), Art. 1 sec. 1 para. 109.

[184] Lavazza and Pizzetti (2020), p. 11.

[185] See, e.g., the contributions of Greely et al. (2007); Karpowicz et al. (2004) ; Lackermair (2017) ; German Ethics Council (2011).

[186] Schröder and Taupitz (1991), pp. 69–70; Taupitz (1991), p. 210; Taupitz (2020a), p. 808; but see, Fink (2005), p. 70 et seq.

[187] Schröder and Taupitz (1991), p. 66.

[188] See Taupitz (2020a), p. 809; Wernscheid (2012), p. 229.

[189] See Taupitz (1991), p. 217, and Fink (2005), p. 74 and the references cited therein. But see Halàsz (2004), pp. 203–204.

payment if there is a market for corresponding body substances.[190] That means the researcher requires the donor's consent if he or she wishes to conduct such a transaction. Should a right to monetize one's right of personality not exist, the obligation to obtain the patient's consent nevertheless follows from the contract between physician and patient.[191] The purchaser of the body material must, by the way, abide by the same restrictions on the use of the substances as the physician removing the substances.[192] Thus, the purchaser may also not generate brain organoids from the body material without the donor's consent.

Moreover, physicians must also obtain the donor's consent if they first generate brain organoids from the body material and then proceeds to sell them: While donors cease to be the owners of their cells once they have been processed (either because they transferred them to the physician or the researcher[193] or because their ownership ends pursuant to sec. 950 BGB), their personality rights are not affected.[194] The sale of the organoids thus encroaches on the right to monetize one's right of personality (and constitutes a breach of contract), especially if—due to the rarity of their properties—the cells can be used for the creation of expensive, and therefore lucrative, medicinal products.[195]

Whether the researcher may transfer removed cells for free largely depends on the use which the acquirer has in mind.[196] In our case, consequently, neither the first nor the subsequent user may generate brain organoids without the donor's consent.

7.2.2.3.4 What About "Presumed Consent"?

Let us now turn to the question of "presumed consent." Some scholars argue that the use of bodily substances for research is lawful if we can presume the concerned person's consent. If presumed consent can justify a physical intervention (under sec. 630d para. 1 sent. 4 BGB), the argument goes, it must do the same, a fortiori, for the use of bodily materials—which, after all, do not pose any risk to the patient's body or health.[197]

The recourse to presumed consent is in most cases less helpful than may appear at first glance, however: Where there are no indications of the patient's will whatsoever, the range of presumed consent can only be assessed by considering objective

[190] Thus Halàsz (2004), pp. 123–124.

[191] Schröder and Taupitz (1991), pp. 71–72.

[192] Schröder and Taupitz (1991), p. 77.

[193] On the abandonment of property and the implied transfer of ownership, Halàsz 2004, pp. 258–259; Breithaupt (2012), p. 215 et seq.

[194] Halàsz (2004), p. 39, pp. 65–68.

[195] Taupitz (1991), p. 218, argues that a violation of the right of personality becomes likely if the remuneration is especially high, and that there may only be a breach of contract between the doctor and his or her patient in other cases. See also Schröder and Taupitz (1991), pp. 78–79. But see Halàsz (2004), pp. 260–261, according to whom the donor transferred the right to economic exploitation of the materials to the physician or researcher, at least in cases in which he or she transfers his or her ownership to the latter.

[196] For greater detail, see Schröder and Taupitz (1991), p. 77 et seq.

[197] Taupitz and Schreiber (2016), p. 306.

considerations,[198] the same considerations we applied to determine whether the right of personality is violated. So, unless there are clear indications of the patient's will, we cannot simply presume he or she would have consented if at the same time, we would presume a violation of the right of personality by weighing all the interests at stake.[199] Consequently, the generation of brain organoids cannot rely on the concerned individual's presumed consent if there is no indication of that individual's preferences.

Moreover, sec. 630d para. 1 sent. 4 BGB seeks to protect the patients' own interests: In an emergency, they should be able to receive the medical treatment that is in their best interests even though they are incapable of expressing their consent. In the case at hand, however, the presumption of consent aims to protect the interests of others, that is, of the researcher. It is unclear whether presumed consent applies in these cases.[200] If it does, one should employ it with due care, and only if the will of the person concerned is known.[201] Furthermore, to avoid attempts at circumventing the right to self-determination, research cannot be based on presumed consent if the persons concerned could have been asked for their consent in good time.[202]

7.3 The Generation of Brain Organoids for Autologous or Allogeneic Transplantation Purposes

7.3.1 Autologous Transplantation

7.3.1.1 The Scope of the German Transplantation Act

Before we spell out the informed-consent requirement under the German Transplantation Act ("*Transplantationsgesetz*," TPG), we must first inquire whether the Act covers the removal of cells for the purpose of transferring (parts of) brain organoids that were grown from these cells either to the cell donor or to third parties.

According to sec. 1 para. 2 sent. 1 TPG, the Transplantation Act applies to the donation and removal of human organs and tissues for the purpose of transfer as well as to the transfer of the organs or tissues, including preparatory measures. Sec. 1a no. 4 defines tissues as "all components of the human body consisting of cells which are not organs according to no. 1, including individual human cells."[203] Sec. 1a no. 6 defines "removal" as the extraction of organs and tissue. Finally, sec. 1a no. 7 TPG defines "transfer" as "the use of organs and tissues in or on a human recipient as well as the application in humans outside the body."

---

198 Wagner (2020b), sec. 630d para. 53.

199 von Freier (2005), p. 327.

200 See Freund and Weiss (2004), p. 317 and the references cited therein.

201 Freund and Weiss (2004), p. 317.

202 Freund and Weiss (2004), p. 317; Wagner (2020b), sec. 630d para. 52.

203 The Tissue Directive 2004/23/EC likewise applies to cells. See Art. 2 para. 1, according to which the Directive applies to tissues and cells, and Art. 3 lit. a and b, which defines tissues and cells, respectively. The German legislature then decided, for the sake of simplicity, to use the term "tissue" for both cells and tissues. See Parliamentary Document 16/3146 (2006), p. 24.

It follows from these provisions that the removal of cells generally falls within the scope of the Transplantation Act, since human cells, including one single cell, constitute tissue within the meaning of sec. 1a no. 4 TPG. Furthermore, the separation of these cells from the human body constitutes a removal within the meaning of sec. 1a no. 6 TPG. Sec. 1 para. 2 sent. 1 TPG should then be read to include the removal of tissues and cells that will be transferred not in their original but in a processed state. After all, the wording of the provision is not limited to unmodified and unprocessed uses of the separated tissue on humans.[204] The explanatory memorandum of the *Bundestag* explicitly confirms this reading.[205]

The removal of cells and tissue to produce brain organoids thus falls within the scope of the Transplantation Act. But does the Act also apply to the transfer of the brain organoids themselves? There is no consensus on this matter. Some scholars argue that while the Transplantation Act covers the removal of cells from which artificial organs (and organoids) are generated, it should not apply to the transfer of those organs (or organoids)[206] because sec. 1 para. 2 sent. 1 TPG mentions the transfer of "the" organs and "the" tissues, not "of" organs and "of" tissues; in other words, they believe the Act only covers the transfer of such organs that have already been removed from the body *as organs*.[207]

I believe, however, that the structure of the statute does not allow applying it only to the removal of organs or tissues, but not to their transfer. The law, I submit, links both aspects of a transplantation so closely to one another it is either fully applicable or not applicable at all.

Thus, the permissibility of removal depends on specific requirements for the planned transfer.[208] According to provisions such as sec. 8 para. 1 no. 2 TPG, for instance, the removal of organs and tissues is only permissible if "the transfer of *the* organ or tissue to the intended recipient is suitable, according to medical assessment, to preserve the life of this person or to cure a serious illness in her, to prevent its aggravation or to alleviate its symptoms." If we stipulate that sec. 8 is applicable to the removal of cells intended to be transformed into organoids, we must also argue that the transfer of the organoid itself must constitute the transfer "of *the* organ" (or tissue) as mentioned in this section. For if (brain) organoids cannot be considered "*the* organ or *the* tissue" within the meaning of sec. 8 para. 2 no. 2 TPG, the removal of the cells is either unlawful—because it does not meet the requirements of sec. 8 (*the* removed cells will not be transferred)—or it falls outside the scope of the Transplantation Act, because this Act would apply only to removals with the aim of transferring *the* removed substances. In that case, the specific informed-consent requirements would not apply.

---

[204] Wernscheid (2012), p. 112; Taupitz (2020a), p. 809; Faltus (2021), p. 131.

[205] Parliamentary Document 16/3146 (2006), p. 21; Wernscheid (2012), p. 112; Taupitz (2020a), p. 809.

[206] Gerke (2020), pp. 291, 295; Taupitz (2020a), pp. 809, 811.

[207] Thus, explicitly, Gerke (2020), p. 295.

[208] König (2005), sec. 1 TPG para. 17; Rixen (2013), sec. 1 TPG para. 3 and footnote 8.

Furthermore, we should bear in mind the wording of sec. 1 para. 2 sent. 1, according to which the Transplantation Act applies to the "removal for the purpose of transfer" as well as to the "transfer of the organs and tissues." It stands to reason that the latter refers to the transfer which was intended at the time of the removal. In other words, we should read the provision as a chronological description of the transplantation process: The physician removes an organ or tissue for a specific transfer and then proceeds to transfer "it" as intended. This means that if the Act were not to cover the transfer of brain organoids, it would not cover the removal for the purpose of this kind of transfer either.

More, it seems contradictory to allow all processing-related intentions during the removal in order to apply the Transplantation Act's relevant provisions but to narrow the statute's scope, because of the processing, when it comes to the transfer itself. Once it has been processed, furthermore, no substance can be considered, strictly speaking, "the" organ or "the" tissue it was at the time of removal, and yet nobody suggests that the Transplantation Act does not apply to the transfer of processed tissue or organs;[209] some authors, for instance, even speak of the transfer of "the tissue" when the tissue in question was artificially produced from human cells.[210] Why, then, should the degree of processing be limited and the threshold exceeded (more or less arbitrarily) once tissue becomes an organ? To the contrary, it is possible to read the wording of sec. 1 para. 2 sent. 2 TPG as "transfer of either the organs and tissues directly obtained, or the organs and tissues produced from the tissue or organs removed, as intended at the moment of removal."

The purpose of the Transplantation Act, which aims to treat the transfer of all human organs and tissues equally, regardless of their origin, points in the same direction. For this reason, artificial organs are also recognized as "parts of the human body" within the meaning of sec. 1a no. 1 TPG, and thus as "organs" within the meaning of the Act, if they were created from human materials.[211] This interpretation is supported by the fact that the Transplantation Act is intended to counteract the dangers arising from the unregulated procurement, processing, storage, and distribution of cells and tissues, such as contamination or the transmission of diseases. These risks also exist in the case of artificially produced tissues (and organs).[212]

Sec. 1a no. 1 TPG, furthermore, does not indicate the Act does not apply to the transfer of brain organoids.[213] According to this provision, the term organs includes "tissues of an organ that can be used for the same purpose as the whole organ in the human body while maintaining the requirements of structure and blood vessel supply, with the exception of such tissues that are intended for the production of advanced therapy medicinal products within the meaning of sec. 4 para. 9 AMG." In other words, tissue that will normally count as an organ loses this organ property if the researcher aims to process it into an ATMP (advanced therapy medicinal prod-

[209] See also Faltus (2021), p. 131.

[210] Gerke (2020), p. 295

[211] Taupitz (2020a), p. 811.

[212] Gerke (2020), p. 291; Taupitz (2020a), p. 811.

[213] But see Taupitz (2020a), p. 811.

uct). This tissue does not lose its organ status once it becomes an ATMP. Rather, it is not removed as an organ in the first place, but "only" as tissue within the meaning of sec. 1a no. 4 TPG.[214] However, of course, the Transplantation Act also applies to the transfer of tissue.

The purpose of sec. 1a no. 1 TPG is to make sure this kind of tissue and the substance resulting from its processing fall under the regulations on medicinal products. Thus, the provision exists in its current form because Reg. (EC) 1394/2007 mandated a narrower definition of organs. According to sec. 2 para. 3 no. 8 AMG, organs within the meaning of sec. 1a no. 1 TPG are not medicinal products,[215] but Reg. (EC) 1394/2007 provided for the medicinal product status of certain substances that did fall under the organ definition of the Transplantation Act, such as cells and tissue from organs. Consequently, single cells were completely excluded from the organ definition in 2009, while tissues were excluded to the extent they are intended for the manufacture of ATMPs.[216] In 2012, the definition of organs was further restricted: now, tissue from an organ only constitutes an organ if the requirements for structure and blood vessel supply mirror that of the whole organ—as is the case, for instance, with split liver donations.[217]

It follows that this provision does not directly cover the case examined here. It is already doubtful whether the production of brain organoids even requires tissue—as opposed to single cells—within the meaning of sec. 1a no. 1 TPG. If the researchers only remove single cells, the tissue they remove never constituted an organ in the first place; in other words, it does not qualify as the object which sec. 1a no. 1 TPG sought to cover.

It is more likely, instead, that the legislature did not have the removal of cells to produce brain organoids in mind at all: Today, after all, we can create something like an organ from a substance that initially did not constitute an organ but may do so after being processed, and at the same time qualifies as an ATMP.[218] The question, then, is which property should prevail. Given sec. 1a no. 1 TPG, it makes more sense to prioritize the ATMP property. This entails that sec. 13 of the Medicinal Products Act, which requires a manufacturing authorization, becomes applicable to the production of brain organoids. This permits reviewing whether the production

---

[214] Pühler et al. (2010), p. 25.

[215] In this context, sec. 17 para. 1 p. 2 no. 2 TPG is very misleading: It refers to medicinal products that are "manufactured from or using organs," which seems to suggest that organs within the meaning of the Transplantation Act can constitute medicinal products in some form after all. To avoid contradicting sec. 2 para. 3 no. 8 of the Medicinal Products Act, sec. 17 para. 2 no. 2 TPG must be read to cover only medicinal products that stem from processed organs which themselves are no longer organs within the meaning of the Transplantation Act—because they no longer form a "functional unit"—and which are also no longer tissues as parts of organs within the meaning of sec. 1a no. 1 TPG.

[216] Parlamentary Document 16/12256 (2009), p. 58 and p. 26; Document of the Federal Council 171/09 (2009), p. 50; Parlamentary Document 16/13428 (2009), pp. 46–47 and p. 75; Federal Law Gazette (2009), p. 2009.

[217] Parliamentary Document 17/7376 (2011), p. 17.

[218] For the medicinal properties of brain organoids, see below, Sect. 7.3.1.3.1.

complies with the requisite technical standards.[219] Furthermore, pharmaceutical regulations on preclinical and clinical testing can be used as a prerequisite for a marketing authorization.[220]

This, it bears emphasizing, would not be the first time that sec. 1a no. 1 TPG would be applied to cases beyond its wording. Thus, the provision also comes into play when a whole organ is removed for the purpose of creating an ATMP, even though the provision states that *tissue from organs* destined to become an ATMP loses its organ property; in this case, then, the whole organ loses its organ property and becomes tissue, as which it is then removed.[221]

To make a long story short, the Transplantation Act applies to the removal of cells to produce brain organoids as well as to their transfer. The reason for that is that organoids constitute tissue within the meaning of the Transplantation Act.[222] They do not count as organs, since they acquire the properties of an ATMP through processing and the Transplantation Act states that tissue ceases to be an organ in case of doubt; but the Act likewise applies to the transfer of tissue.

The removal thus gives rise to specific informed-consent requirements under the Transplantation Act, in particular under secs. 8 et seq. TPG. The Transplantation Act does not regulate the admissibility of transfer of tissue.[223] By subjecting the removal of tissue from living donors to certain conditions, in particular recipient-related criteria (sec. 8 para. 1 no. 2), it does, however, at least indirectly, restrict the possibilities of transfer.

In interpreting both the general regime that governs curative treatments or attempts[224] and the Medicinal Products Act,[225] we must always bear these provisions of the Transplantation Act in mind. In particular, secs. 40 et seq. of the Medicinal Products Act, which regulate the prerequisites of clinical trials, do not constitute a lex specialis compared to the donor regulations of the Transplantation Act; accordingly, the provisions of the Medicinal Products Act cannot loosen the restrictions on permissible donation. First, the reason why secs. 40 et seq. of the Medicinal Products Act are not a lex specialis is that they seek to protect the drug recipient, whereas the secs. 8 et seq. TPG serve to protect the donor. However, the two are not necessarily identical. Second, had the legislator wished to allow autologous transplantations for experimental purposes such as in clinical trials, he could have done so expressly in sec. 8c TPG—a provision that was adopted in 2007[226], i.e., long after the adoption of the provisions on clinical trials.

---

[219] Faltus (2021), p. 131.

[220] Faltus (2021), p. 131.

[221] Thus for the removal of pancreata Pühler et al. (2010), p. 25.

[222] Gerke does not address tissue property and concludes that the transfer of artificial organs does not fall under the Transplantation Act: Gerke (2020), p. 295.

[223] König (2005), sec. 1 TPG para. 17; Rixen (2013), sec. 1 TPG para. 3, also fn 8.

[224] König (2005), sec. 1 TPG para. 17.

[225] See Sects. 7.3.1.3.2 and 7.3.1.3.3, for autologous transplantation, and Sect. 7.3.2.3 for allogeneic transplantation.

[226] Federal Law Gazette (2007), p. 1580.

It bears emphasizing that the Transplantation Act applies irrespective of whether a donation or an autologous transplantation of the brain organoid is planned. According to sec. 1 para. 3 TPG, the Act does not apply to tissues that are removed from a person during the same surgical procedure in order to be transferred back to that person without changing their substantial condition (no. 1). This exception is irrelevant for our purposes, however, because the production of brain organoids requires a change in the substance of cells.[227]

Finally, according to sec. 1 para. 3 no. 2 TPG, the Act does not apply to the removal of blood, which means that the Act does not cover the collection of nuclei-containing blood cells for further donation purposes.[228]

7.3.1.2 Requirements for Informed Consent

Section 8c TPG deals with the informed constent for autogologous transplantations and reflects the general principles that all curative interventions must be medically indicated and require informed consent[229]: It provides that the removal for the purpose of retransfer—i.e., for an autologous transplantation—is only permissible if the person is capable of giving consent, has been informed in accordance with sec. 8 para. 2 sent. 1 and 2 TPG, and has consented to the removal and retransfer (sec. 8c para. 1 no. 1 lit. a) and lit. b)).

The information must cover the purpose of the measure, the prospects of success, possible consequences, and any circumstances that the person concerned evidently considers important; that is also the novelty of the procedure. The same, incidentally, follows from the general principles for new curative methods as well as, generally speaking, from sec. 630e para. 1 sent. 2 BGB.[230] The content of the information and the donor's declaration of consent must also be included in a transcript signed by the persons providing the information as well as the donor (sec. 8 para. 2 sent. 4, 8c para. 4 TPG). Consent may be revoked in writing, electronically or verbally (sec. 8 para. 2 sent. 6, 8c para. 5 TPG).

Crucially, sec. 8c para. 1 no. 2 TPG provides that the removal and retransplantation of the organ or tissue must take place within the context of a *medical treatment* and be necessary—according to scientific consensus—for that treatment. Thus, an autologous transplantation may be used for a curative treatment, and probably also for a curative attempt.[231] It may not be used, conversely, for non-indicated medical interventions,[232] including interventions for research purposes. Researchers must

---

[227] Thus generally for hiPS cell therapies, Gerke (2020), p. 296. For the requirements under sec. 1 para. 3 no. 1 TPG, see Rixen (2013), sec. 1 TPG para. 10.

[228] Gerke (2020), p. 298. See Sect. 7.3.3.

[229] Schmidt-Recla (2013), sec. 8c TPG para. 3.

[230] On the duty to inform about the prospects of success of an organ transplantation, see Müller (2013), p. 167, and below, Sect. 7.3.2.3.

[231] König (2005), sec. 1 TPG para. 17, mentions the possibility of a curative attempt in the context of transplantations (albeit before sec. 8c was adopted).

[232] Schmidt-Recla (2013), sec. 8c TPG para. 6.

bear this in mind when they plan an experiment that falls under the provisions of pharmaceutical law on clinical trials.[233]

In the case of incapacitated persons, a removal for the purpose of retransfer is lawful under sec. 8c para. 2 BGB once the representative has given his or her informed consent, provided the measure does not threaten the welfare of the incapacitated person.[234] To account for the donor's best interests, the representative must give his or her consent if the tissue removal for the purpose of retransfer constitutes a medically indicated and standard treatment. This holds for minors[235] as well as for adults under custodianship. The custodian, moreover, must comply with the wishes of the incapacitated adult; in other words, the wishes specify what counts as the incapacitated person's welfare, provided the wishes do not jeopardize the person's higher-ranking legal interests and do not significantly worsen her overall situation (sec. 1901 para. 3 sent. 1 BGB).[236] In case of doubt, however, the representative must respect the incapacitated adult's self-determination.[237] If the person under custodianship refuses to undergo curative treatment, a coercive treatment is only permissible under the conditions of sec. 1906a BGB.

Curative attempts, that is, treatments that have not yet become the medical standard, are also not automatically incompatible with the best interests of the minor or incapacitated adult.[238] Therefore, deciding whether they are permissible requires a risk-benefit assessment.

7.3.1.3 Can a Retransfer Constitute a Clinical Trial?

7.3.1.3.1 The Medicinal Properties of Brain Organoids

The provisions on clinical trials that subject informed consent to specific requirements are only applicable to the autologous transplantation of brain organoids if the latter qualify as medicinal products. In the following, I will explain why I believe that to be the case.

According to sec. 2 para. 1 AMG, medicinal products are "any substance or combination of substances 1. intended for use in or on the human or animal body and presented as having properties for treating or alleviating or preventing disease in human beings or animals or 2. which may be used in or on the human or animal body or administered to a human or an animal either with a view a) to restoring, correcting or modifying physiological functions by exerting a pharmacological, immunological or metabolic action, or b) to making a medical diagnosis." This definition corresponds (apart from the reference to animals) to that in Art. 1 no. 2 Directive 2001/83/EC, as amended by Art. 1 of the Directive 2004/27/EC.

---

233 See Sects. 7.3.1.3.2 and 7.3.1.3.3, for autologous transplantations.

234 For incapacitated minors, see sec. 1627 BGB. For incapacitated adults under custodianship, see sec. 1901 para. 2 and 3 BGB.

235 Schmidt-Recla (2013), sec. 8c TPG para. 12.

236 Schneider (2020), sec. 1901 BGB para. 11.

237 Schneider (2020), sec. 1901 BGB para. 15.

238 Lipp (2021c), XIII.D. para. 36, 38; Deutsch and Spickhoff (2014), para. 1138-139, 1334.

First of all, organoids that are transplanted for therapeutic reasons are medicinal products, as they are intended to cure human diseases and can be used in or administered to human beings in order to restore, correct or modify physiological functions by exerting a pharmacological, immunological, or metabolic action (Art. 1 no. 2 lit. (a) and lit. (b) of Directive 2001/83/EC as well as sec. 2 para. 1 no. 1 and 2 AMG).[239] The materials generated from hiPS cells are also "substances" as defined in Art. 2 para. 1 no. 3 of Directive 2001/83/EC (and sec. 2 para. 1, 3 no. 3 AMG), that is, any matter, irrespective of its origin (which may be human).

There is also the category of the so-called advanced therapy medicinal products (ATMP), which, according to Art. 4 para. 9 AMG, include gene therapy medicinal products, somatic cell therapy medicinal products, or tissue engineered products in accordance with Art. 2 para. 1 lit. (a) of the Regulation (EC) No 1394/2007 of the European Parliament and of the Council of 13 November 2007 on advanced therapy medicinal products and amending Directive 2001/83/EC and Regulation (EC) No 726/2004. Art. 2 para. 1 lit. (a) of the Reg. (EC) No 1394/2007 defines ATMP as "any of the following medicinal products for human use: a gene therapy medicinal product as defined in Part IV of Annex I to Directive 2001/83/EC, a somatic cell therapy medicinal product as defined in Part IV of Annex I to Directive 2001/83/EC, a tissue engineered product as defined in lit. (b)."

HiPS-cell-based therapeutics such as organoids are ATMPs within the meaning of sec. 4 para. 9 AMG or Art. 2 para. 1 lit. (a) Reg. (EC) No 1394/2007.[240]

Because of the processing steps that they undergo, extracted cells constitute the so-called tissue engineered products as defined in Art. 2 para. 1 lit. (b) Reg. (EC) 1394/2007 which reads as follows: "[t]issue engineered product means a product that contains or consists of engineered cells or tissues, and is presented as having properties for, or is used in or administered to human beings with a view to regenerating, repairing or replacing a human tissue. A tissue engineered product may contain cells or tissues ofhuman or animal origin, or both." According to Art. 2 para. 1 lit. (c), "cells or tissues shall be considered 'engineered' if they have been subject to substantial manipulation, so that biological characteristics, physiological functions or structural properties relevant for the intended regeneration, repair or replacement are achieved (the manipulations listed in Annex I, in particular, shall not be considered as substantial manipulations) or the cells or tissues are not intended to be used for the same essential function or functions in the recipient as in the donor."

The cells from which the brain organoids are generated must first be reprogrammed into hiPS cells and are then differentiated into brain cells. This process constitutes "engineering" within the meaning of the regulation because it achieves biological characteristics, physiological functions, or structural properties that are relevant for the intended regeneration, repair, or replacement.[241] Finally, brain

[239] Gerke (2020), p. 254; Taupitz (2020a), p. 811.

[240] For greater detail, see Gerke (2020), pp. 254 et seq.; Taupitz (2020a), pp. 811–812.

[241] For hiPS cell-based therapeutics in general, see Gerke (2020), pp. 254–255. For organoids, see Taupitz (2020a), pp. 811–812.

organoids also have the purpose—at least in the context of transplantations—of being used in or administered to human beings with a view to regenerating, repairing, or replacing a human tissue.

A question we do not have to answer is whether brain organoids also fall under the definition of a somatic cell therapy medicinal product (Para. 2.2 of Part IV Annex I Dir. 2001/83/EC), since Art. 2 para. 4 Reg. (EC) No 1394/2007 provides that somatic cell therapy medicinal products which also constitute tissue engineered products shall only be considered the latter.

Furthermore, genetic changes to the cells or the brain organoid may cause the latter to become a gene therapy medicinal product. An ATMP which is both a gene therapy medicinal product and a tissue engineered product shall be considered a gene therapy medicinal product (Art. 2 sec. 5 Reg. (EC) No 1394/2007). The definition of gene therapy medicinal products is layed down in Part IV of annex I of the Dir. 2001/83/EC, as amended by the Dir. 2009/120/EC: "Gene therapy medicinal product means a biological medicinal product which has the following characteristics: (a) it contains an active substance which contains or consists of a recombinant nucleic acid used in or administered to human beings with a view to regulating, repairing, replacing, adding, or deleting a genetic sequence; (b) its therapeutic, prophylactic, or diagnostic effect relates directly to the recombinant nucleic acid sequence it contains, or to the product of genetic expression of this sequence."

It is hard to assess in the abstract whether a brain organoid with genetically modified genetic information constitutes a gene-therapeutical medicinal product. Whether it contains a recombinant nucleic acid—so that its therapeutic, prophylactic, or diagnostic effect relates directly to that recombinant nucleic acid sequence it contains, or to the product of genetic expression of this sequence—depends on the method used.[242]

7.3.1.3.2 The Admissibility of Clinical Trials Under the German Medicinal Products Act in its Version Prior to 27 January 2022

Since brain organoids are medicinal products, secs. 40 et seq. AMG apply if these organoids or parts thereof are transferred back to the donor in the context of clinical trials. To decide whether the autologous transfer of brain organoids designed as a clinical trial is permissible, we must also bear in mind the requirements of the Transplantation Act. This yields the following observations.

An autologous transplantation cannot be carried out as a clinical trial on *healthy adults capable of giving consent*. Although clinical trials on these persons are permissible in principle (sec. 40 para. 1 sent. 3 AMG), the Transplantation Act requires that the removal and transfer of autologous tissue must take place in the context of *medical treatment* (sec. 8c para. 1 no. 2 TPG). This means that all non-indicated

[242] It is doubtful, for instance, whether the CRISPR/Cas method, which can change the genetic information of a cell through self-repair mechanisms, creates a gene-therapeutical medicinal product. For an analysis under the German Gene Technology Act, see Deuring (2020), pp. 379 et seq.

medical interventions are excluded,[243] including pure research interventions that fall under the Medicinal Products Act. The same applies to healthy minors.

However, this also means that *adults suffering from a disease who are capable of giving consent can* participate in corresponding clinical trials (sec. 41 para. 1 AMG), provided the use of the medicinal product is indicated, according to scientific consensus, in order to save the life of these persons, restore their health, or alleviate their suffering.

In principle, it is also permissible to conduct clinical trials without direct benefit for the participant if they are at least associated with a direct benefit for the group of patients suffering from the same disease as this person. Due to sec. 8c para. 1 no. 2 TPG, a group benefit is not sufficient, however, since it means the transplantation no longer takes place in the context of a medical treatment and is not necessary for "*the* treatment." It is doubtful, moreover, whether a transplantation without medical necessity would be medically justifiable and therefore admissible as an admissible clinical trial in the first place.

If, however, the clinical trial on adults suffering from a disease and capable of giving consent is permissible in the case of self-benefit, they themselves consent to their participation in writing (sec. 40 para. 1 sent. 3 no. 3 lit. (b) AMG), provided they have been informed according to sec. 40 para. 2 AMG. According to that provision, the persons concerned must be informed about the nature, significance, risks, and implications of the clinical trial as well as about their right to terminate their participation at any time; moreover, they must also receive the information in written form. According to sec. 40 para. 2a sent. 1 AMG, finally, the individuals must be informed about the purpose and scope of the processing of their personal data, in particular their health data.

Sec. 41 para. 2 AMG and sec. 41 para. 3 AMG provide that minors and incapacitated adults who suffer from a disease may participate in corresponding studies. If the conditions stated therein are met, we can assume that the treatments are not contrary to the best interests of these individuals (sec. 1627, 1901 para. 2 and 3 BGB).

This means that *minors who suffer from a disease* may participate if the use of the medicinal product is indicated, according to scientific consensus, in order to save the life of these minors, restore their health, or alleviate their suffering (sec. 41 para. 2 sent. 1 no. 1 AMG). Alternatively, there must be a direct benefit for the group of patients suffering from the same disease (no. 2 lit. (a)), and the research may cause only a minimal risk and burden for the person concerned (no. 2 lit. (d)). At this point, we must recall the requirements of sec. 8c para. 1 no. 2 TPG, however, according to which a group benefit does not suffice.

If the trial is admissible, the researcher must ask the legal representatives for their consent (sec. 40 para. 4 no. 3 sent. 1 AMG) after informing them in accordance with para. 2. The trial must honor the presumed will of the minors to the extent it can be identified (para. 4 no. 3 sent. 2). The minors must also be informed about the trial, the risks, and the benefits if they have the requisite age and mental maturity. If the minors declare that they do not wish to participate in the clinical trial, or express

[243] Schmidt-Recla (2013), sec. 8c TPG para. 6.

their dislike in some other manner, this will must be taken into account (sent. 3). If the minors are capable of understanding the nature, significance, and implications of the clinical trial and of acting accordingly, and is thus capable of giving consent, their consent is also required (sent. 4).

For *adults who are incapable of giving consent and suffer from a disease*, by and large the same principles apply. According to sec. 41 para. 3 no. 1 AMG, the use of the medicinal product to be tested must be indicated, according to scientific consensus, in order to save the life of the persons concerned, restore their health or alleviate their suffering; in addition, such research must relate directly to a life-threatening or highly debilitating clinical condition in which the persons concerned find themselves, and the clinical trial must be associated with the least possible burden and other foreseeable risks for these persons; both the degree of burden and risk must be specifically defined in the trial protocol and constantly reviewed by the investigator. The clinical trial may only be conducted if there is a reasonable expectation that the benefits of using the investigational medicinal product for the persons concerned outweigh the risks or that there are no risks. Consent is given by the legal representatives after they have been informed in accordance with sec. 40 para. 2. Sec. 40 para. 4 no. 3 sent. 2, 3, and 5 apply accordingly. Under German law, then, only direct self-benefit justifies research on persons incapable of giving consent; a mere group benefit does not.

7.3.1.3.3 The Admissibility of Clinical Trials Under Regulation (EU) 536/2014 and the German Medicinal Products Act in its Amended Version as of 27 January 2022

Now, Reg. (EU) No 536/2014 governs questions of consent to clinical trials. Its Art. 28 para. 1 lit. (b), for example, stipulates that subjects or—if a subject is unable to give informed consent—his or her legal representative must be informed in accordance with Art. 29 paras. 2–6. Art. 28 para. 1 lit. (c) states that the subject or, if a subject is unable to give informed consent, his or her legal representative shall give informed consent in accordance with Art. 29 para. 1, 7, and 8.

As regards the admissibility of clinical trials, the Regulation distinguishes between capacitated adults, incapacitated subjects (Art. 31), and minors (Art. 32).

If the criteria of sec. 8c TPG are met, *adults who can give consent* may participate in clinical trials with brain organoids. The Regulation, in other words, changes nothing in this respect.

The law on incapacitated subjects remains largely the same as well. According to Art. 31 and Art. 32 of the Regulation, clinical trials on *incapacitated subjects and minors* are only permitted under certain conditions, as researchers should prioritize trials on subjects who are capable of giving consent (Art. 31 para. 1 lit. (e); Art. 32 para. 1 lit. (f)). Thus, the clinical trial must relate directly to a medical condition from which the subject suffers (Art. 31 para. 1 lit. (f), Art. 32 para. 1 lit. (f)) and must either produce a direct benefit for the subject itself or at least for the population represented by the subject. In the latter case, the clinical trial must pose only minimal risk to, and will impose minimal burden on, the subject concerned in comparison with the standard treatment of the subject's condition (Art. 31 para. 1 lit. (g); Art. 32 para. 1 lit. (g)).

Art. 31 para. 2 points out that more stringent rules prohibiting the conduct of clinical trials on incapacitated subjects remain permissible. German law has made use of this authorization, excluding group-beneficial research on incapacitated adults. The new version of sec. 40b para. 4 sent. 3 AMG, in principle, renders impermissible research projects without direct personal benefit. Thus, the representative is not able to consent to such research projects. Such projects are only permissible insofar as the persons concerned, as persons of legal age who are capable of giving consent have stipulated in writing, upon receiving medical information, that they consent to certain group-beneficial clinical trials in the event that they will be incapable of giving consent. Again, of course, we must bear in mind the conflict with sec. 8c para. 1 no. 2 TPG, pursuant to which clinical trials consisting in the transfer of brain organoids may only take place if there is an intrinsic benefit as defined in this regulation.

Furthermore, both incapacitated subjects and minors must themselves receive the information to which Art. 29 para. 2 refers, and they must do so in a manner that reflects their capacity to understand it (Art. 31 sec. 1 lit. (b); Art. 32 sec. 1 lit. (b)). If they can form an opinion and assess the information they have received, their wish not to take part or to withdraw must be respected (Art. 31 sec. 1 lit. (c); Art. 32 sec. 1 lit. (c)). The new version of sec. 40b para. 3 sent. 2 and para. 4 sent. 2 AMG goes even further, however, providing that the researcher must respect any form of dismissive attitude. All in all, Art. 31 para. 3 and Art. 32 para. 2 stress that the subjects shall take part in the informed consent procedure as much as possible. The new version of sec. 40b para. 3 sent. 1 AMG, incidentally, continues to provide for the co-consensual solution in cases where minors can give consent. Art. 29 para. 8 of the Regulation expressly allows the Member States to retain the co-consensual solution.

7.3.2 Allogeneic Transplantation

7.3.2.1 The Donor's Consent to the Collection of Cells and the Transfer of the Organoid

The removal of cells for the purpose of donation interferes with the physical integrity of the donor and therefore requires his or her consent.[244] This consent is regulated in sec. 8 TPG. Since the collection is always carried out for a specific purpose, the donor simultaneously declares his or her consent to the specific further use, i.e., to the transplantation envisaged at the time of the collection. Although the Transplantation Act only mentions the "consent to removal" and does not touch upon the consent to further uses,[245] the latter remains necessary. But, as just seen,

[244] Ulsenheimer (2019), chapter 24 § 152 IV.2. para. 31; Lipp (2021a), VI.A.I.6.a) para. 30.

[245] In contrast to the Transplantation Act, sec. 6 para. 1 sent. 3 TFG does provide for separate consent.

donors may imply they consent when they consent to the removal *for the purpose of donation*.

The removal of tissue or organs for the purpose of donation is only permissible if the person is of age and capable of giving consent, has been informed in accordance with sec. 8 para. 2 sent. 1 and 2 TPG[246] and has consented to the removal (sec. 8 para. 1 sent. 1 no. 1 lit. (a) and (b) TPG). In addition, the person must be suitable as a donor and may not be endangered beyond the risk of the operation or seriously impaired beyond the immediate consequences of the collection (lit. (c)). According to no. 2, the transfer of the organ or tissue to the intended recipient must be suitable, according to medical assessment, to preserve the life of this person or to cure a serious illness from which he or she suffers, to prevent its aggravation or to alleviate its symptoms.

Furthermore, compared to autologous transplants, the information about allogeneic transplants has to fulfill an additional criterion: It must be provided in the presence of another physician (sec. 8 para. 2 sent. 3).

We can also find many of these elements—particularly the requirements of informed consent and that the donation must have a therapeutic purpose—in the Additional Protocol to the Convention on Human Rights and Biomedicine concerning Transplantation of Organs and Tissues of Human Origin. Art. 14 sec. 1 states that the removal of organs or tissue may not be carried out on persons incapable of consent. Exceptionally, regenerative tissue can be removed, especially if the recipient is a sibling of the donor and the donation has the potential to save the recipient's life.[247] German law, which is much more restrictive in this respect, does not even permit the donation of single cells by persons incapable of giving consent; the only exceptions are the donation of blood cells within the scope of the Transfusion Act and, pursuant to sec. 8a TPG, of bone marrow.

7.3.2.2 Donor Consent to Transfer of Organoids Under the Special Conditions of Sec. 8b TPG

If organs or tissues have been removed from a living person as part of a medical treatment of that person, their transfer is only permissible pursuant to sec. 8b para. 1 sent. 1 TPG if the person has the capacity to consent, has been informed in accordance with sec. 8 para. 2 sent. 1 and 2 TPG, and has consented to the transfer.

7.3.2.3 The Consent of the Recipient to the Transfer of the Organoids

The Transplantation Act does not contain regulations on the information and consent of the recipient. We can find the relevant law, therefore, in secs. 630d, 630e BGB. According to sec. 630d para. 2 BGB, effective consent requires that the patient or the person entitled to consent[248] has been informed in accordance with

---

[246] On the scope of the information, see Sect. 7.3.1.2.

[247] See Art. 14 sec. 2, also with regard to further conditions.

[248] That is, the legal representative of minors, sec. 1626 et seq. BGB, the custodian of incapacitated adults, sec. 1896 et seq. BGB, or the authorized representative, sec. 1901c BGB.

sec. 630e para. 1 to 4. According to sec. 630e para. 1 sent. 1 BGB, the physician is obliged to inform the patient of all circumstances essential to his or her consent. According to para. 2, this includes, above all, the nature, extent, implementation, expected consequences and risks of the measure as well as its necessity, urgency, suitability, and prospects of success regarding the diagnosis or therapy. In the context of allogeneic transplantations of brain organoids or parts thereof, this includes information about the artificiality of the transplanted cells. Both this and the novelty of the procedure increase the requirements which the information must fulfill. Thus, the recipient must be comprehensively informed about the uncertainties of the treatment as well as the unknown chances and risks.[249] The right of personality of minors and incapacitated persons requires, moreover, that they too be provided with the essential information, not only their legal representative.

If the transplantation is carried out as a clinical trial, the provisions of medicinal products law apply, namely, secs. 40 et seq. AMG and Reg. (EU) No 536/2014. Moreover, we must bear in mind what we established, in Sects. 7.3.1.3.2 and 7.3.1.3.3, on the interaction between the Medicinal Products Act and the Transplantation Act: Since sec. 8 para. 1 sent. 1 no. 2 TPG only permits the collection of organs and tissues if the recipient exhibits certain characteristics, a donation for clinical trials is only lawful if the trial excludes participants who do not fulfill these criteria. This means that only persons who suffer from a severe disease can participate in the trials, provided the transplantation promises a benefit described in more detail in sec. 8 para. 1 no. 2 TPG (read together with the provisions of pharmaceutical law, which likewise specify the benefit that must be achieved if the trial involves persons suffering from a disease).

There are less problems, conversely, when the physician transfers brain organoids, within the context of a clinical trial, that were produced from tissue removed for purposes other than donation, sec. 8b TPG. In this case, the recipient must only fulfill the requirements that arise under medicinal products law; sec. 8b TPG does not add any conditions.[250] As already mentioned, it remains doubtful, however, whether a transplantation that does not respond to a medical necessity can be justified. On that view, we may have to rule out research that exclusively benefits others anyway.

7.3.3 The Collection of Nuclei-Containing Blood Cells

If brain organoids are produced using blood cells that contain nuclei, the procedure to collect the blood cells is covered by the Transfusion Act ("*Transfusionsgesetz,*" TFG). According to sec. 28 TFG, the Transfusion Act "does not apply to the collection of a minor amount of blood for diagnostic purposes, to homeopathic autologous

[249] Müller (2013), p. 166. Generally on the scope of the obligation to provide information in the case of novel medical methods, Lipp (2021c), XIII.D. para. 32; Deutsch and Spickhoff (2014), para. 1333.

[250] See Sects. 7.3.1.3.2 and 7.3.1.3.3, for the admissibility of clinical trials.

blood products [or to] autologous blood for the production of tissue engineered products"; this exception, however, covers collections with the purpose of propagation or the processing of other autologous body cells, not the collection of blood cells to produce hiPS cells.[251]

Sec. 6 TFG contains special requirements for informed consent to the collection of a "blood donation." They are relevant for our purposes if the collection of blood cells with the aim of cultivating brain organoids constitutes a "donation" within the meaning of sec. 2 no. 1 TFG. According to this provision, a "donation is the quantity of blood or blood components removed from humans which is an active substance or medicinal product or is intended for the production of active substances or medicinal products and other products for use in humans." The term "donation," then, does not describe the act of transferring a blood sample to another person. Rather, it refers to the collected blood itself, to the result of the blood extraction.

The collection of blood cells with the aim of cultivating brain organoids meets the conditions of this definition since brain organoids are medicinal products and are intended for use in humans. True, both the hiPS cells produced from the collected blood cells and the brain organoids that result from the hiPS cells are themselves no longer subject to the Transfusion Act: they are tissue within the meaning of sec. 1a no. 4 TPG, not blood or blood components.[252] The Transfusion Act remains applicable to the removal of the cells, however, because sec. 2 no. 1 extends the Act's coverage to any blood donation that will be used for medicinal products and other products in general. It is irrelevant whether the product for which the blood cells were removed falls under the Transfusion Act itself (e.g., because it is a blood product). Finally, the Transfusion Act provisions on the collection of blood cells apply regardless of whether the blood cells—or the products derived from them—are intended for an allogeneic or autologous use. After all, sec. 2 no. 1 TFG only mention the "use on humans."

Sec. 6 TFG is thus applicable to the collection of blood cells in order to produce brain organoids that will then be used for autologous or allogeneic transplantations. It provides that the persons concerned must be competently informed about the nature, significance and performance of the removal and the associated examinations. They must also declare that the donation may, in fact, be used. The consent must be confirmed in writing.

While sec. 8 para. 1 no. 2 TPG specifies who may receive allogeneic transplants, sec. 6 TFG does not contain comparable restrictions on the group of people eligible to receive blood donations. Since the Transfusion Act regulates the removal, but not the transfer itself, and the Transplantation Act, despite the tissue properties of brain organoids, does not directly regulate the transfer of tissue, no law limits the eligibility of recipients of brain organoids generated from blood cells. Unlike in the case of organoids produced from other cells, then, secs. 40 et seq. AMG are, in principle,

[251] Faltus (2016), p. 643; Gerke (2020), p. 298 ; Tag (2017), sec. 28 TFG para. 1.

[252] Thus for hiPS cells Gerke (2020), p. 299. For a definition of blood and blood components, see Gerke (2020), p. 297.

fully applicable. The legislature's decision only to regulate the requirements for recipients of cell collections that fall under the Transplantation Act suggests we should not transfer these requirements to the recipients of organoids produced from blood cells. Evidently, the purpose of sec. 8 para. 1 no. 2 TPG was to restrict the removal of certain substances, not to restrict the transfer itself.[253] Whether it is legitimate, from a medical perspective, to include healthy individuals, or whether the Medicinal Products Act precludes their participation, is a different question, of course.

The Transfusion Act does not contain any explicit provisions on the collection of blood from minors or incapacitated adults. There is no reason, however, why the representative or custodian should not be able to consent, according to the principles that generally apply to curative treatments and new curative methods, to the collection of blood cells to produce brain organoids for the purpose of retransplantation.[254] Extracting blood for the purpose of transplanting the resulting organoid to another person, however, may not be in the "best interests" of the person concerned, or conducive to his or her "welfare" (sec. 1627, 1901 para. 2 and 3 BGB), since he or she does not stand to benefit personally.[255] However, each case will have to be assessed individually. Thus, the terms "welfare" or "best interest" can also be understood more broadly (e.g., to enable the healing of a parent if the intervention is negligible). The welfare of persons under custodianship, moreover, partly depends on their personal wishes anyway (sec. 1901 para. 2 and 3 BGB).

7.3.4 Interim Conclusion

The Transplantation Act covers the collection of cells other than nuclei-containing blood cells for the autologous and allogeneic transplantation of brain organoids. Consequently, the informed-consent requirements prior to collection result from secs. 8 et seq. TPG. The transplantation act does not specifically regulate the admissibility of the transfer of organoids, which constitute tissue within the meaning of the act. However, the requisite requirements follow implicitly from the regulations on tissue removal.

Sec. 8c TPG applies to autologous transplants. This means, first of all, that the person concerned or, if he or she is incapable of doing so, his or her legal representative must consent to the procedure. It also follows that autologous transplantation may only take place in the context of medical treatment. That notably excludes experimental transplantations that "only" aim to benefit the population represented by the person concerned. This is particularly important in the context of clinical trials.

According to sec. 8 TPG, allogeneic transplants are only permissible if the donor is of age *and* capable of giving consent. In addition, the transplant must be suitable for

[253] See also Rixen (2013), sec. 1 para. 3 footnote 8.

[254] Thus for autologous blood transfer Tag (2017), sec. 6 TFG para. 10.

[255] Tag (2017), sec. 6 TGG para. 6.

preserving the life of this person or for curing a serious illness, for preventing its aggravation or alleviating its symptoms. This limits the effect of the provisions regarding clinical trials, as such trials are, because of the Transplantation Act, only permissible if the person concerned will benefit personally. The restriction does not apply, however, if the physician uses body materials that were removed for other purposes (sec. 8b TPG); in that case, the admissibility of clinical trials solely depends on the requirements established in the Medicinal Products Act (and Reg. (EU) 536/2014).

In the case of the collection of nuclei-containing blood cells for the autologous or allogeneic transfer of brain organoids generated from them, the informed-consent requirements follow from sec. 6 TFG. Unlike the Transplantation Act, the Transfusion Act does not impose any restrictions on who may receive the transfer.

7.4 Questions of Data Protection

Research with brain organoids generates data, specifically genetic data, at least if the procedure or the planned investigations involve the analysis of the genome. This brings the regulations of data protection law into play.

To begin with, most scholars do not consider the separated body substance itself—i.e., the removed cells and the organoid created from them—data: According to Art. 4 no. 1 of the Regulation (EU) No 2016/679 on the protection of natural persons with regard to the processing of personal data and on the free movement of such data, "'personal data' means any information relating to an identified or identifiable natural person ('data subject') (...)."[256] According to Art. 4 no. 13, "genetic data" means "personal data relating to the inherited or acquired genetic characteristics of a natural person which give unique information about the physiology or the health of that natural person and which result, in particular, from an analysis of a biological sample from the natural person in question."

Body substances, in other words, constitute the "biological sample" mentioned in Art. 4 no. 13, but not the data itself.[257] Classifying bodily substances as data would have the unacceptable consequence, moreover, that the removal of bodily material, which would then be considered data collection, would at times be possible without the consent of the person concerned.[258]

The analysis of substances, at least of identifiable and not (de facto) anonymized ones,[259] falls within the scope of Reg. (EU) No 2016/679, however, as it can

[256] See recital 26 for the criteria that determine whether a person is "identifiable." It remains disputed whether a pseudonymization (Art. 4 no. 5 Reg. (EU) No 2016/769) qualifies, at least from the perspective of the data processing agents, as anonymization. See Spindler and Dalby (2019a), Art. 4 Reg. (EU) No 2016/769 paras. 14 et seq.; Taupitz (2020b), pp. 606 et seq.

[257] See Schreiber (2019), p. 105, and the references cited therein. Breyer (2004), p. 660, Breithaupt (2012), p. 240, and Fink (2005), p. 60, also reject classifying body substances as data.

[258] von Freier (2005), p. 324; Schreiber (2019), p. 105.

[259] Schreiber (2019), pp. 130–131; for greater detail on the meaning of anonymization and de facto anonymization, especially in the context of genetic data, see Taupitz (2020b), p. 605 et seq.

generate genetic data.[260] It constitutes a collection of data within the meaning of Art. 4 no. 2, which defines "processing of data." Genetic data are particularly protected, as they constitute, according to Art. 9, a special category of data. For that reason, they may only be processed if the following specific requirements are met.

According to Art. 9 para. 2 lit. a), data such as genetic data may be processed "if the data subject has given explicit consent to the processing of those personal data for one or more specified purposes (...)." Consent is defined in Art. 4 no. 11 as "any freely given, specific, informed and unambiguous indication of the data subject's wishes by which he or she, by a statement or by a clear affirmative action, signifies agreement to the processing of personal data relating to him or her." It is important to note that consent must be explicit; implied consent is not sufficient.[261] Written consent, however, is not required.[262] The information requirements, moreover, follow from Arts. 13 et seq. The requisite information includes the purpose of and responsibility for the processing, contact details, the intended data transfer—in particular to countries outside the EU—as well the rights of the data subjects. The latter must include, among other things, the right to withdraw consent and the right to information.[263]

However, Art. 9 para. 2 lit. (j) also permits the processing of particularly sensitive data without consent if "processing is necessary for archiving purposes in the public interest, scientific or historical research purposes or statistical purposes in accordance with Article 89(1) based on Union or Member State law which shall be proportionate to the aim pursued, respect the essence of the right to data protection and provide for suitable and specific measures to safeguard the fundamental rights and the interests of the data subject." According to Art. 89 para. 1, the provisions at national or EU level that implement Art. 9 para. 2 lit. (j) must provide "appropriate safeguards for the rights and freedoms of the data subject." The German legislature, for its part, has implemented Art. 9 para. 2 lit. (j) in the new Federal Data Protection Act ("*Bundesdatenschutzgesetz*," BDSG).[264] Art. 27 para. 1 sent. 1 BDSG provides that the processing of sensitive data is also permissible absent consent for *scientific or historical research purposes or for statistical purposes, if* this is necessary to achieve the stated purposes and the interests of the responsible person in the processing significantly outweigh the interest of the data subject who my refuse any processing of his or her data.[265] Some argue that we should read Art. 27 narrowly

---

260 Art. 4 no. 1; Schreiber (2019), p. 106; Fink (2005), p. 61.

261 Schreiber (2019), p. 109; Albers and Veit (2020), Art. 9 Regulation (EU) No 2016/679 para. 50-51.

262 Schreiber (2019), p. 109.

263 Schaar (2017), p. 215.

264 The BDSG applies to public bodies of the federal level as well as to private persons. The data protection laws of the *Länder* will not be dealt with separately in this chapter.

265 Spindler and Dalby (2019b), Art. 9 Regulation (EU) No 2016/679 para. 25; Schreiber (2019), p. 116. For greater detail regarding the balancing of interests, especially the possibility of obtaining consent, see Taupitz (2020b), p. 621 et seq. For criticism, see Fleischer (2018), p. 302.

and demand that the research in question cannot be carried out in any other way than by processing the data in question.[266]

In any event, data processing is only permissible pursuant to Art. 27 para. 1 sent. 1 BDSG if the responsible person takes appropriate and specific measures to protect the interests of the data subject in accordance with Art. 22 para. 2 sent. 2 BDSG, which provides a ten-point example catalogue of protective measures.[267] This provision aims to fulfill the requirements for protective measures to safeguard the rights of the data subjects as provided in Art. 9 para. 2 lit. (j) and Art. 89 of the EU-Regulation.[268]

In the context of data collection on the basis of such "research clauses" as Art. 27 BDSG, children, who enjoy special protection under the General Data Protection Regulation, must be paid particular attention to. Thus, it follows from Art. 6 para. 1 lit. (f), at least for "normal" data, that processing data of children without the consent of their parents is only permissible following a particularly careful consideration of their interests.[269]

It remains to be seen, however, whether "research clauses" such as sec. 27 para. 1 sent. 1 BDSG should even apply to genetic data. Most scholars consider the latter part of the core of fundamental personality rights. For that reason, they argue that the processing of (certain) genetic data should not be permissible absent explicit consent.[270] At least according to the German Federal Constitutional Court, however, the analysis of *non-coding* gene segments does not implicate the core of the right of personality: examining non-coding areas, the Court argued, does not allow any conclusions about personality-relevant characteristics such as hereditary dispositions, character traits or diseases of the person concerned. It does not, in other words, permit the creation of a personality profile.[271]

The Court did not explicitly comment on examinations of coding areas. We can infer from the decision, however, that the creation of a personality profile through genetic examinations does indeed affect the core of the right of

---

[266] Schreiber (2019), p. 116; Fleischer (2018), p. 302.

[267] Schreiber (2019), p. 117.

[268] Schreiber (2019), p. 117.

[269] Schreiber (2019), pp. 119–120.

[270] See, e.g., Goerdeler and Laubach (2002), p. 117; Keller (1989), p. 2292. Fink (2005), p. 66, argues that the core of the right of personality is not affected if the research merely aims to uncover the (not yet established) connection between genetic predispositions and the development of a disease (i.e., a personality-related characteristic). Taupitz (2020b), p. 613 et seq., argues that the "research clauses" apply to genetic data, as neither Art. 9 nor sec. 27 differentiate between genetic data and other sensitive data. Yet, Art. 9 para. 4 allows the member states to introduce further conditions, including limitations, with regard to the processing of genetic data. This allows narrowing the scope of sec. 27, if the consideration of fundamental rights requires it, without coming into conflict with EU law. See Schreiber (2019), p. 122.

[271] Federal Constitutional Court (2000), p. 32.

personality.[272] I suggest we go further and consider, in principle, any examination that yields information on personal characteristics an intrusion on the core of the right of personality, regardless of whether it suffices to establish an entire profile of the person concerned. After all, it makes little sense to exclude a characteristic that the person concerned considers particularly sensitive and intimate—an assessment that, incidentally, cannot be reviewed against an objective standard. Furthermore, a single point of data can be particularly sensitive as well: If it falls into the wrong hands (e.g., insurance companies, employers, etc.), the person concerned may run the risk of considerable disadvantages. Alternatively, we should regard at least some genetic data as part of the core of the fundamental right of personality, viz., data that, if it falls into the wrong hands (e.g., employers, insurance companies, etc.), may create considerable disadvantages for the person concerned.[273]

It follows, then, that research involving the examination of coding gene segments cannot in principle be based on "research clauses" but requires the consent of the person concerned or their representative. This finding adds to our previous ones: We have already established that the generation of brain organoids from substances that have been separated for other purposes requires the consent of the donor. Now we know this applies even more if the research conducted with the organoids involves the investigation of coding gene segments.

If the genetic data was collected upon the donor's consent, however, the question arises whether the data may subsequently be processed for purposes other than those for which they were originally collected. Data protection law provides for such further use within narrow limits. In principle, Art. 5 para. 1 lit. (b) Reg. (EU) No 2016/679 makes clear, data may only be used and thus further processed for the purpose for which they were collected. This applies not only to the first user but also to subsequent users.[274] Further processing for other purposes is only permissible if the new purpose is compatible with the original one. This compatibility test requires an evaluative judgment,[275] although further processing for other research purposes should always, in accordance with Art. 5 para. 1 lit. (b), be considered compatible,

[272] Fink (2005), p. 66. But see Schreiber (2019), p. 123 et seq., who argues the purpose of the use should also be relevant. The objection to this argument, however, is that it eviscerates the notion of a fundamental right's core protection. In particular, it does not follow from the Federal Constitutional Court's "Diary Decision" that the purpose of use alters the personal and intimate nature of the data concerned. Rather, the Court argued (in a questionable manner) that the diary entries could be used for criminal investigations because they "inherently affect"—by providing information about the cause and background of the criminal offense—"the sphere of others or the interests of the community" (Federal Constitutional Court 1989, p. 379). Crucially, the information embodied in one's genes does not "inherently affect" the interests of others or the community.

[273] See Taupitz (2020b), p. 609.

[274] Schreiber (2019), p. 280; Taupitz (2020b), p. 619.

[275] Schreiber (2019), p. 281.

provided the requirements of Art. 89 para. 1 Regulation (EU) No 2016/679 are met.[276]

In the case of further use for purposes other than research, the compatibility test shall be based on Art. 6 para. 4 Regulation (EU) No 2016/679, which contains a non-exhaustive list of criteria.[277]

Moreover, contrary to recital 50—which is considered a drafting error—a separate legal basis is required for each further use.[278] As a result, further processing of particularly sensitive personal data must fulfill one of the conditions of Art. 9 para. 2 Regulation (EU) No 2016/679.[279] Art. 27 para. 1 sent. 1 BDSG, which complements the opening clause of Art. 9 para. 2 lit. (j) Regulation (EU) No 2016/679, can (again) be used as a legal basis in this regard.[280] As we have seen above, however, the research clause does not apply to genetic data, at least not to data that results from coding areas; further use of such data without the consent of the data subject is therefore impermissible.

This means that genetic data collected in the context of (consensual) research with and on brain organoids cannot be reused for other research purposes without explicit consent. It also means that the non-consensual disclosure of such data to third parties—which, if the third parties want to carry out research projects with the data, is covered, in principle, by Art. 9 para. 2 lit. (j) Regulation (EU) No 2016/679 in conjunction with Art. 27 BDSG[281]—is unlawful, as Art. 27 BDSG does not apply to genetic data. If the person disclosing the data is a physician, it should also be noted that he or she is bound to medical confidentiality under sec. 203 of the Criminal Code.[282]

Moreover, data protection law also provides for the possibility of "broad consent." As a result, the data subject's consent may refer to the processing of data not only in a specific research project but more generally in different research areas. Consider, for instance, recital 33: "It is often not possible to fully identify the purpose of personal data processing for scientific research purposes at the time of data collection. Therefore, data subjects should be allowed to give their consent to certain areas of scientific research when in keeping with recognized ethical standards for scientific research."[283]

[276] Schreiber (2019), p. 281.

[277] Schreiber (2019), p. 281.

[278] Schantz (2016), p. 1844. See also Schantz (2020), Art. 5 Regulation (EU) No 2016/679 para. 22; Spindler and Dalby (2019b), Art. 9 Regulation (EU) No 2016/679 para. 23; Schreiber (2019), pp. 282–283; Fleischer (2018), p. 294 et seq. For a contrasting opinion, see Schlösser-Rost (2020), sec. 27 BDSG para. 13.

[279] Weichert (2017), p. 540; Schreiber (2019), p. 283.

[280] Fleischer (2018), p. 301; Greve (2020), sec. 27 BDSG para. 15.

[281] Albers and Veit (2020), Art. 9 Regulation (EU) No 2016/679 para. 88; also in favor of disclosure being a form of processing, Taupitz (2020b), p. 618 et seq.

[282] On this provision, Fleischer (2018), pp. 308–309.

[283] Schreiber (2019), p. 111; Fleischer (2018), p. 296.

7.5 Conclusion

Because they have a right to self-determination, adults capable of giving consent can consent to research on and with brain organoids generated from cells taken from their bodies. For persons incapable of consent, the right to consent lies with their legal representative: This group of individuals, I have argued, is not generally barred from participating in research project, provided the project in question does not conflict with their best interests and welfare. That is all the more true if the research project involves only minor physical interventions. The only thing that may exclude this group from research projects involving brain organoids is the possibility to conduct the research just as effectively with cells—and, therefore, brain organoids—that originate from persons who *can* give consent. Minors who are capable of giving consent may not consent to research projects on their own, since the projects affect the right to property that persons hold with regard to their separated bodily substances.

The information provided to the person concerned or his or her legal representative must touch upon all relevant circumstances, such as the planned intervention, the associated risks, and the intended use of both the substances and the data, including a possibly intended disclosure to third parties. The person concerned must also be informed of his or her right of revocation as well as of what will happen to the substances, the data, and the research results should he or she exercise that right.

In my opinion, it is impermissible to use, without the concerned person's consent, substances that were separated for other ends in order to generate brain organoids for research purposes. First, this results from the fact that such research may be considered controversial from an ethical perspective; moreover, the patient's right to informational self-determination is violated if the substances are not anonymized. Second, the research may involve the collection of genetic data, which, according to data protection regulations, requires consent—at least in the case of identifiable materials. (Admittedly, this point is disputed.) Nor is the non-consensual disclosure of genetic data relating to indetifiable subjects to third parties lawful, even if the third parties merely wish to use the data for their research.

Both in terms of data protection law and in general, however, the so-called broad consent is permissible. The person concerned, therefore, may consent to the removal and disposal of bodily substances for research in general. In principle, this also applies to individuals who cannot give consent. The information provided should, however, indicate that the individual has the right to exclude those types of research of which he or she does not approve.

If brain organoids are to be (re)transplanted, the Transplantation Act and the Transfusion Act become relevant, depending on the cell type to be removed from a person's body. Moreover, the Transplantation Act and the regulations on medicinal products (the German Medicinal Products Act and Reg. (EU) 536/2914) must be read together and complement each other whenever brain organoids are to be transferred in the context of clinical trials. This means that transplantations—be they autologous or allogeneic—may only be carried out in the context of clinical trials if the person concerned can expect a direct benefit. A benefit for the population he or she represents is not sufficient. The same result, incidentally, may already follow from a risk–benefit assessment of the transplantation. It is doubtful whether a transplantation for research purposes is lawful absent medical necessity.

The informed-consent requirements regarding the donor follow from the Transplantation Act and the Transfusion Act; regarding the recipient, they follow from the Medicinal Products Act (Reg. (EU) 536/2014) if the transplantion occurs within the context of a clinical trial. If the transplantation does not occur during a clinical trial, informed consent regarding the recipient is subject to the general principles established in sec. 630d and 630e of the Civil Code.

References

Albers M, Veit R-D (2020) art. 9 Regulation (EU) No 2016/679. In: Brink S, Wolff HA (eds) BeckOK Datenschutzrecht, 34th edn. C.H. Beck, München
Balaban E, Teillet M-A, Le Douarin N (1988) Application of the quail-chick chimera system to the study of brain development and behavior. Science 241(4871):1339–1342
Bartfeld S, Clevers H (2018) 3.5 Aus stammzellen abgeleitete organide und ihre bedeutung für die biomedizinische forschung und therapie. In: Zenke M, Marx-Stölting L, Schickl H (eds) Stammzellforschung. Aktuelle wissenschaftliche und gesellschaftliche Entwicklung, 1st edn. Nomos, Baden-Baden, pp 90–95
Baston-Vogt M (1997) Der sachliche Schutzbereich des zivilrechtlichen allgemeinen Persönlichkeitsrechts. Mohr Siebeck, Tübingen
Bender D (2005) Heilversuch oder klinische Prüfung? Annäherung an eine diffuse Grenze. MedR 23:511–516
Breithaupt J (2012) Rechte an Körpersubstanzen und deren Auswirkungen auf die Forschung mit abgetrennten Körpersubstanzen. Nomos, Baden-Baden
Breyer P (2004) Der datentschutzrechtliche Schutz von Körpersubstanzen, die Patienten zu Analysezwecken entnommen werden. MedR 22:660–667
Buchanan A, Brock DW, Daniels N et al (2012) From chance to choice. Cambridge University Press, Cambridge
Central Ethics Committee (2003) Die (Weiter-)Verwendung von menschlichen Körpermaterialien für Zwecke medizinischer Forschung (20.02.2003). https://www.zentrale-ethikkommission.de/stellungnahmen/koerpermaterialien/. Accessed 15 Jan 2021
Chen HI, Wolf JA, Blue R et al (2019) Transplantation of human brain organoids: revisiting the science and ethics of brain chimeras. Cell Stem Cell 25(4):462–472
Daviaud N, Friedel RH, Zou H (2018) Vascularization and engraftment of transplanted human cerebral organoids in mouse cortex. eNeuro 5(6):1–18
Dederer H-G (2009) Die Garantie der Menschenwürde (Art. 1 Abs. 1 GG). Dogmatische Grundfragen auf dem Stand der Wissenschaft. In: Häberle P (ed) Jahrbuch des öffentlichen rechts, vol 57. Tübingen, Mohr Siebeck, pp 91–124
Dettmeyer R, Madea B (2004) Aufklärung und Einwilligung bei der Beschaffung und Verarbeitung von menschlichen Zellen und Geweben. Rechtsmedizin 14(2):85–93
Deuring S (2020) Rechtliche Herausforderungen moderner Verfahren der Intervention in die menschliche Keimbahn. Springer, Heidelberg
Deutsch E, Spickhoff A (2014) Medizinrecht. Arztrecht, arzneimittelrecht, medizinprodukterecht und transfusionsrecht, 7th edn. Springer, Berlin
Document of the Federal Council 171/09 (2009). http://dipbt.bundestag.de/dip21/brd/2009/0171-09.pdf. Accessed 14 Jan 2021
Dreier H (2013) Art. 1 para. 1 GG. In: Dreier H (ed) Grundgesetz-Kommentar, 3rd edn. Mohr Siebeck, Tübingen

Faltus T (2016) Stammzellenreprogrammierung. Der rechtliche Status und die rechtliche Handhabung sowie die rechtssystematische Bedeutung reprogrammierter Stammzellen. Nomos, Baden-Baden
Faltus T (2021) Organoide, assembloide, gastruloide, SHEEFs & organ-on-chip. MedR 39:125–134
Farahany NA, Greely HT, Hyman S et al (2018) The ethics of experimenting with human brain tissue. Nature 556:429–432
Federal Constitutional Court (1970) Decision of 15 Dec 1970 – 2 BvF 1/69, 2 BvR 629/68, 2 BvR 308/69. In: Official collection of decisions, vol 30 ("*BVerfGE* 30"), pp 1–47
Federal Constitutional Court (1977) Decision of 21 Dec 1977 – 1 BvL 14/76. In: Official collection of decisions, vol 45 ("*BVerfGE* 45"), pp 187–271
Federal Constitutional Court (1989) Decision of 26 Jul 1989 – 1 BvR 685/89. In: Official collection of decisions, vol 80 ("*BVerfGE 80*"), pp 367–383
Federal Constitutional Court (2000) Decision of 14 Dec 2000 – 2 BvR 1741/99, 2 BvR 276/00, 2 BvR 2061/00. In: Official collection of decisions, vol 103 ("*BVerfGE 103*), pp 21–41
Federal Court of Justice (1953) Decision of 25 Sept 1952 - III ZR 322/51. NJW 1953:700–701
Federal Court of Justice (1976) Decision of 29 Jun 1976 - VI ZR 68/75. NJW 1976:1790–1792
Federal Court of Justice (1994) Decision of 09 Nov 1993 - VI ZR 62/93. NJW 1994:127–128
Federal Court of Justice (2004) Decision of 26 May 2004 - 2 StR 505/03. NJW 2004:2458–2461
Federal Court of Justice (2007) Decision of 10 Oct 2006 - VI ZR 74/05. NJW 2007:217–220
Federal Court of Justice (2014) Decision of 20 May 2014 – VI ZR 381/13. NJW 2014:2190–2192
Federal Court of Justice (2017) Decision of 30 May 2017 – VI ZR 203/16. NJW 2017:2685–2686
Federal Law Gazette (2007) Part I No 35. www.bgbl.de. Accesses 09 Mar 2021
Federal Law Gazette (2009) Part I No 43. www.bgbl.de. Accessed 14 Jan 2021
Fink S (2005) Einwilligung und vertragliche Regelungen zur Entnahme von Körpersubstanzen, deren Aufbewahrung und Verwendung in Biobanken. Mannheim
Fleischer H (2018) Rechtliche Aspekte der Systemmedizin. LIT, Münster
Forsberg SL, Ilieva M, Michel TM (2018) Epigenetics and cerebral organoids: promising directions in autism spectrum disorders. Transl Psychiatry 8(1):1–11
Förster C (2020) § 823 BGB. In: Hau W, Poseck R (eds) BeckOK BGB, 55th edn. C.H. Beck, München
Franken A (2020) § 12 Klinische Prüfung von Arzneimitteln. In: Fuhrmann S, Klein B, Fleischfresser A (eds) Arzneimittelrecht. Handbuch für die Rechtspraxis, 3rd edn. Nomos, Baden-Baden
Frankenberg G (2000) Die Würde des Klons und die Krise des Rechts. Kritische Justiz 33:325–332
Freund G, Weiss N (2004) Zur Zulässigkeit der Verwendung menschlichen Körpermaterials für Forschungs- und andere Zwecke. MedR 22:315–319
Gerke S (2020) Die klinische translation von hiPS-Zellen in Deutschland. In: Gerke S, Taupitz J, Wiesemann C et al (eds) Die klinische Anwendung von humanen induzierten pluripotenten Stammzellen. Springer, New York, pp 243–328
German Ethics Council (2011) Mensch-Tier-Mischwesen in der Forschung. Berlin. https://www.ethikrat.org/fileadmin/Publikationen/Stellungnahmen/deutsch/DER_StnMischwesen_Deu_Online.pdf. Accessed 15 Jan 2021
Goerdeler J, Laubach B (2002) Im Datendschungel - Zur Notwendigkeit der gesetzlichen Regelung von genetischen Untersuchungen. ZRP 2002:115–119
Greely HT, Cho MK, Hogle LF et al (2007) Thinking about the human neuron mouse. Am J Bioeth 7(5):27–40
Greve H (2020) § 27 BDSG. In: Eßer M, Kramer P, von Lewinski K (eds) Auernhammer. DSGVO/BDSG, 7th edn. Carl Heymanns, Köln
Halàsz C (2004) Das Recht auf bio-materielle Selbstbestimmung. Grenzen und Möglichkeiten der Weiterverwendung von Körpersubstanzen. Springer, Berlin
Hardtung B (2017) § 228 StGB. In: Joecks W, Miebach K (eds) Münchener Kommentar zum StGB, 4th edn. C.H. Beck, München
Higher Regional Court Hamm (2020) Decision of 29 Nov 2019 – 12 UF 236/19. NJW 2020:1373–1376

Karpowicz P, Cohen CB, van der Kooy D (2004) Is it ethical to transplant human stem cells into nonhuman embryos. Nat Med 10(4):331–335
Katzenmeier C (2020) § 630d BGB. In: Hau W, Poseck R (eds) BeckOK BGB, 55th edn. C.H. Beck, München
Keller R (1989) Die Genomanalyse im Strafverfahren. NJW 38:2289–2296
Kern B-R (1994) Fremdbestimmung bei der Einwilligung in ärztliche Eingriffe. NJW 47:753–759
Kern B-R (2019a) § 4 Berufsethik: Schutz der Prsönlichkeit des Kranken. In: Laufs A, Kern B-R, Rehborn M (eds) Handbuch des Arztrechts, 5th edn. C.H. Beck, München
Kern B-R (2019b) § 131 Die medizinische Forschung. In: Laufs A, Kern B-R, Rehborn M (eds) Handbuch des Arztrechts, 5th edn. C.H. Beck, München
Kern B-R, Rehborn M (2019) § 42 Der Arztvertrag. In: Laufs L, Kern B-R, Rehborn M (eds) Handbuch des Arztrechts, 5th edn. C.H. Beck, München
Kersten J (2004) Das Klonen von Menschen. Eine verfassungs-, europa- und völkerrechtliche Kritik. Mohr-Siebeck, Tübingen
Klumpp S (2017) Vorbemerkungen zu §§ 104 ff. BGB. In: Staudinger BGB. Buch 1. Allgemeiner Teil, §§ 90-124; §§ 130-133. Sellier/de Gruyter, Berlin
König P (2005) § 1 TPG. In: Schroth U, König P, Gutmann T et al (eds) Transplantationsgesetz. C.H. Beck, München
Lackermair M (2017) Hybride und Chimären. Die Forschung an Mensch-Tier-Mischwesen aus verfassungsrechtlicher Sicht. Mohr Siebeck, Tübingen
Lancaster MA, Renner M, Martin C-A et al (2013) Cerebral organoids model human brain development and microcephaly. Nature 501:373–390
Laufs A, Reiling E (1994) Schmerzensgeld wegen schuldhafter Vernichtung deponierten Spermas? NJW 1994:775–776
Lavazza A, Massimini M (2018) Cerebral organoids: ethical issues and consciousness assessment. J Med Ethics 44:606–610
Lavazza A, Pizzetti FG (2020) Human cerebral organoids as a new legal and ethical challenge. J Law Biosci 2020:1–22
Li R, Le S, Fang A et al (2017) Recapitulating cortical development with organoid culture in vitro and modeling abnormal spindle-like (ASPM related primary) microcephaly disease. Protein Cell 8(11):823–833
Lipp V (2021a) VI. Rechtsfragen der transplantation, transfusion, sektion und intensivmedizin. In: Laufs A, Katzenmeier C, Lipp V (eds) Arztrecht, 8th edn. C.H. Beck, München
Lipp V (2021b) VII. Sterilisation, schwangerschaft und sexualmedizin. In: Laufs A, Katzenmeier C, Lipp V (eds) Arztrecht, 8th edn. C.H. Beck, München
Lipp V (2021c) XIII. Heilversuch und medizinische forschung. In: Laufs A, Katzenmeier C, Lipp V (eds) Arztrecht, 8th edn. C.H. Beck, München
Lippert H-D (2001) Forschung an und mit Körpersubstanzen - wann ist die Einwilligung des ehemaligen Trägers erforderlich? MedR 2001:406–410
Mansour AA, Gonçalves JT, Bloyd CW et al (2018) An in vivo model of functional and vascularized human brain organoids. Nat Biotechnol 36(5):432–441
Müller S (2013) Die Aufklärung des Organspendeempfängers über Herkunft und Qualität des zu transplantierenden Organs. Ärztliche Pflichten im Spannungsfeld zwischen Standardbehandlung und Neulandmedizin. Peter Lang, Frankfurt
Muotri AR, Nakashima K, Toni N et al (2005) Development of functional human embryonic stem cell-derived neurons in mouse brain. Proc Natl Acad Sci U S A 102(51):18644–18648
National Ethics Council (2004) Biobanken für die Forschung. Berlin. https://www.ethikrat.org/fileadmin/Publikationen/Stellungnahmen/Archiv/NER_Stellungnahme_Biobanken.pdf. Accessed 15 Jan 2021
Nitz G, Dierks C (2002) Nochmals: Forschung an und mit Körpersubstanzen - wann ist die Einwilligung des ehemaligen Trägers erforderlich? MedR 2002:400–403
Ohly A (2002) "Volenti non fit iniuria" Die Einwilligung im Privatrecht. Mohr Siebeck, Tübingen
Parliamentary Document 16/12256 (2009). http://dipbt.bundestag.de/dip21/btd/16/122/1612256.pdf. Accessed 15 Jan 2021

Parliamentary Document 16/13428 (2009). http://dipbt.bundestag.de/dip21/btd/16/134/1613428.pdf. Accessed 15 Jan 2021

Parliamentary Document 16/3146 (2006). http://dip21.bundestag.de/dip21/btd/16/031/1603146.pdf. Accessed 15 Jan 2021

Parliamentary Document 16/5374 (2007). http://dipbt.bundestag.de/doc/btd/16/053/1605374.pdf. Accessed 15 Jan 2021

Parliamentary Document 17/7376 (2011). https://dip21.bundestag.de/dip21/btd/17/073/1707376.pdf. Accessed 15 Jan 2021

Prütting J, Merrem M-T (2019) § 630d BGB. In: Prütting D (ed) Medizinrecht Kommentar, 5th edn. Hermann Luchterhand, Köln

Pühler W, Middel C-D, Hübner M (2010) Pankreasinseln. Was tun, wenn ein Organ zum Gewebe wird? MedR 2010:23–27

Qian X, Nguyen HN, Song MM et al (2016) Brain-region-specific organoids using mini-bioreactors for modeling ZIKV exposure. Cell 165(5):1238–1254

Qian X, Jacob F, Song MM et al (2018) Generation of human brain region-specific organoids using a miniaturized spinning bioreactor. Nat Protoc 13(3):565–580

Quadrato G, Nguyen T, Macosko EZ et al (2017) Cell diversity and network dynamics in photosensitive human brain organoids. Nature 545(7652):48–53

Rehmann WA (2020) Arzneimittelgesetz (AMG), 5th edn. C.H. Beck, München

Rixen S (2013) § 1 TPG. In: Höfling W (ed) Transplantationsgesetz, 2nd edn. Schmidt, Berlin

Schaar K (2017) Anpassung von Einwilligungserklärungen für wissenschaftliche Forschungsprojekte. ZD 5:213–220

Schantz P (2016) Die Datenschutz-Grundverordnung - Beginn einer neuen Zeitrechnung im Datenschutzrecht. NJW 26:1841–1847

Schantz P (2020) Art. 5 regulation (EU) No 2016/679. In: Brink S, Wolff HA (eds) BeckOK Datenschutzrecht, 34th edn. C.H. Beck, München

Schlösser-Rost J (2020) § 27 BDSG. In: Brink S, Wolff HA (eds) BeckOK Datenschutzrecht, 34th edn. C.H. Beck, München

Schmidt-Recla A (2013) § 8c TPG. In: Höfling W (ed) Transplantationsgesetz, 2nd edn. Schmidt, Berlin

Schneider A (2020) § 1901 BGB. In: Säcker FJ, Rixecker R, Oetker H et al (eds) Münchener Kommentar zum BGB, vol 10, 8th ed. C.H. Beck, München

Schreiber M (2019) Die medizinische Forschung mit abgetrennten Körpersubstanzen Minderjähriger. LIT, Münster

Schröder M, Taupitz J (1991) Menschliches Blut. Verwendbar nach Belieben des Arztes? Zu den Formen erlaubter Nutzung menschlicher Körpersubstanzen ohne Kenntnis des Betroffenen. Enke, Stuttgart

Schroth U (2009) Die Einwilligung in eine nicht-indizierte Körperbeeinträchtigung zur Selbstverwirklichung - insbesondere die Einwilligung in Lebendspende, Schönheitsoperation und Piercing. In: Hassemer W (ed) In dubio pro libertate. Festschrift für Klaus Volk zum 65. C.H. Beck, München, pp 719–741

Schünemann H (1985) Die Rechte am menschlichen Körper. Lang, Frankfurt

Spickhoff A (2008) Autonomie und heteronomie im alter. AcP 2008:345–415

Spickhoff A (2018a) § 107 BGB. In: Säcker FJ, Rixecker R, Oetker H et al (eds) Münchener kommentar zum bürgerlichen gesetzbuch, vol 1, 8th edn. C.H. Beck, München

Spickhoff A (2018b) § 630d BGB. In: Spickhoff A (ed) Medizinrecht, 3rd edn. C.H. Beck, München

Spindler G (2020) § 823 BGB. In: Gsell B, Krüger W, Lorenz S et al (eds) BeckOGK BGB. C.H. Beck, München

Spindler G, Dalby L (2019a) Art. 4 regulation (EU) No 2016/679. In: Spindler G, Schuster F (eds) Recht der elektronischen Medien, 4th edn. C.H. Beck, München

Spindler G, Dalby L (2019b) Art. 9 regulation (EU) No 2016/679. In: Spindler G, Schuster F (eds) Recht der elektronischen Medien, 4th edn. C.H. Beck, München

Spranger TM (2001) Fremdnützige Forschung an Einwilligungsunfähigen, Bioethik und klinische Arzneimittelprüfung. MedR 2001:238–247

Sternberg-Lieben D (2019) § 228 StGB. In: Schönke A, Schröder H (eds) Strafgesetzbuch, 30th edn. C.H. Beck, München
Stock C (2009) Ist die Verbesserung des Menschen rechtsmissbräuchlich? In: Wienke A, Eberbach WH, Janke K et al (eds) Die Verbesserung des Menschen. Tatsächliche und rechtliche Aspekte der wunscherfüllenden Medizin. Springer, Berlin, pp 145–162
Suhr K (2016) Der medizinisch nicht indizierte Eingriff zur kognitiven Leistungssteigerung aus rechtlicher Sicht. Springer, Wiesbaden
Tag B (2017) § 6 TFG, § 28 TFG. In: Joecks W, Miebach K (eds) Münchener Kommentar zum StGB, 3rd edn. C.H. Beck, München
Taupitz J (1991) Wem gebührt der Schatz im menschlichen Körper? Zur Beteiligung des Patienten an der kommerziellen Nutzung seiner Körpersubstanzen. AcP 1991:201–246
Taupitz J (1998) Das recht auf nichtwissen. In: Hanau P (ed) Festschrift für Günther Wiese zum 70. Luchterhand, Neuwied, pp 583–602
Taupitz J (2000) Empfehlen sich zivilrechtliche Regelungen zur Absicherung der Patientenautonomie am Lebensende? Gutachten A für den 63. Deutschen Juristentag. In: Ständige Deputation des Deutschen Juristentages (ed) Verhandlungen des 63. Deutschen Juristentages. C.H. Beck, München, pp A41–A130
Taupitz J (2012) Medizinische Forschung an jungen und alten Patienten. MedR 30:583–588
Taupitz J (2020a) Organoide. MedR 30:805–813
Taupitz J (2020b) Biobanken: Forschungsprivilegien im Datenschutzrecht. In: Katzenmeier C (ed) Festschrift für Dieter Hart. Medizin – Recht – Wissenschaft. Springer, Berlin, pp 603–625
Taupitz J, Schreiber M (2016) Biobanken - zwischen Forschungs- und Spenderinteressen. BGesB 1:304–310
Ulsenheimer K (2019) § 152 Strafrechtliche aspekte der organtransplantation. In: Laufs A, Kern B-R, Rehborn M (eds) Handbuch des arztrechts, 5th edn. C.H. Beck, München
von Freier F (2005) Getrennte Körperteile in der Forschung zwischen leiblicher Selbstverfügung und Gemeinbesitz. MedR 2005:321–410
Wachenhausen H (2016a) § 4 para. 23 AMG. In: Kügel JW, Müller R-G, Hofmann H-P (eds) Arzneimittelgesetz, 2nd edn. C.H. Beck, München
Wachenhausen H (2016b) § 40 AMG. In: Kügel JW, Müller R-G, Hofmann H-P (eds) Arzneimittelgesetz, 2nd edn. C.H. Beck, München
Wagner G (2020a) § 630d BGB. In: Säcker FJ, Rixecker R, Oetker H et al (eds) Münchener Kommentar zum BGB, vol 5, 8th edn. C.H. Beck, München
Wagner G (2020b) Vor §§ 630a BGB. In: Säcker FJ, Rixecker R, Oetker H et al (eds) Münchener Kommentar zum BGB, vol 5, 8th edn. C.H. Beck, München
Watanabe M, Buth JE, Vishlaghi N et al (2017) Self-organized cerebral organoids with human-specific features predict effective drugs to combat zika virus infection. Cell Rep 21(2):517–532
Weichert T (2017) "Sensitive Daten" revisited. DuD 2017:538–543
Wenzel F (2019) Chapter 4. Das Recht der medizinischen Behandlung. In: Wenzel F (ed) Handbuch des Fachanwalts Medizinrecht, 4th edn. Hermann Luchterhand, Köln
Wernscheid V (2012) Tissue Engineering - Rechtliche Grenzen und Voraussetzungen. Universitätsverlag Göttingen, Göttingen
Winnands S (2016) § 22 AMG. In: Kügel JW, Müller R-G, Hofmann H-P (eds) Arzneimittelgesetz, 2nd edn. C.H. Beck, München

Part IV

The Chimera Issue

Moral Permissibility of Transplantation of Human Brain Organoids into Animals

8

Insoo Hyun

8.1 Introduction

This chapter considers the moral permissibility of transplanting human brain organoids into laboratory animals. In secular ethics, an activity is morally permissible if it is neither morally obligatory nor morally forbidden. To be morally permissible is to be a matter of individual choice—to do or not to do, with no moral obligation one way or another. For the matter at hand, which involves a research technique aimed at the advancement of human brain organoid research and disease modeling in laboratory animals, the moral grounds for permissible choice lie at the intersection of two ethically sensitive areas of science. Understandably, given the ethical heat already present around both human brain organoid research and human-to-animal interspecies chimera research, people's thinking about the permissibility of transplanting human brain organoids into the brains of laboratory animals is liable to be murky, at least initially.

We can turn down the heat and increase the light at the juncture of organoid and chimera research if we consider separately what makes brain organoids and chimeras each so ethically sensitive and then consider whether the combination of these two scientific pursuits raises any additional ethical concerns that must be addressed in a new way.

I. Hyun (✉)
Department of Bioethics, Case Western Reserve University, School of Medicine, Cleveland, OH, USA

Center for Bioethics, Harvard Medical School, Boston, MA, USA
e-mail: insoo.hyun@case.edu; insoo_hyun@hms.harvard.edu

H.-G. Dederer, D. Hamburger (eds.), *Brain Organoids in Research and Therapy*, Advances in Neuroethics, https://doi.org/10.1007/978-3-030-97641-5_8

8.2 Ethical Sensitivities Surrounding Brain Organoids and Chimeras

Unlike other human organoids generated in the lab, such as gut organoids or liver organoids, brain organoids are of special ethical concern because the human brain itself is so intimately tied to people's subjective sense of personal identity. One might say that the natural human brain is distinct from other organs because of its role in supporting persons' experiences, memories, agency, creativity, and all those other mental properties that make human lives distinctive. The brain can easily be viewed as the seat of an individual's subjective sense of identity, so much so that a sudden and severe brain injury might cast doubt on whether the "same person" will survive, even as the body otherwise remains intact. Irreversible damage to other organs, even if it results in the need for a whole organ transplant, does not raise the same type of concerns about the continuity of a person's identity.

Furthermore, the brain itself is central for the possibility of personhood in Western philosophy. That is to say, human beings are classified as *persons* due to their capacity for rational agency. According to one popular version of the concept of rational agency—what contemporary philosophers call personal autonomy—an individual's rational agency consists in his or her ability to act thoughtfully on those motivations, appetites, or desires that he or she approves of having on a higher cognitive level of self-reflection.[1] To put the point another way, an autonomous person acts deliberately in accordance with his or her own values.[2] Certainly, neither personhood nor autonomy would be possible without the complex cognitive functions supported by the brain, the sum of which makes having distinctively human lives possible and gives human existence its felt coherence. John Locke once wrote that a person is "a thinking intelligent being, that has reason and reflection, and can consider itself as itself, the same thinking thing, in different times and places."[3]

It may be argued, therefore, that the moral significance of the brain derives from its role in supporting personhood, rational agency, personal identity, and personal interactions, all of which are crucial for grounding our everyday judgments of moral approbation and blame. In light of these important philosophical connections, it is easy to see why people may be much more concerned about human brain organoids than other types of organoid models. If the natural human brain is morally significant, then some might reason that organoid models of the human brain could also be morally significant, especially if they are capable of exhibiting or supporting the types of cognitive human traits mentioned above. Even if it is extremely unlikely or impossible for human brain organoids in a dish to ever exhibit agency on the level of personal autonomy in humans with intact brains, which would require the ability to interact with the world and each other, concern over the possible emergence of some basic level of humanlike "consciousness" might suffice to motivate a cautious approach to human brain organoid research.

---

[1] Dworkin (1970) and Frankfurt (1971).

[2] Hyun (2001).

[3] Locke (1694/1975), 2.27.9.

A similar concern underlies people's apprehensions about human-to-animal chimera research, in particular the possibility of acute neurological chimerism in laboratory animals generated through the transfer of human stem cells or their direct neuronal derivatives. Here the chief worry appears to be less about whether chimeric animals could gain "consciousness" in the form of conscious access to sensory stimulation, or wakefulness, vigilance, focal attention, or sentience. Host animal species such as laboratory rodents already possess all of these mental capacities without the addition of human neural cells. Rather, the ethical concern with both neurological chimeras and human brain organoids is that organoids maintained in a dish (in vitro) or human-to-animal chimeras could somehow gain the morally significant characteristic of humanlike consciousness in the form of *subjective self-awareness*: i.e., conscious awareness of oneself as a temporally extended being with experiences, beliefs, and interests, all of which can be mentally reflected upon by oneself.

Is an ethical concern about the emergence of humanlike consciousness a realistic one for either chimeras or brain organoids? Would the transfer of brain organoids into animal models heighten the concerns around neurological chimerism that already exist in the stem cell field? To answer these questions we must first consider (a) what science has revealed about the possibilities of human stem cell-based chimera research (and thus what can be said of its moral permissibility in general) and (b) whether human brain organoids can support ethical concerns about the emergence of humanlike consciousness through their developmental capacity in vitro and/or upon transplantation into laboratory animals' brains—that is, within the context of brain organoid-generated chimeric animals.

8.3 The Promises and Limitations of Chimera Research

In ancient Greek mythology the Chimera was a monster composed of three different animals—a lion's head, a goat's body, and a serpent's tail. In contemporary biomedical science, research chimeras are entities that contain functionally integrated populations of cells from at least two zygotes of the same or different species. Experimental chimeras composed of cells from two individuals, particularly in the mouse, are used in everyday biomedical research for generating transgenic animals. More recently, advances in the generation of chimera-competent pluripotent stem cells and interspecies chimera research are opening new paths for applications of chimeras for basic biology and regenerative medicine. Generating human-to-animal chimeras using reprogrammed patient cells called induced pluripotent stem cells (iPS cells) might create an in vivo setting to study human disease and to generate transplantable human organs inside livestock animals.

Generally speaking, chimera research is not new. It pre-dates the advent of human pluripotent stem cell research. In biomedical research, the transfer of human somatic cells into animal hosts has become commonplace over the past several decades, in large part because of the scientific advantages it offers over non-chimeric animal research. Non-chimeric laboratory animals (typically rodents) are generated to mimic human diseases via selective breeding, genetic engineering, or by physical or chemical means, after which they are used to assess the effectiveness of new drug interventions and other novel therapies. However, these purpose-built laboratory animals usually do not closely replicate human biology. For this reason, non-chimeric animal models of human disease do not always provide the surest means to aid the development of new therapeutic protocols.

To overcome these limitations, human-to-animal chimera research aims to introduce localized human cellular and biological characteristics into laboratory animals. Animal models of human disease composed specifically of localized human tissues of investigational interest can be studied for their human-specific biological processes without experimentation on human subjects during very early stages of translational research.[4] In essence, the overarching purpose of human-to-animal chimera research is to *biologically humanize* research animals in order to study human processes without using living human subjects.

Stem cell scientists join this ongoing scientific tradition of chimera research in several ways. In basic stem cell research, human-to-animal chimera experiments can illuminate on how human stem cells and their derivatives behave in a living organism and integrate into complex organ systems. In translational stem cell research, chimera experiments take place when multipotent human stem cells or the derivatives of human pluripotent stem cells are transferred into laboratory animals to assess the safety and efficacy of new stem cell-based interventions. In fact, the United States Food and Drug Administration (FDA) recommends preclinical proof-of-principle studies using at least two different animal models for all stem cell-based biological product developments.[5] Therefore, human-to-animal chimera research is a pathway for stem cell research toward clinical applications in humans. Whether for basic or for translational stem cell research, chimeric animals can be utilized to help broaden our understanding of stem cell behavior beyond the confines of the culture dish, but before stem cells are studied in humans. Assuming all animal research standards are ethically upheld in the process, it is the long-range goal of expanding knowledge and promoting clinical translation that makes stem cell-based chimera research morally permissible today.

For some observers, the scientific and social value of human-to-animal chimera research might not be enough to justify it. Specifically, some may worry that, in the process of biologically humanizing animals, scientists may inadvertently humanize animals in a *moral* sense. In pursuing stem cell-based human-to-animal chimera research, a fear is that researchers might end up creating new creatures with full or near human moral status sufficient to make experimenting on them ethically problematic.[6] A strong version of this fear might correspond to the emergence of "rational agency" of the type implied by a theory of moral personhood discussed in Sect. 8.2. A weaker version might require only that the chimeric animal exhibits some new, yet-to-be explained humanlike cognitive capability. On either interpretation,

[4] Behringer (2007).

[5] Halme and Kessler (2006).

[6] Streiffer (2005).

this concern goes beyond some people's more general objections to animal research. Indeed, for many, this is a separate concern—namely that chimera research may be ethically undesirable even if one accepts that animal research is ethically permissible in other biomedical areas. This difference in attitude could be based on the belief that stem cell-based chimerism has the potential to radically humanize the biology of laboratory animals, depending on the type and number of human stem cells transplanted, the species and developmental stage of the host animal, and the anatomical location of the animal host where the human stem cells are transferred. When human stem cells are transplanted into a postnatal animal, it is unlikely these cells will integrate significantly into the animal's existing biological structures. But if human stem cells are introduced into an embryonic or fetal animal host that is then gestated, then the percentage of differentiating human cells and the degree of human physiological integration in the developing chimeric animal may turn out to be high, especially if there is less evolutionary distance between humans and the animal species used. The worry therefore is that, in the process of biologically humanizing a research animal, scientists may end up also morally humanizing the resulting chimera, especially if there is acute chimerism of the central nervous system.

Is the potential for acute neurological chimerism a real possibility? A recent comprehensive literature review of human–animal neurological chimera experiments suggests we are scientifically far from realizing this fear.[7] This review analyzed 150 peer-reviewed scientific publications involving the transfer of human stem cells or their direct derivatives into the central nervous systems of mice, rats, and nonhuman primates. None of these studies showed any evidence that the resulting chimeric animals gained altered cognitive or behavioral traits that would make them more "humanlike." Indeed, few of these studies which involved the transfer of human cells into diseased or injured animal models restored the resulting chimeras' cognitive or motor functions to the same level of healthy control animals.

Perhaps the limited chimerism evidenced in these studies could be explained in part by the short time periods in which the chimeric embryos and fetuses were allowed to develop and by the differences in developmental timing between human and nonhuman cells. Regardless of the specific reason, an important qualification to consider is that researchers did not produce acute neurological chimeras despite the fact that they transferred human cells into embryonic and fetal animal hosts—one of the methodologies most feared by chimera research critics.

8.4 Chimera Successes: Transplanting Human Glial Progenitor Cells and Brain Organoids

There have been, however, two exceptional research strategies to date that have resulted in significantly higher levels of neurological chimerism than the 150 studies mentioned above. These experiments deserve closer examination here.

[7] Crane et al. (2019).

The first is the neurological chimera work of the Goldman lab at Rochester Medical Center in New York. In a widely publicized study, Goldman and colleagues reported that human glial progenitor cells (GPCs)—which are technically not stem cells, but a little more developed—can successfully integrate into the brains of neonatal immunodeficient mice, where they generate high levels of human glial progenitors and astrocytes.[8] Not only do the transplanted human cells mature in vivo to adulthood, but these cells also retain the size and unique structural complexity of human astrocytes and even appear to serve their normal functions of regulating synaptic transmission, plasticity, and learning. Indeed, the experimental outcome that drew the most public attention was the team's claim that their human glial chimeric mice outperformed control mice in four different learning tasks: auditory and contextual fear conditioning; Barnes maze; and novel object-location. Importantly, on the other hand, there was no evidence that neurological chimerization had any effect on how these chimeric mice interacted with control mice and littermates. Their "sociability" was not affected in any discernable way. This is an important point we shall return to shortly.

Goldman's human glial-chimeras provoke intriguing questions about the role of human GPCs in cognition. Do human glia influence neural network function in a species-specific manner? Since human astrocytes possess greater fiber complexity than those of non-primate mammals, can human glial-chimera models inform questions about the role of human-specific GPC in human cognitive evolution? As Goldman and colleagues write, the ability to generate high degrees of human glial chimerization in mice "should permit us to address these questions, by rigorously evaluating the in vivo contributions of both human astrocytes and their progenitor cells to neural network activity, and hence their respective roles in human cognition."[9]

These long-term research ambitions may not be so far-fetched, for another major finding of the Goldman lab was that transferred human GPCs tend to thrive in their mouse neural environments—so much so that they can developmentally outcompete their hosts' resident GPCs. By the time the chimeric mice reached adulthood, very large proportions of their forebrain glia were comprised of human cells. The remarkable competitive advantage of human GPCs was also shown in some of the Goldman lab's earlier work. Nine months after transplantation, nearly all of the mouse glial progenitors were replaced by human GPCs.[10]

This ability to generate neurologically chimeric mice containing large populations of aggressively expanding human glial cells opens up exciting new scientific possibilities. For one, this makes it feasible to explore the role glial cells might play in hereditary human neurological disorders, as the contribution of these cells to neuropsychiatric pathologies is very challenging to define.

In the case of studying childhood-onset schizophrenia, human glial chimeric mice engrafted with GPCs from patient-derived induced pluripotent stem cells were found to develop abnormal astrocytic morphology, hypomyelination, and

---

[8] Han et al. (2013)

[9] Goldman et al. (2015).

[10] Windrem et al. (2014).

behavioral and sleep abnormalities.[11] These results suggest again a strong causal contribution of cell-autonomous glial pathology to the development of neurological disease. The chimeric mice's behavioral abnormalities—increased anxiety, antisocial traits, and disturbed sleep—suggest it is impaired glial function itself that may be causing these abnormal patterns. Recall that in Goldman's earlier work the chimeric mice produced from healthy human GPCs exhibited none of these unusual behaviors. Goldman's healthy glial chimeric mice could learn faster, but they were not "antisocial."

As one can see from the disease modeling work of the Goldman lab, even chimeric animals that have large amounts of *disease-specific* human neural cells are not cognitively enhanced above species-typical levels, much less so to justify people's worries about "moral humanization." Instead, neurologically chimeric human disease model animals are much more likely to experience functional deficits that call into action the typical issues around animal welfare in animal research. As long as such research is scientifically justified and conducted humanely—with appropriate standards for interventional euthanasia approved by animal research committees and the veterinarian staff overseeing the work—then it should be regarded as morally permissible.

Besides the transfer of human stem cells or their direct neuronal derivatives in disaggregated form into animals, are there any scientifically justified reasons for transplanting whole human brain organoids into the brains of laboratory animals? And would such experiments raise additional ethical issues not found in current forms of disaggregated human neural cell chimeric transplantation? The answer seems to be "yes" to the first question and "no" to the second.

The first transfer of human brain organoids into the brains of laboratory animals was reported by Fred Gage's team at the Salk Institute in 2018.[12] Since brain organoids lack the vasculature, microenvironment, and neuronal circuits that exist in vivo, researchers engrafted 40 to 50-day old human brain organoids into immunodeficient mice and observed them for 0.5–8 months to see if any of these missing aspects could be established. The organoid grafts showed good integration, vascularization, and survival in their in vivo environment. Gage and colleagues further demonstrated that human brain organoids could integrate and form progressive neuronal differentiation, maturation, gliogenesis, integration of microglia, and axon growth into multiple regions of the mouse host brain. Optogenetic control of the grafts suggested that synaptic connectivity was established between the organoids and their host brains. Finally, the team assessed the spatial learning abilities of the grafted mice in comparison to ungrafted mice using the Barnes maze. There were no observed differences between the two groups, although the grafted mice did not perform as well as their controls when tested for spatial memory. There seemed to be no other observed ill effects (or any benefits) conferred to experimental mice by human brain organoid engraftment.

[11] Windrem et al. (2017).

[12] Mansour et al. (2018).

The overarching scientific rationale for this brain organoid engraftment study was to enable the eventual study of the pathogenesis of neurodevelopmental, neuropsychiatric, and neurodegenerative disorders (and perhaps preclinical drug testing) under physiological conditions of the host animal using human brain organoids derived from patient-specific iPS cells. Again, like the Goldman studies, the scientific and translational value of transplanting human brain organoids into animals should make this research methodology morally permissible, assuming that animal welfare standards for biomedical research are upheld.

It is important to note however that, unlike the transfer of disaggregated human neuronal cells into animal models, the transplantation of human brain organoids faces two limitations not usually associated with other forms of neurological chimera research. First, given the limitations imposed by an animal's small skull size, pieces of the animal's brain have to be removed prior to transplantation to allow room for the human brain organoid. This fact alone drastically limits both the size and number of human brain organoids that can be transferred into a single animal, since removing too much brain tissue would injure the animal beyond what would be ethically acceptable for animal research and beyond what would be beneficial for the study itself (a gravely injured or dead animal holds little scientific value for furthering one's research aims). Second, human brain organoids cannot be transplanted just anywhere the researcher would like, since it would be detrimental to remove existing tissue from crucial areas of the brain necessary for the animal's survival and function. These two types of limitations further constrain how much human neural matter can actually be transferred into an animal via human brain organoids and, consequently, how much of an impact human brain organoids are likely to have on an animal's cognitive capacities.

Finally, it is worth mentioning that if, for some unexpected reason, a human brain organoid could go beyond integration, vascularization, and survival in an animal's brain to actually play a role in improving the chimeric animal's cognitive functioning, then the clinical implications of this discovery for stroke patients and other people with brain injuries would be enormous. This would be such a significant experimental result, in fact, that this discovery—that transplanted human brain organoids could rescue cognitive function in mammalian brains that have had tissue removed and engrafted with organoids—would far eclipse ethical concerns that chimeric animals in this proof-of-concept study might have experienced a cognitive gain-of-function beyond control species levels.

8.5 Going Forward: Chimera Research Oversight

Going forward, research involving the transfer of human brain organoids into laboratory animals will have to be reviewed and approved through a process of scientific and ethics oversight before it can be deemed to be ethically permissible. What would this review process look like?

Review will most likely be in accordance with the ethical standards already put into place by the International Society for Stem Cell Research (ISSCR) for stem cell-based human-to-animal chimera research.[13] Because human brain organoids are themselves derived from human stem cells, the ISSCR standards for chimera research will be directly relevant for human brain organoid engraftment studies like the one performed by the Gage team.

According to the ISSCR, any time human stem cells or their direct derivatives are integrated into the central nervous systems of laboratory animals, stem cell specific review must take place to oversee chimera research. This review should build upon and remain consistent with animal welfare principles, but with added stem cell expertise to consider the further developmental effects on animal welfare of human-to-animal chimerism.

Past experience with genetically altered laboratory animals has shown that reasonable caution is warranted if genetic changes carry the potential to produce new behaviors and especially new defects and deficits. Best practices dictate that research involving genetically modified animals must involve the following: (1) the establishment of baseline animal data; (2) ongoing data collection during research concerning any deviation from the norms of species-typical animals; (3) the use of small pilot studies to ascertain any welfare changes in modified animals; and (4) ongoing monitoring and reporting to oversight committees authorized to decide the need for protocol changes and the withdrawal of animal subjects.

In addition to adopting these standards, researchers must also justify why a particular species of host is necessary for their experiments. For the time being, it appears that the transfer of human brain organoids into animal models can be done using rodents; thus, researchers who wish to use larger animal species will have to explain why. In principle, the use of other laboratory animal species commonly used for neurological research is potentially permissible, including nonhuman primates, except great apes and lesser apes (i.e., except chimpanzees, gorillas, orangutans, bonobos, gibbons, and siamangs). The use of great and lesser apes is excluded for two main reasons. First, it is a widespread international research restriction that apes cannot be used for invasive biomedical research. Second, the use of great and lesser apes for human organoid chimera research would not be justified as long as other "lower" NHP species that are more evolutionarily distant to humans are available and routinely used for neurological research. As long as researchers can scientifically justify why a particular host species is necessary and that there are no adequate alternatives available (which alone would restrict how often non-rodent species might reasonably be used), then the main ethical issues to consider for the moral permissibility of this research are essentially animal welfare considerations appropriate for animal research.

Some might object that this primary focus on animal welfare misses the mark when it comes to what may be most ethically worrisome about brain organoid transplantation and other forms of neurological chimerism—namely the concern that researchers might create a morally ambiguous research animal in the process. As alluded to above, I believe this concern has a tendency to run too far ahead of the actual science, and that it erroneously conflates higher degrees of *biological*

[13] Hyun et al. (2021) and ISSCR (2021).

structural humanization with greater *moral* humanization. In the strong version of this moral concern, moral humanization would involve the emergence of humanlike cognitive capacities such as higher-order intellectual processing capabilities and thought, and of self-consciousness. Such complex mental traits are not biologically assured even in infant brains that are 100% human, without the social and nurturing conditions of child-rearing over many years.[14] Since the social support and language-use conditions necessary to support human consciousness in this most robust sense are absent from the laboratory conditions within which neurological chimeras are created and maintained, the threat of conscious self-awareness does not appear to be a serious ethical challenge for biomedical research employing the transfer of human brain organoids.

8.6 Concluding Thoughts

The incremental ethical approach outlined above for chimera research attempts to avoid giving undue influence to unsupported, imagined possibilities and strives to be grounded in observable animal behaviors and reasonable inferences. A concern over the possible emergence of humanlike consciousness seems to motivate a cautious approach to advancing human brain organoid and neurological chimera research. However, I wish to conclude by acknowledging that humanlike consciousness—that which forms the very basis of the moral life of humans—can only be realized in normally functioning human brains starting from infancy within nurturing social environments and through the acquisition of language that would enable one to have reflective beliefs about one's own beliefs and experiences.

Perhaps ultimately people's concerns around human brain organoid and chimera research reflect a broader unease about interfering with the natural (and implicitly normative) order of the world. However, such a "natural law" framework—broadly speaking—for determining moral right and wrong is not an easy fit for modern scientific pursuits such as those being considered in this chapter. In order for natural law to provide guidance for research, many, if not most, scientific techniques would have to be abandoned. According to the implicit norms of the natural law tradition, scientists must only passively observe and record the "natural order" of the world; they must never disrupt this order during their act of observation. However, organoid and chimera research—like most other forms of biomedical research and technological advancement—violates this observation constraint. For example, in the process of studying the full potential of human stem cells, researchers must cultivate or form new biological entities (e.g., bioengineered organoids or chimeric animals) that do not have direct natural analogues in the developed human body or in the animal kingdom. The "unnatural" is an anathema for the natural law tradition in ethics. But all branches of the modern biological sciences must proceed through the performance of unnatural acts.

---

[14] Hyun (2016).

Secular ethics, such as the research ethics framework sketched out in the chapter on chimera research, at least allows for the possibility that unnatural acts may be ethically permissible. If unnatural acts are an ineliminable part of modern science and biomedical research, then research ethics must provide room for the performance of unnatural acts. Ultimately, the important distinction for research ethics is not between the *natural* and the *unnatural*; rather, it is between the *ethically* unnatural and the *unethically* unnatural. The transfer of human brain organoids into animal models is an ethically unnatural act, but it is ethically permissible according to secular ethics.

References

Behringer R (2007) Human-animal chimeras in biomedical research. Cell Stem Cell 1:259–262

Crane A et al (2019) Concise review: human-animal neurological chimeras: humanized animals or human cells in an animal? Stem Cells 37:444–452

Dworkin G (1970) Acting freely. Noûs 4:367–383

Frankfurt H (1971) Freedom of the will and the concept of a person. J Philos 68:5–20

Goldman SA, Nedergaard M, Windrem MS (2015) Modeling cognition and disease using human glial chimeric mice. Glia 63:1483–1493

Halme DG, Kessler DA (2006) FDA regulation of stem-cell-based therapies. NEJM 355:1730–1735

Han X et al (2013) Forebrain engraftment by human glial progenitor cells enhances synaptic plasticity and learning in adult mice. Cell Stem Cell 12:342–353

Hyun I (2001) Authentic values and individual autonomy. J Value Inq 35:195–208

Hyun I (2016) Illusory fears must not stifle chimaera research. Nature 537:281

Hyun I et al (2021) ISSCR guidelines for the transfer of human pluripotent stem cells and their direct derivatives into animal hosts. Stem Cell Rep 16:1409–1415

ISSCR (2021) Guidelines for stem cell research and clinical translation. https://www.isscr.org. Accessed 1 March 2021

Locke J (1694/1975) An essay concerning human understanding. Clarendon Press, Oxford

Mansour AA et al (2018) An in vivo model of functional and vascularized human brain organoids. Nat Biotechnol 36:432–441

Streiffer R (2005) At the edge of humanity: human stem cells, chimeras, and moral status. KIEJ 15:347–370

Windrem MS et al (2014) A competitive advantage by neonatally engrafted human glial progenitors yields mice whose brains are chimeric for human glia. J Neurosci 34:16153–16161

Windrem MS et al (2017) Human iPSC glial mouse chimeras reveal glial contributions to schizophrenia. Cell Stem Cell 21:195–208

Transplantation of Human Brain Organoids into Animals: The Legal Issues 9

Nils Hoppe, Maria Lorenz, and Johannes Teller

9.1 Introduction

Other authors in this book are more qualified to provide a scientifically sound introduction into the scientific mechanisms underlying brain organoids than we are (see, in particular, Chap. 2). We will therefore not rehearse these aspects but rely on two claims for the purposes of this chapter: brain organoids are three-dimensional functional cell constructs that mimic at least some functions of a human brain, and they are created from human cellular source material. In addition, it is vital to focus our analysis on a specific set of contextual circumstances, and we do so by working along the guard rails of a case study which we outline in the next section. This is necessary because, otherwise, the legal analysis of the transplantation of brain organoids into animals would either fray into a chaotic free-for-all (quite likely) or (less likely) systematically grow into an entire book of its own (which would be well outside the scope of this chapter). Another challenge we faced when writing was that it did not seem desirable to focus on one specific jurisdiction. Instead, we wanted to use this opportunity to highlight those normative challenges that would lend themselves to resolution through the instruments provided by law, independent of a specific realm of domestic law. Where appropriate, we provide examples from different jurisdictional contexts but we explicitly do not claim to provide an exhaustive overview of the legal issues in relation to one domestic law (the use of which, in the context of predominantly cross-border biomedical research, would be questionable at best in any case).

Biomedical research has a proven track record of producing knotty normative challenges and of translating a technology from wildly experimental to routinely

N. Hoppe (✉) · M. Lorenz · J. Teller
Leibniz Universität Hannover, Hannover, Germany
e-mail: nils.hoppe@cells.uni-hannover.de

H.-G. Dederer, D. Hamburger (eds.), *Brain Organoids in Research and Therapy*, Advances in Neuroethics, https://doi.org/10.1007/978-3-030-97641-5_9

available at great speed. An often-cited example of this phenomenon is the breakneck progress made in the context of transplantation in just a few decades,[1] particularly with the acceleration in developments provided by advances in vascular surgery and the clinical introduction of cyclosporine in the 1980s. It is therefore not breaking news to say that a novel technology swiftly antagonises regulatory frameworks and tests their suitability for purpose. At the same time, we are confronted by the perennial dilemma of technology regulation: at the point in time where we are in a position to regulate easily, we do not understand a technology sufficiently to do so; when we fully understand the extent to which regulation is necessary, it has become incredibly hard to do so. As Collingridge famously put it:

> This is the dilemma of control. When change is easy, the need for it cannot be foreseen; when the need for change is apparent, change has become expensive, difficult and time consuming.[2]

To compound this dilemma, regulatory frameworks in technology and science have additional normative gymnastics to perform: on the one hand, they are relied upon to ensure that a technology is reasonably safe to use before it is allowed into regular use. On the other hand, we demand sufficient regulatory elasticity to ensure that innovation is not stifled and that appropriate risks can be taken to develop novel approaches.

Organoids as a technology are not exempt from these challenges, and the notion of *human brain organoids* in particular activates, in a reasonable reader, connotations of consciousness and sentience-in-a-dish that produce intuitive uneasiness. To some extent, this may be an indictment of the scientific community having hastily given a three-dimensional heap of cells such an evocative label. In any case, it was inevitable that the connections would be made in scientific literature and normative scholarship. Farahany et al. write:

> [...T]he closer the proxy gets to a functioning human brain, the more ethically problematic it becomes.[3]

That is surely the case when thinking of the scientific development of brain organoids as residing, at any given time, somewhere on an epistemic trajectory from the first hypothesis on paper to a fully functional and brain-equivalent artefact in a jar. Our chapter, however, addresses a normatively significant junction on that track in that we not only have to bear in mind the ever-increasing ethical and legal complexity of brain organoids as they become better at mimicking an entire human brain. We specifically address the legal aspects of the departure of that organoid from the dish and into an organism. To that extent, our chapter's premise is leapfrogging a vast section of Farahany's trajectory. Whilst the brain organoids may not have become more like a functioning human brain, they are embedded in an organism

---

[1] For an historical overview, see Stehlik et al. (2018).

[2] Collingridge (1980), p. 11.

[3] Farahany et al. (2018), p. 429.

which provides the milieu in which it may no longer be fanciful to expect additional functional effects, however difficult these may be to predict. Indeed, if availing of such effects is not exactly the point of the exercise, it is difficult to envisage (and justify) the transplantation into the organism in the first place.

In order to ensure that our discussion of the legal aspects of the transplantation of a human brain organoid into an animal does not deteriorate into a wildly meandering thought experiment, it is important to ground the analysis by way of a real-life reference point. To this end, we have selected a case study which provides a plausible backdrop to our analysis (and we outline this in more detail in the next section). After that, we will deal with the cell donors' interests in the use of their cells to produce an organoid which is subsequently implanted in an animal. Others in this book have dealt in greater depth with the consent requirements surrounding the use of human cells for the generation of the right type of source material for creating an organoid, such as induced pluripotent stem cells (iPSC) (see Chaps. 6 and 7). We do not propose to repeat these here but seek to address an important partial aspect of that consent in our specific context, the foreseeability of the use of cells. We then turn our attention in more detail to Collingridge's dilemma in a section entitled Technology Regulation Challenges and will seek to unpack the regulatory conundrum of the technology's blurry risk profile in a legislative arena which often favours erring on the side of safety. Then, our focus becomes that of animal welfare in biomedical research, before we turn our attention to whether the potential of sentience or consciousness should make any normative difference in legal terms.

The velocity of technological developments in this field produces a very specific risk for chapters such as this one. By the time it is published, it may well be the case that phenomena that we anticipate have been proven to be non-existent, or that public perceptions which we posit to be likely have been shown to be different. We bear these aspects in mind when we outline our exploration of this field and try to provide analysis which survives impact with reality, at least for a whilst.

9.2 Background and Context

Biomedical law and ethics often encounter contexts in which an unprepared observer's first reaction might be an incredulous 'Why would you do that?' Many processes underlying biomedical innovations of the last 100 years must have initially evoked that response: organ and tissue transplantation, pacemakers and other implants, cloning and assisted reproduction. The creation of human organoids in a petri dish (to put it with rather reduced complexity) is such an activity which challenges fundamental notions of human forms and existence—these things do not look like our idea of an organ, yet they still evoke that association with organ transplantation, the gift of life, and the many connected ethical and legal questions. Human *brain* organoids take this a step further; the concept seems to propose that the very part of us that puts the person into the body can be replicated and kept alive in a glass vessel. Notwithstanding the fact that the technology strives to move along that trajectory which brings the proxy ever close to the fully functional human brain, it is clear from even the most cursory literature review that, currently, these organoids do not look like brains and do not do what we think our brains mainly do.[4] Nonetheless, accepting the possibility of modelling not only entire organs in a lab, but also of creating something like a human brain in a dish, we are presented with yet another step in this cascade—putting a human brain organoid into a live animal. *Why* would you do that? Before we can systematically identify and discuss the legal issues in this kind of case, it is worth answering this question.

For the purposes of this chapter, we will do so on the basis of a case study that we have chosen which is a real-life application of human brain organoids in animal models. Picking a case study and addressing the associated legal issues can sometimes be a sign of intellectual laziness, of unwillingness (or inability) to give a comprehensive overview of the normative challenges of a particular case. In the case of this chapter, it is the only way to be able to provide an overview without losing sight of the issues. The reason for this is that *context matters*, and the content of normative analyses shifts in myriad ways with every tiny tweak in that context: is the source material from a patient, or is it from a commercial cell line? The legal consequences are vastly different. Is this basic research or clinical care? The legal frameworks diverge. Is the animal a rodent or a primate? The permissibility goes from *go* to *no*. In order to get a handle on the normative complexity, we have therefore decided to work our way along a case study which we will outline in all brevity here. Afterwards, we will discuss the legal issues that our case study gives rise to, clustered into a small number of categories.

For our case study, we have chosen a process which was recently described by Xu et al.[5] the modelling of human Down syndrome in mouse models. In this case, previously produced human pluripotent stem cells are processed in vitro before implantation via craniotomy into neonatal mice. The mice undergo behavioural tests and EEG measurement, before being killed at predetermined points in time for subsequent immunohistochemical and microscopic analyses. The case study establishes a number of boundaries which are important to guide our analysis of the legal aspects of transplanting brain organoids into animals.

1. The context in which are working is one of basic research (rather than, say preclinical trials or even a clinical application).
2. The material does not come from specific patients by way of a biopsy, but from an existing cell line.
3. The species, number of animals, the method of insertion, the experiments performed, and means of sacrifice are well-described.

This last point is particularly important as the assessment of the justifiability of an experiment using animals depends on many moving parts. We will discuss the

[4] See, e.g. Paşca et al. (2015), p. 671.

[5] Xu et al. (2019).

described process below and, where appropriate, we will point out legal aspects that branch off but are not captured by our case study without making them the subject of detailed discussion.

9.3 Cell Donors' Interests

The production of brain organoids necessitates the use of raw material of human origin. In this case, adult human somatic cells that are induced (or reprogrammed) to become pluripotent stem cells (iPSC) through the introduction of four specific transcription factors. The resulting iPSC are then introduced to a bioreactor and induced to form the appropriate type of cells which lead to the development of a neural structure. When viewing this part of the process from the perspective of the original cell donor's interests, it quickly becomes clear that there are at least three subsets of legal issues that are in play here: the actual physical artefacts that are the donor's cells, informational issues related to the original donor, and autonomy and personality related aspects.

Both in cases where the source of the material is a specific patient and where a commercial cell line is first produced, a biopsy is performed to harvest cells for later reprogramming. This gives rise to a fundamental property question to be resolved: in some jurisdictions, the biopsy is capable of being the subject of property rights and this is usually dealt with as part of the information and consent procedure for the biopsy, which tends to include a transfer of ownership clause. In these cases, ownership in the sample vests in the researcher or research organisation taking control of the sample. Occasionally, out of date or incomplete contractual arrangements at this point have significant knock-on effects for research projects. It is worth closely examining these processes before using biopsies for reprogramming. In some jurisdictions, there is lingering doubt in relation to the possibility to exert property rights over unchanged human biological material, with significant jurisprudential contortions being performed to find remedies for donors in cases of loss.[6] However, jurisdictions that fail to provide a property-based remedy to donors at the point of the biopsy are also not likely to interfere with the de facto right of possession that rests with the individual taking the biopsy, likely rendering this question moot for the purposes of our discussion.

In addition, the process we have outlined necessitates the complete loss of the original cell biopsy. The whole point of the reprogramming exercise is to replace the less-useful original adult somatic cells with newly generated and more-useful pluripotent cells as fuel for the subsequent bioreactor work. This process therefore means that the original material (which may or may not be the subject of a property interest by one or more persons) is no longer in existence. In terms of proprietary rights, this would ordinarily mean that the original material donor has no further

[6] See, e.g. the debate surrounding the loss of human sperm samples in Yearworth and others v. North Bristol NHS Trust [2010] QB 1, and the less than generous views expressed by the court in relation to the approach taken by the German BGH in a case with indistinguishable facts.

interest in an existing, tangible artefact but would only be able to enforce restitutional interests in very specific circumstances. That means if one sought to establish a tangible proprietary continuum between the original biopsy at the outset and the brain organoids at the end, many legal systems would require an element of dishonesty or fault to preserve the original donor's tangible property interests throughout the conversion.[7] This is relatively unlikely in commercial cell line activities. Where a researcher, or biotechnology company, takes a sample in good faith and converts (in the legal sense) the cells to a new material, it is an uphill struggle to argue against the severance of the original donor's proprietary interests. In other words, even where the parties have failed to adequately discharge the original donor's proprietary interests (if any), it would usually be difficult to establish a tangible reach-through property right in the absence of dishonesty or negligence.

A significantly more interesting context is that of the donor's interests in the intangible. The original biopsy is, of course, more than a simple heap of cells devoid of any other personally attributable characteristics. A somatic cell contains the DNA of the original donor, and with it additional information points about the donor and their family. The reprogramming of that cell into an iPSC does not remove this information nor does it create new information which can be said to have been successfully severed from the original donor's. Whilst the cells can therefore be tangible artefacts that can be destroyed and reproduced by different possessors and owners, the replicated data remains, by and large, that of the original source. This assessment would be meaningless in a legal sense if the data were just a pointless jumble of As, Cs, Gs, and Ts. We therefore need to establish a particularly protected characteristic of the data, with a respective rights holder. Genetic data are powerful in their potential to reveal information about an individual's family relations, inherited or acquired genetic characteristics such as predispositions to disease or drug responses, or physical traits. It is for this reason that they fall squarely into the category of particularly sensitive personal data and take a prominent place in data protection instruments: whereas the EU's now obsolete data protection directive (95/46/EC) only concerned itself with health-related data (quite clearly also including genetic and genomic data), the General Data Protection Regulation is explicit in establishing a normative exceptionalism for genetic data, with Art. 9(1) firmly closing the door on any processing absent of any of the grounds contained in Art. 9(2). Shabani and Borry[8] write 'Sharing identifiable genomic data is a form of processing of personal data, and as such would fall within the scope of data protection laws'. We would agree and add that it is difficult to envisage a context in which the kind of genomic or genetic data subject to our discussion would not be identifiable. We do not therefore share the view of some legal commentators that anonymisation or pseudonymisation is a sustainable mechanism for rendering genetic data non-personal data from a data protection perspective. Instead, we would endorse the

[7] It is worth noting that even in cases of proven dishonesty, the courts in some jurisdictions have still been reluctant to act in the interests of the donor. Moore v. Regents of the University of California 51 Cal. 3d 120; 271 Cal. Rptr. 146; 793 P.2d 479.

[8] Shabani and Borry (2018), p. 149.

view expressed by Hudson and Collins[9] that 'non-identifiability is increasingly illusory'.

A third, and final, aspect that needs to be addressed in this part of our analysis is one that cannot be readily delineated from property interests of rights of informational self-determination: an individual cell donor's autonomy rights. These rights certainly underpin the donor's other subsequent legitimate interests, but they also apply much earlier. Respecting an individual's right to make choices about participating in research (or other activities) can be distilled from a range of different norms, depending on which kind of jurisdiction is in play. In common law jurisdictions, principles such as the one settled, inter alia, in professional self-regulation, as well as general criminal and civil law can be best summarised by the dictum in Schloendorff v. Society of New York Hospital (105 N.E. 92 (N.Y. 1914)): 'Every human being of adult years and sound mind has a right to determine what shall be done with his own body'. In some civil law jurisdictions, this can be found in constitutional values and associated personality rights. The essence of these types of norms is this: unless subject to a statutory derogation, an individual with capacity must be allowed to freely decide how to participate in an activity, clinical, or otherwise. A manifestation of this entitlement can be found in the requirement to seek informed consent to the procurement, storage, and use of cells and we want to briefly reflect on a partial aspect of consent to the donation of cells which will subsequently be used to create a brain organoid (see Chaps. 6 and 7 for a more detailed exploration of consent issues). What we are concerned with is the specific use of the brain organoids in our case study: implantation into an animal. This specific use raises the question of the range of the original consent that was sought for the cell donation: in most cases, this will be a relatively broad consent for use in biomedical research but without individual specifics. The reasons for this are clear—where a commercial cell line is produced from donor cells, it is all but impossible to predict the exact uses for these cell lines; indeed, it would go against the grain of innovation in science to restrict their use to contexts which are known in advance. It is for this reason, amongst others, that there is a general acceptance of broad consent for biomedical research—which is—we argue—not without limits, however. We agree with analyses such as that of Sheehan[10] that broad consent can amount to a proper exercise of, and respect for, an individual's autonomy, and that it is important to take into account future decision-making mechanisms to come to any sort of meaningful conclusion. At the same time we are concerned that there may be some unenvisaged uses of a donor's cells that are fatal to this account of decision-making in biomedical research. Not unlike in the fundamental principles of tort law, we submit that there needs to be a foreseeability of harm; in our case, even the most tenuous of links of foreseeability between the ultimate use of the cells and the moment of consent.

Using iPSC derived from a donor's cells to create an embryo that is then implanted and brought to term would, we argue, already leave the scope of what

[9] Hudson and Collins (2013), p. 142.

[10] Sheehan (2011).

could reasonably be consented to through a broad consent procedure. This type of activity would clearly create legal and social consequences that would require a more specific reflection, understanding, and agreement from the original donor of the cells. Borrowing from tort one more time, we would ask whether a donor would have consented to the use of their cells but for the lack of knowledge about the intended application. The creation of a brain organoid from the cells may not pierce that boundary of broad and unspecific consent just yet, but the introduction of these cells into a living being with all the possible intended and unintended (and unmeasurable) effects does, we argue, leave a sizeable dent in the general justification of broad consent for biomedical research. It would therefore be necessary to ensure that the cells used were procured with a clear contemplation of being used in, or with a view to the creation of, another organism.

9.4 Technology Regulation Challenges

The core concern with regulating innovative technological contexts is premised on the assumption that *sui generis* regulation is the gold standard. The general rules of rights, freedoms, and obligations provided for in all areas of law usually also encompass technological innovations, so any assertion that an innovation takes place in a legal vacuum is rarely correct and almost certainly never correct in jurisdictions that have constitutionally settled fundamental rights and freedoms. At the same time, it may very well be the case that these general rules only poorly address the regulatory requirements of an innovation. In these cases, the concern is focused on the creation of *sui generis* regulation for the technology. This is where the Collingridge Dilemma, outlined at the beginning of this chapter, becomes visible: where the general rules poorly capture the technology, and *sui generis* regulation is desirable, the exact content and mechanism of this regulation are not sufficiently certain. The reason for this is that there is no empirical knowledge about the societal effects of the technology as yet; and by the time that the effects can be determined, there may already have been a detrimental impact. Fenwick et al.[11] suggest that this leaves the regulator in the unenviable position of having to choose between recklessness and paralysis in regulation.

The procurement and processing of cells prior to the creation of organoids is, as we have seen, already subject to specific regulation: depending on whether the cells are destined for research or for clinical application, a clear regulatory framework is in place. The creation of organoids in the lab produces no specific additional legal problems. However, the introduction of these functional cell clusters into an animal may give rise to as yet unresolved technology regulation issues. The reason, of course, why they are unresolved is that we are unfamiliar with the exact regulatory target—we simply cannot know the realistic risks until they have manifested. Absent a protected non-public environment in which they can manifest safely and be documented, a regulator would ordinarily tend to err on the side of caution, particularly

[11] Fenwick et al. (2017), p. 590.

where a possible risk is one which may significantly impact society (such as the unwitting production of sentient mice or the inadvertent release of modified germ lines into nature). This type of preventive regulation can come in the form of a blanket prior ban on the technology or process, which is the course of action usually taken when the balance of risks is fairly straightforward (i.e. no balancing to be done at all) and the expected outcome is likely so detrimental that a complete prohibition is proportionate. Alternatively, the regulatory response could be one of controlling the emergence of the technology tightly and restrictively. A normative concept often deployed in this context is that of the precautionary principle, which justifies strict prohibitions in situations where there is insufficient scientific knowledge about the likely effects of technology, but enough information to reasonably apprehend a risk to human, animal, or plant health or the environment. The EU currently favours this approach to technology regulation under conditions of uncertainty (Art 191 TFEU; COM(2000)1), though this is likely to be of little interest in the context of experimentally introducing human brain organoids to mice (unless the context is expanded to seek to introduce such mice to the market as a product).

Instead, it is worth asking the question whether a reactive technology regulation approach is the right one in our case. This hinges to a great extent on the answer to the question whether there are obvious risks that should proportionately be addressed by a regulator, or whether the current state of knowledge favours non-intervention. One aspect to be born in mind here is that whilst it is clear that the societal risks are difficult to identify and quantify—and the resultant uncertainty may create an intuition that regulation is a sensible response—there is a flip side of uncertain benefits inherent in the scientific process. It is no doubt contentious to suggest that uncertain risks and uncertain benefits cancel each other out. It is less contentious to assert that this is the bread and butter of technological innovation and scientific process. We would propose, then, that—absent a change in circumstances which, for example, means that our scientists are seeking to develop a commercial human brain organoid mouse line for introduction to the market—the current regulatory regime is sufficient in addressing existing and arising challenges, and that no prohibitive sui generis regulation is called for.

9.5 Animal Welfare

The use of animals in biomedical research—and in general—is rightly subject to strict regulation. Nonetheless, significant numbers of animals are killed every year in the interests of scientific progress. The European Commission reports nearly 10m animals were used for research, testing, routine production, and training purposes in 2017.[12] Regulators' and scientists' firm commitment to the three R (*Reduce, Replace, Refine*; see Russell and Burch[13]) when contemplating the use of animals is a reflection of societal desire to converge to a state of affairs in which we use as few

[12] ALURES section 1 query for 2017, 20 June 2021.

[13] Russell and Burch (1959).

animals as possible in research. At the same time, scientific practice is hampered in its efforts to forego animal experimentation because of a number of interconnected factors: the inertia of established scientific processes and traditions, the lack of a clear regulatory framework for validating alternative methods, and the desire to produce the greatest possible amount of epistemic certainty before exposing a human to an experimental—and possibly risky—treatment.

In the context of implanting a human brain organoid into an animal organism as described in our case study, a number of the usual discussions on animal welfare are more or less devoid of meaning. The entire point of the exercise is to place the organoid into specific surroundings to test its interaction with the host organism, which excludes the possibility that there may be a non-animal alternative model to use (bar that of an actual human research participant, of course). Instead, the discussion on animal welfare has to focus on whether the type of experiment we are contemplating is in line with the requirements as set down by the salient legislation. At EU level, Directive 2010/63 EU sets the framework for protecting animals used for scientific purposes and the Directive's measures have percolated into the domestic law of all EU Member States. Over and beyond rather technical stipulations in relation to the care and accommodation of animals, the Directive explicitly endorses Russell and Burch's[14] 3R approach and limits the purposes for which an experiment may be conducted. At domestic level, this is usually expressed in limiting experimental use of animals to purposes for which this is absolutely necessary and cannot be achieved by other means. The number of animals used has to be the minimum, and harm and suffering must also be minimised as much as possible. In addition, there are usually strict requirements for staff to have appropriate qualifications and training prior to performing the experiment.

Overall, the regulatory framework for animal welfare is a reasonably restrictive one. It is clear from the language found in the different legislative instruments that animals are not viewed as a consumable which can be used and disposed of arbitrarily. Indeed, some jurisdictions have given animal welfare constitutional protection (e.g. in Germany: Art. 20a of the Basic Law) and have thereby conclusively removed animals from the category of commodity and provided a basis for additional restrictions on their use. When contemplating the genesis of the change in the German constitution, the decisive characteristics that underpin this *sui generis* categorisation become clear: there is a moral responsibility to treat animals appropriately and there is an ethical obligation to take into consideration animals' ability to suffer—it follows from this that avoidable harm must indeed be avoided.[15]

In terms of implanting human brain organoids into mice for experimental purposes, it therefore makes sense to first explore the nature of the proposed experiment. Art. 5 of the Directive 2010/63 enumerates in a rather broad fashion all types of scientific activity, including basic research. The balance that the norm giver is seeking to achieve here is clearly one of providing a restrictive corset for the use of animals in science whilst at the same time not stifling science's occasionally

[14] Russell and Burch (1959).

[15] BT-Drs. 14/8860 (2002), p. 3.

unpredictable innovative pathfinding. It seems clear that the more specific the experimental design and its objectives are, the more likely it is that a robust justification for the use of animals can be postulated. The epistemic goal of the experiment in the case study is to gain a better understanding of the neurodevelopmental phenotypes expressed in the context of Down Syndrome. It can therefore be posited that the planned experiment using mice has a clearly delineated objective, related to an identifiable and serious human genetic condition. Indeed, searching for the use of mouse models in Down Syndrome research on databases such as PubMed yields thousands of results from the last 20 years. If we accept that the use of mice in experimental work on Down Syndrome does not immediately raise any red flags, we need to explore whether the type of experiment proposed here triggers any of the regulatory safeguards of animal welfare.

The organoids are injected into the young animals using a needle through the skull. Before the cell transplantation is performed, the animals are anaesthetised by putting them on ice for five minutes. The experimental endpoint in our case study lies beyond the animals' death, with a significant amount of cell-based work performed after the mice have been sacrificed. It is part of the experiments' design that the animals have to be killed prior to further analysis, and this is done by way of decapitation after anaesthetisation.[16] Whilst it is clear that the latter part of the procedure falls squarely into the severity category 'no recovery', it is beyond our expertise to state whether the former part is mild, moderate, or severe. It would be unduly anthropocentric to substitute one's own expectations of the discomfort expected when injected into the brain with a needle, and it serves to illustrate the epistemic conundrum that is requiring researchers to perform a species-and-experiment-specific severity assessment when there is little dependable knowledge available in relation to actually measuring suffering in experimental animals.

9.6 Additional Issues

Moving away from the more readily identifiable legal problems surrounding the use of cells, or the creation of a chimeric mouse for research purposes, we have to address—in all brevity—a more difficult to define normative issue. The transplantation of brain organoids into mice is an interference with a bright line species boundary which is used to determine categorical belonging. One effect of blurring the species boundaries in this way, and of making mice more human, is that a specific parameter that delineates animal experimentation from human experimentation is called into question. If we accept that the introduction of human brain organoids into a mouse has made the mouse more human, then we must at some point address the consequences of the experiment crossing from animal experiment into a preliminary limbo of nearly human experimentation (which would, under the current guidance, have to be immediately stopped for obvious reasons—sacrificing humanesque research participants to look at their brain cells would be unacceptable). This seems

[16] Xu et al. (2019), p. e4.

to be prima facie rather far-fetched, but does bear contemplation for the purposes of this analysis.

An often-cited objection to the apprehension of introducing some sort of human consciousness into an experimental animal is that primate brains take significantly longer to develop than other species.[17] But our context is different in a material way—the concurrent, but too slow, development of neurons that lead to some form of consciousness is not an undesirable side effect of chimera research as outlined by Karpowicz et al. In our case, we are introducing brain cells into the animal specifically to test their function in that setting whilst also bearing in mind that what we are actually trying to better understand is the functioning (or lack thereof) of human cells in an host organism (rather than specifically how a neural chimera might come into existence and what this would mean). Nonetheless the pacifying logic availed of by Karpowicz et al. of suggesting that by the time we are done with the animal, no meaningful human consciousness could be developed is simply not readily available in our case study. In addition, Bourret et al.[18] report that human brain cells can outperform mouse brain cells,[19] thereby adding credence to the concern that a fully humanised mouse brain is at least a possibility. At the same time, the significantly higher complexity (and size) of the human brain (with the associated capacity to develop more complex faculties) is a compelling argument that takes some of the force out of these concerns.

This chapter is concerned with the legal consequences that would flow from such an effect, but it is difficult to envisage precisely how a legal system would react to the creation, or the risk of creation, of such an unusual entity. It is just as difficult, if not more so, to envisage how the law would grapple with notoriously uncertain concepts such as consciousness, and how to quantify such consciousness to give meaning to a normative balancing exercise between conflicting rights and obligations. Existing law deals in discrete categories of entities and an artefact created as part of a scientific experiment cannot easily straddle two such normative categories. The assessment must therefore either be that an artefact is an animal—in which case the rules in relation to regulating animal experimentation apply—or that the artefact has sufficiently crossed the species boundary to be deemed a human. In the latter case, the law would have to be applied in the same way as though the experiment would be performed on human participants. One particularly interesting, but somewhat far-fetched, aspect of this would be that (by design) the entity would be an animal at the outset and a human at the endpoint of the experiment. This, yet again, presents the bright-line-desiring practice of law with another spectrum in the middle of which ontologies become untenable, not unlike the normative challenges of defining the moral and legal status of budding life from the gamete, via fertilisation and implantation to late term pregnancy and birth. In addition, some concern must surely lie in the apprehension that the introduction of the human cells may give rise

---

[17] Karpowicz et al. (2004), p. 334.

[18] Bourret et al. (2016), p. 87.

[19] Windrem et al. (2014), p. 16153.

to chimerism which can in some way find its way into the animal's germ line, though we accept that this is a most unlikely scenario.

It seems clear that there is a sliding scale of the weight of the status of the entity in question,[20] but the exact points in time, and the normative effect, depend greatly on very specific cultural legal and moral circumstances. In essence, therefore, an analysis of the legal aspects of this particular part of the experiment is rendered not particularly helpful unless it is underpinned by some kind of scientific and societal agreement about the status of the entity in question. This, however, brings our discussion full circle back to Collingridge: we cannot agree that categorisation until we have seen it in action, and once we have seen it in action it may be too late to regulate it.

9.7 Conclusions

The discussion of legal issues arising from a highly experimental undertaking is a difficult task. This is particularly the case where the context produces novel entities that pierce established normative boundaries. At the same time, the law has always shown itself to be particularly good at capturing new challenges—first in its general framework of rights and obligations within societies, and later (where needed) through specialised norms that better address the legal requirements of an innovative technology or new process. The same is true when we discuss transplanting human brain organoids into an animal model: existing normative frameworks will swing into action, discrepancies will become apparent, and a discussion will take place about additional safeguards (or, indeed, relaxations of safeguards) being needed. Exceptional new challenges produced by scientific progress are therefore not confounders of law, but desirable antagonists, leading to an evolution of norms in technology regulation. For this reason, it is also clear that the legal frameworks surrounding innovative scientific processes must retain a significant amount of elasticity (at the expense of accepting risks) in order not to stifle innovative progress. The dilemma of technology regulation is, we argue, therefore not a flaw in the system, but an inevitable phenomenon that occurs when society encounters a new scientific product. At the very starting point of this encounter, overriding rights and obligations (such as the informational and personal rights of the original material donors') are adequately respected by the legal mechanisms that are already well-established. They are sufficient to provide a lever that would stop the process if they are realised adequately. Our only concern in this area is that the current practice of broad consent to material donations for research may lead to a lack of an ability to fully give effect to the donor's actual preference where the ultimate use of the cells is entirely unforeseeable. It is at this point that we would advocate strongly for a further debate around developing frameworks which allow the seeking, and obtaining, of fresh consent for particularly innovative and unexpected uses.

[20] For a more detailed discussion, see Lavazza and Pizzetti (2020).

Over and beyond this—we accept, rather fundamental—concern about the availability of donors' cells for all uses, we argue that the risks inherent in transplanting human brain organoids into animals do not pose a novel challenge to the legal frameworks that are usually in play. The inability to adequately foresee the downstream risks of a novel technology create a duty of regulatory vigilance. This public duty is underpinned by the scientific community's inherent processes of testing results, data and assumptions, with undesirable and risky effects of a technology inevitably being part of the scientific discourse. The balancing exercise between risks and benefits, that initially seems such a conundrum in our context, is a routine undertaking in biomedical law. Indeed, we encounter it again when we discussed animal welfare and the need to identify the severity of the experiment, and the harm likely to be caused to the animals, in order to come to a robust conclusion as to the experiment's justifiability. There is an inevitable epistemic uncertainty in these exercises that makes it impossible to achieve the certainty we would desire as a society—this is where the law translates the exercise of balancing uncertainties into a legal certainty which makes the intended act permissible, and where a risk materialises, designates the right parties owing redress.

Finally, we struggled with the question how the law would approach an entity that metamorphoses from one kind of protected entity (cells) via another (animal) to a third (humanesque animal). All analyses of such a scenario must, ultimately, sound like an attempt at using a normative crystal ball. Aside from pre-emptive mechanisms grounded in some variety of the precautionary principle, it is difficult to see how a predominantly reactive legal system would encapsulate a changed moral status. Analogies from changes to the way animals have been protected in law suggest that it may well be the case that such a reaction may initially be symbolic in nature: mice with human brains are not animals. But they can be treated like animals unless some law or other says otherwise. What remains are two overarching conclusions: the need to continue the legal discussion at both ends of the spectrum opened up by our reference experiment. At the beginning, we need to address concerns in relation to the foreseeability of the possible uses of human cells in the context of broad consent. At the end, we need to contemplate innovative legal approaches to categorising entities that are deserving of legal protection. Not everything will always be a thing, an animal, or a human.[21]

References

Bourret R, Martinez E, Vialla F et al (2016) Human-animal chimeras: ethical issues about farming chimeric animals bearing human organs. Stem Cell Res Ther 7(1):87. https://doi.org/10.1186/s13287-016-0345-9

Collingridge D (1980) The social control of technology. Pinter, London

Farahany NA, Greely HT, Hyman S et al (2018) The ethics of experimenting with human brain tissue. Nature 556(7702):429–432. https://doi.org/10.1038/d41586-018-04813-x

[21] Lavazza and Pizzetti (2020).

Fenwick MD, Kaal WA, Vermeulen EPM (2017) Regulation tomorrow: what happens when technology is faster than the law? Am Univ Bus Law Rev 6(3):561–594

Hudson KL, Collins FS (2013) Biospecimen policy: family matters. Nature 500(7461):141–142. https://doi.org/10.1038/500141a

Karpowicz P, Cohen CB, van der Kooy D (2004) It is ethical to transplant human stem cells into nonhuman embryos. Nat Med 10(4):331–335. https://doi.org/10.1038/nm0404-331

Lavazza A, Pizzetti G (2020) Human cerebral organoids as a new legal and ethical challenge. J Law Biosci 7(1):5. https://doi.org/10.1093/jlb/lsaa005

Paşca AM, Sloan SA, Clarke LE et al (2015) Functional cortical neurons and astrocytes from human pluripotent stem cells in 3D culture. Nat Methods 12(7):671–678. https://doi.org/10.1038/nmeth.3415

Russell WMS, Burch RL (1959) The principles of humane experimental technique. Methuen, London

Shabani M, Borry P (2018) Rules for processing genetic data for research purposes in view of the new EU General Data Protection Regulation. Eur J Hum Genet 26(2):149–156. https://doi.org/10.1038/s41431-017-0045-7

Sheehan M (2011) Can broad consent be informed consent? Publ Health Ethics 4(3):226–235. https://doi.org/10.1093/phe/phr020

Stehlik J, Kobashigawa J, Hunt SA et al (2018) Honoring 50 years of clinical heart transplantation in circulation: in-depth state-of-the-art review. Circulation 137(1):71–87. https://doi.org/10.1161/CIRCULATIONAHA.117.029753

Windrem MS, Schanz SJ, Morrow C et al (2014) A competitive advantage by neonatally engrafted human glial progenitors yields mice whose brains are chimeric for human glia. J Neurosci 34(48):16153–16161. https://doi.org/10.1523/JNEUROSCI.1510-14.2014

Xu R, Brawner AT, Li S et al (2019) OLIG2 drives abnormal neurodevelopmental phenotypes in human iPSC-based organoid and chimeric mouse models of down syndrome. Cell Stem Cell 24(6):908–926. https://doi.org/10.1016/j.stem.2019.04.014

Part V

The Enhancement Conundrum

Building a Better Beast: Enhancing the Minds of Animals

10

Eric Schneider and Veljko Dubljević

10.1 Introduction

For centuries, nonhuman animals have been a testing bed for novel science and technology. Ancient Greeks studied animal forms and developed Hippocratic medicine, structuring their knowledge around the ideas that blood, phlegm, yellow bile, and black bile were the constituents of the body and that their improper balancing led to illness.[1] Islamic scholars examined animal bodies to discover that blood circulates throughout the body via a closed vasculature in humans.[2] Dolly the sheep clone was born in 1996 from somatic cells, heralding the ability to clone entire eukaryotic beings for developmental, epigenetic, and other forms of research.[3] Animals in this century are used for a vast range of applications from psychological experimentation to pharmaceutical testing.[4] With the broadening of researchers' horizons and more intricacies learned of the natural world, it is hardly a leap of the imagination that the variety in how humans and animals interact also changes. Enter brain organoids. Also called cerebroids, these are agglomerations of cells derived from a progenitor with the capacity to develop and organize themselves into near-faithful models of brain tissues.

Creating high-quality models of living systems is valuable for research because the way such systems respond to a set of variables guides future decisions about how to develop technology that could improve human lives. Brain organoids

[1] Dey et al. (2010).

[2] Soubani and Khan (1995).

[3] Weintraub (2016).

[4] Grayson et al. (2016) and Hajar (2011).

E. Schneider · V. Dubljević (✉)
NeuroComputational Ethics Research Group, NC State University, Raleigh, NC, USA
e-mail: veljko_dubljevic@ncsu.edu

H.-G. Dederer, D. Hamburger (eds.), *Brain Organoids in Research and Therapy*, Advances in Neuroethics, https://doi.org/10.1007/978-3-030-97641-5_10

allow researchers to study tissues almost as they are in nature without the exact same ethical and technical challenges of investigating a real person's brain. This is a technology with the potential to enhance, or at least shift, the cognitive ability of sentient beings[5] outside of what nature has thus far granted. When a human cerebroid is implanted into an animal, it creates what is called a chimera; a single organism that possesses the phenotypes of two separate organisms.[6] While such a thing may seem to lie in the realm of science fiction, humanity is no stranger to enhancing the minds and bodies of animals in pursuit of a better companion, meal, or subject. Dogs have been artificially bred from wolves to fill specialized roles as companions, guards, detectors, or even guides in the case of seeing eye dogs,[7] chickens have been genetically and endocrinologically altered to produce higher quantities of more nutritious meat,[8] and rats have been intensively inbred to express genotypes for reliable scientific experimentation.[9] Personal care for dogs can reach levels only surpassed by human children and the USA even protects animals from certain mistreatment via legislation.[10] As enhanced animals, dogs are further subject to much higher social-behavioral expectations than their wolf counterparts in the wild. The cognitive and behavioral enhancement of animals in these ways can then enhance their moral status in the eyes of humans on both individual and institutional levels. Considering the potential for a brain organoid to enhance the cognition of nonhuman animals by way of increased neuron counts or brain wrinkling, for example, it stands to reason that enhanced thinking confers an enhancement in moral status and the proto-moral agency of an animal itself. Such changes in the moral relevance of an animal via grafting of cerebroid tissue may come to elevate moral decision-making in animals to complicate what is permissible, encouraged, or even prohibited. Institutions have the power to exert influence over human behavior and thus answer these questions of permission and prohibition. Therefore, those with influence must be well informed of the benefits that may come from cerebroid research involving chimerization in addition to its drawbacks. Benefits include modeling human tissue development, while drawbacks may lead to the mistreatment of morally relevant beings. In this chapter, we explore some of the social implications and the paradigm shift in the growth of knowledge that could be foreseen given the development and proliferation of brain organoid technology. We describe more nuanced models of describing how human institutions may manipulate the proliferation of a morally concerning technology. Also, we address the problem of media hype, which could lead both to outsized hope and wholesale fear of the technology, which ought to be preempted.

[5] Lancaster et al. (2013) and Wang et al. (2020).

[6] Rogers (2018).

[7] Larson and Bradley (2014).

[8] Zuidhof et al. (2014).

[9] Vandenbergh (2000).

[10] 89th Congress (1966).

10.2 The Basics of Brain Organoid Science and Technology

To understand the ethical dilemmas and potential consequences of chimerization with human brain organoids, the cerebroids must be understood at least on a basic foundational level. Further, one should appreciate that these models of life can be as complex in themselves as the life they are supposed to mimic. Cerebroids exist in an intermediate stage of complexity between that of tissue and that of an entire organ.[11] They are three-dimensional cultures of tissue that pose different cultivation challenges than simpler two-dimensional ones grown on, say, an agar plate. Cells in 3D cultures often have more heterogeneous access to growth media, cells may receive signaling via extracellular matrix (ECM) producing unexpected outcomes, and it remains an open question how realistically organized the cells in the subject culture are.[12] When a pluripotent cell is subject to a niche (a particular set of environmental conditions), it will proliferate while it consumes media and differentiates into different specialized cell types. These cell types will communicate with each other to organize over time into a series of transition states along a path towards more complex, intermediately functional units. It is these functional amalgams of cells, organoids, which can approximate both form and function of more complex systems to varying extents. How strong the approximation is depends on a litany of factors, but the better approximation of nature in an organoid model does improve the connection between controlled experiment and practical application. In effect, these are a sort of miniature fractions of a brain that have valuable applications in modeling how viral infections affect brain tissue; such as, the Zika virus,[13] or exploring potential relationships between neural tissue structure and neurological disorders such as schizophrenia or Alzheimer's.[14]

10.3 Chimerism: How Does It Work?

While fear for the challenge chimerism can pose to our ideas of nature or order is not new, as it is seen in literature[15] and digital media,[16] the availability of technology to permit the cohabitation of human and animal cells in an individual organism is a novel breakthrough of the twentieth and twenty-first centuries. A major boon to this technology is the ability to reprogram differentiated somatic cells into effectively naive pluripotent stem cells. Pluripotency, the ability of a cell type to proliferate and develop into a profile of multiple functional cell types, may be induced in somatic cells via treatment with protein factors. This has allowed for the indirect derivation

[11] Mansour et al. (2018).

[12] Edmondson et al. (2014).

[13] Garcez et al. (2016).

[14] Quadrato et al. (2016) and Amin and Paşca (2018).

[15] Shelley (1998).

[16] Biller (1996).

of a human brain organoid from virtually any cell in the body.[17] Despite the potential of this technology, blood or epithelial cells are typical and simple sources for harvesting from a consenting donor. These human-induced pluripotent stem cells (hiPSCs) are themselves not controversial to the public (in the USA and globally) and are used in both research and biomanufacturing to develop complex cell or tissue cultures.[18] However, their use in developing brain organoids has generated friction among researchers, ethicists, and other attentive stakeholders in the future of neurological research.[19] There have been rapid improvements in the sophistication of models of morally relevant structures. Concerns over this have drawn attention to how their synthesis with nonhuman beings will affect those animals' moral status. For if in Europe and the USA it is either illegal or highly discouraged to perform experiments on great ape primates,[20] then it is not a serious stretch of the imagination to consider some degree or kind of cognitive enhancement in another nonhuman species could justify the extension of some of society's protections to those nonhuman individuals.

10.4 Enhancement with Brain Organoids?

While no brain organoid is currently thought to have been complex enough to warrant being considered a conscious being, the fact that vascularized brain organoid tissues ever more strongly resemble structures that do has given some researchers pause.[21] Chen and colleagues have recently argued that the enhancement of animals is a salient ethical challenge we stand to face in the near future. The urgency of this ethical challenge depends then on the level of functional enhancement conferred to the animal, ranging from low when the animal's enhancement is minor and only in the degree to higher when the chimera appears to display behavior unseen in unaltered individuals (see Table 10.1).

As these "mini-brains" grow more sophisticated and larger in scale, such as being able to model multiple regions of a brain,[22] their application in nonhuman animals will pose questions to their validity as valid subjects of experimentation. According to Chen and colleagues, there are several viable avenues of research that could take advantage of the sheer volume of neurons in human brains as well as the computational capacities and structural organization unique to human brains. These avenues could be put to good use with the aid of human brain organoids for the express purpose of enhancing animals (see Table 10.2).

Though the mere mention of human experimentation may raise concerns, direct experience gained via research on human subjects can be an impactful tool for

[17] Takahashi and Yamanaka (2006) and Lancaster et al. (2013).

[18] Loh et al. (2009) and Zhou et al. (2011).

[19] Hyun et al. (2020).

[20] Project R and R (2015).

[21] Hyun et al. (2020).

[22] Bagley et al. (2017).

Table 10.1 Significance and ethical importance of brain function enhancement in chimeric animals

Functional enhancement level	Functional enhancement expression	The relative degree to which enhancement influences perceived moral status
1. Basic neurological functions – Movement/sensory perception	Chimera develops faster reaction times and finer visual perception	Low: Would be considered a novelty and may provoke broad ethical concerns about the brain function enhancement
2. Cognitive functions – Attention/memory	Chimera develops the capacity for faster learning, quicker decision-making, and memory with higher recall accuracy	Medium: Will be perceived with greater scrutiny and create concerns among a wider range of disciplines
3. Self-awareness – Sentience	Chimera develops metacognition (i.e., thinking about thinking) as well as self-awareness of mental states	High: Discussion will expand from concerns centered primarily around brain enhancement to concerns of perceived "humanization" of brain chimeras

Table 10.2 Possible brain enhancements that could result from theoretically successfully replacing a common laboratory rat's brain with human brain organoid cells

Variables potentially enhanced	Standard laboratory rat brain features	Human brain features	Possible results of successful introduction of human neurons
Number of neurons	200 million total neurons (31 million cortical neurons)	86 billion total neurons (16 billion cortical neurons)	(If neurons introduced at an early embryonic stage) cortical expansion would result in 2–3 times larger brain + increase of ~230 million neurons
Computational capacity of individual neurons	I. High cortical synaptic density II. Lack specific cell types such as dopamine interneurons required for specialized neuromodulation	I. Significantly long dendritic arbors in V-pyramidal neurons II. Possess certain cell types such as dopamine interneurons allowing for unique neuromodulation	I. Augmentation of cortical computation through increased electrical compartmentalization and input–output properties II. Increased capacity for more complex neuromodulatory systems where neuromodulatory transmitters such as dopamine can be involved in uniquely human neural cognitive computation (working memory, reasoning, reflective exploratory behavior)

(continued)

Table 10.2 (continued)

Variables potentially enhanced	Standard laboratory rat brain features	Human brain features	Possible results of successful introduction of human neurons
Structural organization and connectivity of neurons	Relatively flat and smooth cerebral cortex	Relatively large cerebral cortex covered in gyri and sulci to increase surface area and form brain divisions	Significant gains in computational capacity would be achieved if the cerebral structure of the hypothetical chimera resembled the human cerebral cortex—still questionable if human brain architecture can be replicated with only 0.5% of cells

improving lives, in particular those suffering from neural pathologies. Though since such research *can* lead to good rather than *necessarily* leading to good outcomes, it thus follows that research on humans *can* also make sets of people worse off. Further, those who volunteer for clinical trials are often those who are "Have-nots" in either financial or medical senses of the term.[23] People who do not have great freedom in rejecting meager financial incentives from organizations conducting experimental trials can then readily be exploited which leads to the exchange lopsidedly benefiting the "Haves." One need only look to experimental economic policies, medical procedures, or infamous psychological experiments to see what damage can come from a combination of improperly gained consent, carelessness for the worst-off among us, poor grounding of methodology, or sheer scientific illiteracy to name but a few.[24]

10.5 Who Even Counts? The Sliding Scales of Moral Weight

In the mind of a typical researcher that studies and categorizes living things (professionally or otherwise), there may reasonably exist a hierarchy of moral relevance that tends to flirt with an upward trend in complexity. Humans are considered most complex with certain other mammals such as dogs, great apes, dolphins, and whales standing below but near humans in importance. Those in the Western world have organized to avoid scientific testing on great apes, while many individuals have qualms over the treatment of lab rats bred for experimentation, and many are indifferent about the treatment of or even disgusted by insects or reptiles.[25] Thus it stands to reason that if an organism, say a mouse, were to have grafted onto its brain tissue a cerebroid (synthetic human brain tissue), and if approaching similarity to human biology confers greater moral considerations, then that animal

[23] Parekh and Desai (2020).

[24] Barrett and Carter (2010), Baker et al. (2005), and Kaplan (2004).

[25] Project R and R (2015), University of California Irvine (2013), Janovcová et al. (2019), and College (2016).

or any like it could be elevated in how worthy it is of being considered in moral decision-making via chimerization of that animal with human brain organoid. Its biology, of course not human in full, possesses structures of those who get to be considered people. Further, while the grafted human tissue onto animal tissue[26] confers only a small patch of human biology to a developing animal, the earlier this grafting occurs, perhaps during gestation, the greater a fraction of that individual's cells become human. This human character may provide to that animal serious enough enhancement in cognition, relative to non-augmented animals, that serious changes in the moral status of the animals take place. Such stark enhancement has not yet taken place but it is valuable to keep this in mind as well as the sheer evolutionary distance in structures between humans and, say, horses. Many proteins and genes are rather similar between humans and animals, especially those sharing mammalian heritage, and so there is cause to believe that those life-forms which are often given lower moral status by Western standards may be able to ascend to higher levels of cognition. Such levels of cognition could demand interspecies interactions even more pronounced than the bond between humans and canines. That said, future advances in research have the potential to change the narrative of human self-understanding and thus how humanity interacts with the rest of the animal kingdom, especially those enhanced by human (synthetic tissue) technology.

Whether or not it is possible to enhance the cognitive abilities of an animal in such a way that the animal is perceived to be also enhanced morally is separate from the question how that technology proliferates. There is an intersection by way of the research defining the scope of what is possible, but how actualized that possibility becomes is a question for the political, economic, ethical, legal, religious, and myriad of other institutions individuals constitute. To give an idea of what sort of proliferation, or lack thereof, brain organoid technology's application to animal subjects may see, one can look towards the constellations of thought, ideologies, which help describe the landscape of the world we live in. Ideology, as it is used here, should not be understood as a strict set of internally consistent beliefs that match each other in language. As an example, associations between American political conservatism and what has been called bioconservatism need not necessarily, but may overlap (see Fig. 10.1). While the White House' council on biotechnology from 2003 largely represented voices that may be considered bioconservative in varying ways[27] and were also affiliated with the conservative government of the Bush-Cheney Administration,[28] notable brain surgeon Ben Carson contributed to this panel and has been reported to be associated with a company that sold alleged enhancement technology.[29] Is Dr. Carson conservative? As an actor in American politics, he is, but he is not wholly so in the context of the social phenomenon of enhancement.

---

[26] Chen et al. (2019).

[27] Browne and Clarke (2019).

[28] Kass (2003).

[29] Maremont (2015).

Fig. 10.1 Classical model of biopolitics

Before describing potential associations with outlooks such as biolibertarianism or bioconservatism, it is important to remember that contemporary patterns of association between individuals and ideologically aligned institutions may not hold in the future. The heads of political, economic, or other sorts of institutions' list of essential supporters are often in flux. A conservative now with restrictive views on some matters may switch to holding wildly different views on the matter of elevating the moral status of animal beings as political incentives change.

10.6 Counterproductive Enthusiasm: Incentive and Opinion

Grafting human cells onto an animal so that one can create an organism that shares unique features between sexually incompatible species is a rather exciting subject for those involved with such research. It is cutting-edge technology, and it has fascinating implications for the future of humanity. It also gives great material to write headlines about in the science section of non-academic media. However there often is a strange "game of telephone" played as scientists make their findings and communicate those findings to others and the writers who have to almost translate between the dialect of their audience and the dialect of scientific literature. Outcomes A and B may be found to be associated with conditions C and D which may be reported as such in literature but by the time it reaches commercial media the narrative may alter into something so simple as "C causes A, avoid C in your diet altogether." This may or may not be true, but the initial piece of literature detailing the research conducted likely did not contain such a narrative.

As if this were not challenging enough, science journalists have their own array of incentives pushing and pulling them towards and away from various actions. Information propagates through media, such as digital or print, like proteins and nucleic acids migrate down a gel. An informative and lengthy discussion of the limitations in cerebroid grafting onto animals may be a stellar read, but such pieces rarely come with Tweet-length summaries. Further, they often paint a narrative for the reader of what technology cannot do. In effect, they counter against more salacious pieces and pull the mind away from a fantasy of technology without costs. Having the content to convey complexity is important as is covering the limitations of our reach, but more precaution can be less savory to many news or social media consumers.

While research has shown that media narratives surrounding emerging technologies have the power to distort the knowledge and expectations people hold about those technologies,[30] the media also can improve understanding: to steer

[30] Dubljević et al. (2014).

narratives away from the extremes of utopia versus dystopia and provide a reality-centered discourse. How this occurs in different political landscapes as described above is varied, but there is a shared understanding that institutions which represent the best interests of the people they are accountable to are preferable: institutions that balance the democratic value of broad-based accountability with the restraint afforded by tailored judgment from technocrats. The ratio of these values to each other will fluctuate over time as old systems are replaced with new ones, but the aim of individuals organized in this way must be outcome-oriented to measure where fault and success lie in media narratives around cerebroid research and innovation. Outcomes may include timely implementation of technology where it is demanded, low incidence of lawsuits surrounding alleged misuse of the technology, government funding of research into the ethics and biology of animals augmented with human cerebroids, intersection between authors writing for scholarly and public audiences, and results from research on public sentiments. This list is not exhaustive; many more options exist to creatively measure how and where harm can be reduced as human brain organoid technology blossoms. But the objective must be to holistically reduce harm whether it comes from the technology itself, from the people serving justice to those harmed, or even due to lost accountability in government via regulatory capture.

10.7 The Ideological Runway

Bioconservatism and biolibertarianism are each of two wings along a putative axis which describes the beliefs held by fractions of a population. Broadly, bioconservatives would take a view that some defined set of characteristics of a creature are natural and thus worthy of conservation. A range of norms to neither be brought up nor down from an existing moral range of acceptability. These norms exist to defend and maintain the status quo which serves the extant powers-that-be of a society. Some leeway exists in terms of permitting therapeutic uses for enhancement technology to restore to normal some function that has been lost, such as vision or memory; however, such permissiveness may be more constricted if the subject is not seen as sufficiently human. Further, it of course depends also on the scope of what is to be conserved as defined by the norms of society. Perhaps one with a bioconservative outlook would seek some restoration of mental or visual acuity in an aging and beloved pet, but not necessarily one that enables super-standard communication.

Biolibertarianism, conversely, takes a considerably more Laissez-faire view that permissible means whatever an individual, absent coercion, agrees to. It is individualistic and generally supports looser institutional constraints on individual expressions of desire. If a person wishes their dog, for whom they make the decision, to be enhanced so that it can better perceive certain symbols such as sign language or if a military organization wishes to enhance dolphins to more efficiently sweep for naval mines[31] then that freedom to make those decisions is desirable because it

[31] Thompson (2018).

Fig. 10.2 A more nuanced view of contemporary biopolitics

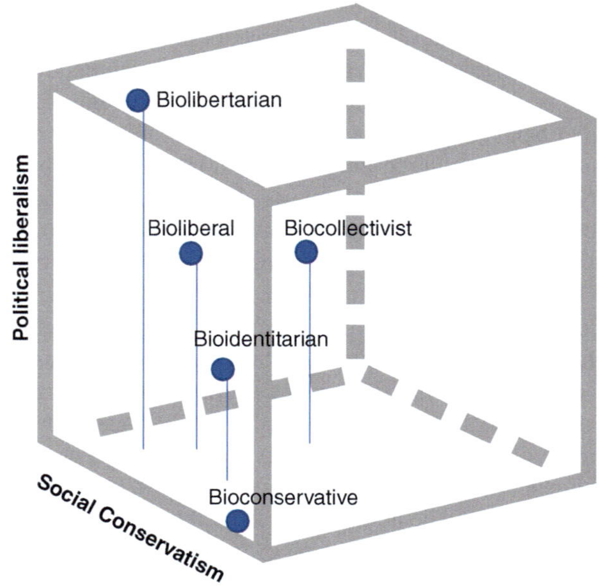

is free from nonaggressive coercion.[32] However, despite modern affinities for describing types of beliefs as falling somewhere along a left-right spectrum, human thought can be a push and pull between many different directions depending on the lived experience in one's niche, and in what follows we will argue that biopolitics should be reimagined along multiple dimensions to respect the unique patterns of pressure systems that contribute to what is attractive and what is repulsive to the relevant stakeholders of societal institutions. In particular, it will be described along with the scales of adherence to more traditional political theories of conservatism, liberalism, and socialism (see Fig. 10.2).

To illustrate some of these clusters of thought that may develop in concert with the biopolitics of animal enhancement, three labels can be used: biocollectivism, bioidentitarianism, and bioliberalism. These labels may be used to describe an individual's leanings, but here they refer more to how the constituents of a political body organize themselves to write the rules of the road; even the most ideological of a system's adherents may act contrary to their dogma if the incentives are correctly arranged. Biocollectivism refers to what can be thought of as a maximalist approach to the democratic decision-making process. It shares some features with biolibertarian ideas, such as avoidance of coercion that some contemporary libertarians colloquially have referred to as the non-aggression principle.[33] A biocollectivist institution may reward beliefs that what it is to be canine has very little essential connection to a current understanding of what makes any animal an animal. In such a system people would concern themselves less with the fact that

[32] Hughes (2002).

[33] Christmas (2017).

what it is to be a particular animal has changed, rather they would be far more sensitive to the exploitation of that animal, or how that animal's experiences fit into a class-based view of society. In a way, the focus is more on the relationship of the thing to the world than the nature of the thing itself. While everyone in such a society's value on the enhancement of non-living things may fall anywhere on the axis of biolibertarian to bioconservative, the pressures of a biocollectivist niche are going to encourage contextualization of libertarian-conservative positions into what kind of class conflict is incentivized by the enhancement of animals. They may look in varying ways to explore how an enhanced animal perceives their experience in the world: whether they can see themselves as participants in society working for themselves or if their nature lends them well to being submitted to exploitation by another. An additional pertinent question would be whether or not an enhanced animal class can act as a sort of antithesis to some power centers in society. It is reminiscent of the broad strokes of socialist thinking and thus one could expect it to perhaps be expressed also by members of organizations affiliated along such lines of thought. Besides the criticism offered from such groups, dissemination and development of animal enhancement in a society with biocollectivist values may lean towards what could be considered hasty development in early stages and comparatively indifferent proliferation in later, more successful stages of history. A reason for this is that opposition to a ruling administration demands coalition building and inclusion of similarly motivated identities, as seen in election cycles where there is an incumbent to unseat. However, if a ruling regime were sufficiently of the people and had done away with unjust exploitation, there would be weak, if any, pushing force to go either for or against further increases in nonhuman moral weightiness from a biocollectivist perspective as an expansion of the body politic is neither beneficial nor hostile to a maximally democratic regime.

Bioidentitarianism is perhaps the most amorphous of the three terms. It is in some ways conservative but not most accurately described as such. Institutions of such a perspective may implement really whatever is palatable to those of the correct identity and that identity need not be one associated with dominant power structures. It is a considerably less rigid outlook and is permissive of nearly any position from biolibertarianism to bioconservatism, but not every position. Bioidentitarian institutions would uphold a certain identity as a standard-bearer for others to aspire to be like. An identity that is perhaps estranged by a mainstream but aligned enough with some powerful institutions to fuel it. This could be the ideal of a pure and unmodified human being who extends this purity to the rest of the natural world if modification becomes prolific in day-to-day life. Alternatively, this could be the ideal of a human being who considers one set of identities to be supreme and all others, even nonhuman, are owed elevation to a greater moral plane (e.g., a transhumanist identity) in a world where modification and enhancement are largely eschewed. Essentially, it relies on a distinction between self, or self-adjacent depending on the individual's tolerance, and non-self where the self is simultaneously stronger and weaker than the non-self. The nature of that interaction may be hostile or friendly, combative or non-combative, exploitative or Laissez-faire, but bioidentitarianism must not be wholly inclusive of those of the out-group and must exclude certain elements from society on some set of rules to build a sense of in-group identity. Otherwise, the biological character of identity is not being upheld. While this perspective is not in itself a necessarily harmful one, it is disruptive to some set of the current norms of a society. Nevertheless, hijacking by bioidentitarian-esque movements of regulatory institutions that exist in a society to empower and constrain other institutions tasked with adjudicating who is and is not a valid member of society as well as what counts for acceptable conduct has historically and currently led to serious abuses of humanity by those who hold absurd and vile biases[34] and thus is a mentality which should not be taken lightly by democratic societies that seek to uphold themselves.

The third of these perspectives, bioliberalism, refers to many of the developments made during the twentieth and twenty-first centuries upon the Laissez-faire style economic liberalism of the nineteenth and early twentieth centuries. Bioliberal institutions are those which generally value individualist approaches to empower decisions to be made by those who do not identify with a collective while acknowledging that there is an advantage in understanding how group identity motivates an individual towards or away from particular changes or sets of changes. Much of what incentivized it to come forth was the structural tensions in the let-it-be governing of bodies that led to the rise in collectivist and identitarian ideologies during the early-mid-twentieth century. In this vein, it comes as a model of society with ancestry in the problems faced and solutions provided by various attendees of the Walter Lippmann Colloquium in 1938.[35] While some adherents espoused views favoring considerably less interactive roles for the state such as Hayek,[36] others such as Milton Friedman, Wilhelm Röpke, or Walter Lippmann suggested a model involving a more interactive state that ensures the market and society remain in a competitive condition of creative destruction to generate wealth that can be allocated following just social policy and respect for human rights.[37] Again, this is distinct from the biolibertarian perspective which holds more staunchly individualistic and market fundamentalist values. This basic idea has been given multiple names by numerous critics and supporters over the decades since its inception, but to evaluate the artificial enhancement of an animal's moral relevancy it will be called bioliberalism. Such a framework leaves latitude to institutions that are tasked with regulating the proliferation and development of the technology, aiming to determine what research practices are acceptable from the input of a broad and inclusive coalition of stakeholders, but tempering some of that popular will with more exclusive institutions to inoculate it against populist tides. Two democratic societies, one with a reverence for felines and canines (e.g., USA)[38]

[34] United States Holocaust Memorial Museum (2010), Libaridian (2017), p. 223, 21st Congress (1830).

[35] Foucault (1979), pp. 131–133.

[36] Hayek (1944), pp. 42–44.

[37] Reinhoudt and Audier (2018), Gregg (2010), pp. 12–13, and Friedman (1951).

[38] 115th Congress (2018).

and the other with one for ruminants (e.g., India), in this way are simultaneously free to pursue separate paths where one has state policies that are loose and conducive to researchers having access to government funds and the other is free to restrict such practices under the spiritual and cultural beliefs of its people. Of course, this example is rather reductive, and international relations are scarcely so simple in the real world, but essentially the bioliberal model on inter- and intranational levels enables a degree of experimentation in including a greater diversity of life forms into the body politics of democratic societies while recognizing that decisions so revolutionary to our existing conceptions of human–nonhuman interaction are not atomic enough to be left to a strictly individualist level.

From mistakes of the past and present, future researchers and members of societies' leadership can learn what rules and incentives lead to fair, science-backed, and dynamically stable systems to support the achievement of both long- and short-term prosperity. Human brain organoid technology stands to be an excellent, albeit limited, model for regions of living brain tissue. Included with it are potential ethical tradeoffs but those tradeoffs do not necessitate outright prohibition of the technology.[39] Banning the technology in advance, or at least starving it of funding to achieve the same effect, shows both a lack of will to find the hard solutions and is harmful to those who stand to gain from discoveries found using chimeras of human and animal tissue.

To identify and empower those with the most direct stakes in what benefits and tradeoffs come from cerebroid research is a daunting task. It will demand collaboration between academics of many fields and professionals of many stripes if the beings, human or otherwise, impacted by human curiosity are to be treated fairly. As Lippmann observed, to execute justice we have to understand not only what raw information exists and is accessible, we also must understand how that information is processed in the minds of those who access it.[40] Senses of wonder and excitement at promising technology are not bad things in themselves and can be strong engines of innovation. But these sentiments have to be tempered with proper messaging via collaborative and countervailing institutions of experts and communicators if their desired outcomes are to be achieved decently and humanely. The shape those institutions take will be greatly molded by the ideological perspectives mentioned above along messy, ideologically impure pathways. The authors of this chapter do not explicitly advocate any one perspective as superior to any other as there is a time and place for all things, but this is not to say the perspectives ought to be considered morally, ethically, or practically equivalent to one another. It will depend on the mistakes that have been recognized in the past, the will to experiment in the present, and what visions of the future appeal to the present. Complex topics such as this are the domain of future scholarship as much as they are of current actors and thus subject to the forces of creative destruction.

---

[39] Di Lullo and Kriegstein (2017).

[40] Lippmann (1922), p. 85.

10.8 Concluding Remarks

Humanity's experience building better animals is a long one full of hybrid and domesticated creatures. However, never before has human experience become so sophisticated that chimeric organisms could be grown and studied as models of human biology, where a human brain organoid can be cultured and grown in the short-term using nonhuman biology as a kind of scaffold for mimicking natural human development. This is a potent technology which many may stand to gain from and as such the way it will be used over the coming years will be impacted by the shape of our civic and political institutions. What principles guide those in charge of such institutions will fall along ideological borders to an extent and may be described using language of values, but will mostly reflect the coalitions institutional leaders see themselves as accountable to. Of course, accountability in this context can be determined in many ways. For bioidentitarianism, accountability comes from how well a system upholds the supremacy and interests of given biology. In the biocollectivist framework, institutions' leaders are or, at least will appear to be, subject to the will of the greatest majority. Contrasting both of those is the bioliberal view that while majority opinion is valuable to uphold, determining the appropriate majority is relevant work and can be done correctly as well as incorrectly. Further, in any system there are institutions that serve to inform those tasked with making decisions. Media organizations, academic or popular, are some of the institutions which have an ability to influence outcomes to the benefit of some groups over others based on the narratives they build and reinforce. What narratives are built or torn down depends on the coalitions an institution's decision-makers view their careers as incumbent upon. Ideological lenses may be used to answer the question of who the right people for a certain set of institutions to be accountable to are; though, it will in reality be an ongoing process of redefining the problems and rewriting answers. A process which may be fraught for the near future with solutions which create no-win scenarios, perfect dilemmas, undue harm, and harsh injustice. These are mistakes which will lay the foundation for more mature solutions and affirm the value of evidence-based policy and cautious decision-making. Solutions which create structures in society which minimize unproductive pain, extract meaningful inputs from the relevant stakeholders, and seek to close out conflict over the correct rules of the road. It will take arrogance, humility, grudges, forgiveness, and all kinds of human failure and success to build worlds that engender informed and legitimate consent: consent from the researchers whose careers may be built on this science, from the patients whose well-being and families' well-being are on the line, from the countless private interests in markets fixing to exploit a new opportunity, and from the interests that can monitor chimeras so as to limit the abuse of what may become a morally relevant class. Thus, consent must be given not only by those who stand to gain the most, but also those who potentially lose the most.

References

115th Congress (2018) An Act to prohibit the slaughter of dogs and cats for human consumption, and for other purposes

21st Congress (1830) An act to provide for an exchange of lands with the Indians residing in any of the states or territories, and for their removal west of the river Mississippi

89th Congress (1966) Animal Welfare Act

Amin N, Paşca S (2018) Building models of brain disorders with three-dimensional organoids. Neuron 100:389–405. https://doi.org/10.1016/j.neuron.2018.10.007

Bagley J, Reumann D, Bian S et al (2017) Fused cerebral organoids model interactions between brain regions. Nat Methods 14:743–751. https://doi.org/10.1038/nmeth.4304

Baker S, Brawley O, Marks LS (2005) Effects of untreated syphilis in the negro male, 1932 to 1972: A closure comes to the Tuskegee study, 2004. Urology 65:1259–1262. https://doi.org/10.1016/j.urology.2004.10.023

Barrett C, Carter M (2010) The power and pitfalls of experiments in development economics: some non-random reflections. Appl Econ Perspect Policy 32:515–548. https://doi.org/10.1093/aepp/ppq023

Biller K (1996) Tuvix. Voyager 2, Star Trek

Browne T, Clarke S (2019) Bioconservatism, bioenhancement and backfiring. J Moral Educ 49:241–256. https://doi.org/10.1080/03057240.2019.1576125

Chen H, Wolf J, Blue R et al (2019) Transplantation of human brain organoids: revisiting the science and ethics of brain chimeras. Cell Stem Cell 25:462–472. https://doi.org/10.1016/j.stem.2019.09.002

Christmas B (2017) Rescuing the libertarian non-aggression principle. Moral Philos Polit 5:305–325. https://doi.org/10.1515/mopp-2017-0007

College W (2016) America's top fears 2016 - Chapman University Survey of American Fears. In: The voice of Wilkinson. https://blogs.chapman.edu/wilkinson/2016/10/11/americas-top-fears-2016/. Accessed 17 Jun 2021

Dey N, De P, Smith B, Leyland-Jones B (2010) Of mice and men: the evolution of animal welfare guidelines for cancer research. Br J Cancer 102:1553–1554. https://doi.org/10.1038/sj.bjc.6605692

Di Lullo E, Kriegstein A (2017) The use of brain organoids to investigate neural development and disease. Nat Rev Neurosci 18:573–584. https://doi.org/10.1038/nrn.2017.107

Dubljević V, Saigle V, Racine E (2014) The rising tide of tDCS in the media and academic literature. Neuron 82:731–736. https://doi.org/10.1016/j.neuron.2014.05.003

Edmondson R, Broglie J, Adcock A, Yang L (2014) Three-dimensional cell culture systems and their applications in drug discovery and cell-based biosensors. Assay Drug Dev Technol 12:207–218. https://doi.org/10.1089/adt.2014.573

Foucault M (1979) The birth of biopolitics lectures at the College de France, 1978-1979. Picador, New York, pp 131–133

Friedman M (1951) Neoliberalism and its prospects. In: Collected works of Milton Friedman – collected works of Milton Friedman. https://miltonfriedman.hoover.org/objects/57816/neoliberalism-and-its-prospects. Accessed 17 Jun 2021

Garcez P, Loiola E, Madeiro da Costa R et al (2016) Zika virus impairs growth in human neurospheres and brain organoids. Science 352:816–818. https://doi.org/10.1126/science.aaf6116

Grayson DS, Bliss-Moreau E, Machado CJ et al (2016) The Rhesus monkey connectome predicts disrupted functional networks resulting from pharmacogenetic inactivation of the Amygdala. Neuron 91:453–466. https://doi.org/10.1016/j.neuron.2016.06.005

Gregg S (2010) Wilhelm Röpke's political economy. Edward Elgar, Cheltenham

Hajar R (2011) Animal testing and medicine. Heart Views 12:42. https://doi.org/10.4103/1995-705x.81548

Hayek F (1944) Road to serfdom. Routledge, London, pp 42–44
Hughes J (2002) The politics of transhumanism. http://www.changesurfer.com/Acad/TranshumPolitics.htm. Accessed 20 Jun 2021
Hyun I, Scharf-Deering J, Lunshof J (2020) Ethical issues related to brain organoid research. Brain Res 1732:146653. https://doi.org/10.1016/j.brainres.2020.146653
Janovcová M, Rádlová S, Polák J et al (2019) Human attitude toward reptiles: a relationship between fear, disgust, and aesthetic preferences. Animals 9:238. https://doi.org/10.3390/ani9050238
Kaplan R (2004) Treatment of homosexuality during apartheid. BMJ 329:1415–1416. https://doi.org/10.1136/bmj.329.7480.1415
Kass L (2003) The President's Council on Bioethics: full document. In: Beyond therapy: biotechnology and the pursuit of happiness. https://bioethicsarchive.georgetown.edu/pcbe/reports/beyondtherapy/fulldoc.html. Accessed 17 Jun 2021
Lancaster M, Renner M, Martin C et al (2013) Cerebral organoids model human brain development and microcephaly. Nature 501:373–379. https://doi.org/10.1038/nature12517
Larson G, Bradley D (2014) How much is that in dog years? PLoS Genetics, The advent of canine population genomics. https://doi.org/10.1371/journal.pgen.1004093
Libaridian G (2017) The ultimate repression: The Genocide of the Armenians, 1915-1917. In: Modern Armenia. Routledge, London, p 223. https://doi.org/10.4324/9781315124667-9
Lippmann W (1922) Public opinion. Harcourt, San Diego
Loh Y, Agarwal S, Park I et al (2009) Generation of induced pluripotent stem cells from human blood. Blood 113:5476–5479. https://doi.org/10.1182/blood-2009-02-204800
Mansour A, Gonçalves J, Bloyd C et al (2018) An in vivo model of functional and vascularized human brain organoids. Nat Biotechnol 36:432–441. https://doi.org/10.1038/nbt.4127
Maremont M (2015) Ben Carson has had ties to dietary supplement firm that faced legal challenge. Wall Street J. https://www.wsj.com/articles/ben-carson-has-had-ties-to-dietary-supplement-firm-that-faced-legal-challenge-1444057743. Accessed 17 Jun 2021
Parekh T, Desai A (2020) Demographic and socioeconomic disparities among cancer survivors in clinical trials participation, USA, 2016–2018. J Cancer Educ. https://doi.org/10.1007/s13187-020-01790-6
Project R and R (2015) International bans: laws: release & restitution for chimpanzees. https://releasechimps.org/laws/international-bans. Accessed 17 Jun 2021
Quadrato G, Brown J, Arlotta P (2016) The promises and challenges of human brain organoids as models of neuropsychiatric disease. Nat Med 22:1220–1228. https://doi.org/10.1038/nm.4214
Reinhoudt J, Audier S (2018) The Walter Lippmann colloquium: the birth of neo-liberalism. Palgrave Macmillan, Cham
Rogers K (2018) Chimera. In: Encyclopædia Britannica. https://www.britannica.com/science/chimera-genetics. Accessed 17 Jun 2021
Shelley M (1998) Frankenstein. Oxford University Press, Oxford
Soubani A, Khan F (1995) The discovery of the pulmonary circulation revisited. Ann Saudi Med 15:185–186. https://doi.org/10.5144/0256-4947.1995.185
Takahashi K, Yamanaka S (2006) Induction of pluripotent stem cells from mouse embryonic and adult fibroblast cultures by defined factors. Cell 126:663–676. https://doi.org/10.1016/j.cell.2006.07.024
Thompson L (2018) U.S. Navy Marine Mammal program performs mine clearance operations during RIMPAC SOCAL 201. In: United States Navy. https://www.navy.mil/Press-Office/News-Stories/Article/2249862/us-navy-marine-mammal-program-performs-mine-clearance-operations-during-rimpac/. Accessed 17 Jun 2021
United States Holocaust Memorial Museum (2010) United States Holocaust Memorial Museum. https://encyclopedia.ushmm.org/content/en/article/defining-the-enemy. Accessed 17 Jun 2021
University of California Irvine (2013) Euthanasia of research animals. https://research.uci.edu/compliance/animalcare-use/research-policies-and-guidance/euthanasia.html. Accessed 17 Jun 2021

Vandenbergh J (2000) Use of house mice in biomedical research. ILAR J 41:133–135. https://doi.org/10.1093/ilar.41.3.133

Wang Z, Wang S, Xu T et al (2020) Cerebral organoids transplantation improves neurological motor function in rat brain injury. CNS Neurosci Therap 26:682–697. https://doi.org/10.1111/cns.13286

Weintraub K (2016) 20 years after dolly the sheep led the way-where is cloning now? Scientific American. https://www.scientificamerican.com/article/20-years-after-dolly-the-sheep-led-the-way-where-is-cloning-now/. Accessed 17 Jun 2021

Zhou T, Benda C, Duzinger S et al (2011) Generation of induced pluripotent stem cells from urine. J Am Soc Nephrol 22:1221–1228. https://doi.org/10.1681/asn.2011010106

Zuidhof M, Schneider B, Carney V et al (2014) Growth, efficiency, and yield of commercial broilers from 1957, 1978, and 2005. Poult Sci 93:2970–2982. https://doi.org/10.3382/ps.2014-04291

Legal Arguments in Favour of and Against Neuroenhancement by Means of Brain Organoids

11

Tade M. Spranger

11.1 Introduction

In essence, enhancement can be understood as a biomedical intervention in the physiological constitution of humans that is aimed at improving their abilities and characteristics and thus—for example, with regard to physical stamina or attractiveness, cognitive abilities or behavioural attributes—goes beyond the restoration and preservation of their health.[1] The commonly discussed forms of enhancement include plastic/cosmetic surgery, sports doping and genetic enhancement. Even this short list makes it clear that it is often impossible to draw a clear distinction between medically indicated therapies on the one hand and non-indicated enhancement on the other. For example, it is possible to understand a cosmetic operation that is exclusively due to aesthetic considerations as enhancement, whereas the same measure as a result of an accident, for example, can be understood as a medical treatment. If, however, on the occasion of a medically indicated treatment, not only the original constitution of the patient is restored, but "on occasion" an "improvement" is brought about, the classification is much more difficult.

A look at the current developments in the so-called St. Nicholas jurisprudence (Nikolaus-Rechtsprechung) in Germany is an example of the fact that a case-by-case consideration is required in this regard, taking into account all circumstances that are relevant for the respective individual case. In its famous decision of 6 December 2005, the Federal Constitutional Court stated that under certain

[1] Fuchs et al. (2002); Runkel and Heinemann (2010), p. 211.

T. M. Spranger (✉)
Faculty of Law, University of Bonn, Bonn, Germany

Rittershaus Rechtsanwälte, Mannheim, Germany
e-mail: spranger@jura.uni-bonn.de

H.-G. Dederer, D. Hamburger (eds.), *Brain Organoids in Research and Therapy*, Advances in Neuroethics, https://doi.org/10.1007/978-3-030-97641-5_11

conditions, policyholders have a claim against the statutory health insurance for the assumption of treatment costs even if the treatment method in question is not established:

> It is not compatible with Art. 2 (1) of the Basic Law ("Grundgesetz") in conjunction with the constitutional principle of the welfare state to subject the individual (...) to compulsory insurance in the statutory health insurance and to legally promise the necessary treatment of illnesses for his contributions, which are based on his economic capacity, but on the other hand, if he suffers from a life-threatening or even regularly fatal illness for which there are no conventional medical methods of treatment, to exclude him from the provision of a certain method of treatment by the health insurance and to refer him to financing of the treatment outside the statutory health insurance. In this case, however, the other treatment method chosen by the insured person must promise a not entirely unlikely prospect of cure or at least a noticeable positive effect on the course of the disease, based on circumstantial evidence. (…) (A non-acceptance of the payments) in the extreme situation of a life-threatening illness is also not compatible with the state's duty to protect life under Article 2 (2) sentence 1 of the Basic Law. If the state takes responsibility for the life and physical integrity of the insured person with the system of statutory health insurance, preventive care in cases of a life-threatening or regularly fatal illness belongs, under the conditions mentioned, to the core area of the obligation to provide benefits and the minimum care required by Article 2 (2) sentence 1 of the Basic Law (...). In such cases, the social courts called upon by the insured person in the case of dispute have to examine, if necessary with expert assistance, whether there are serious indications for the treatment undertaken or intended by the doctor after conscientious professional assessment of a not entirely remote success of the cure or even only of a noticeable positive effect on the course of the disease in the concrete individual case (...).[2]

In consequence of the Nikolaus decision, an extraordinarily lively case law[3] has unfolded, which, surprisingly, in an extremely large number of cases concerns a constellation that hardly anyone would have previously attributed to a "danger to life due to illness". This refers to liposuction, i.e., a procedure in which fat cells are removed from certain areas under the skin using cannulas. Although liposuction is usually performed as a cosmetic operation, it is also increasingly performed for illness-related reasons in the case of the so-called lipedema. The borderline is therefore extremely difficult to draw and only becomes easier in the present example when liposuctioned fat is used for "modelling" the body by means of fat transfers.[4]

The aforementioned definition of "enhancement" is therefore extremely helpful, but must not lead to the false assumption that a generalised or schematic view of the topic is possible. Having mentioned this, the discussion on the so-called neuroenhancement will be presented in the following, before a specific examination of the challenges of enhancement through brain organoids takes place on this basis.

---

[2] Federal Constitutional Court (2005), p. 25 et seq.

[3] Deister (2016), p. 337; Deister (2017), p. 61; Eichberger (2019), p. 217; Kunte and Kostroman (2014), p. 610.

[4] The "Brazilian butt lift", for example, but also the modelling of the so-called washboard bellies in men are well known and have received a lot of media attention. On this: Tiryaki (2016).

11.2 Neuroenhancement

In the course of the general "boom" in neuroscience,[5] the so-called neuroenhancement also became the focus of interest about ten years ago. It has been known for some time that many psychotropic substances lead to an enhancement of cognitive abilities not only in sick people but potentially also in healthy users. However, the formerly massive side effects have been significantly reduced in the meantime,[6] which has led to considerable, medically non-indicated use.[7] In this respect, it is sometimes argued that neuroenhancement always takes place in a non-medical context.[8] As the questions of differentiation presented in the introduction should have shown, such a clear-cut classification is in fact not possible.

Two of the best known (and most controversial) psychotropic drugs that directly affect the neurological system are methylphenidates (brand names: Ritalin, Medikinet) and Modafinil (Vigil). While Ritalin is mainly used for attention deficit hyperactivity disorder (ADHD), Modafinil (Vigil) is a drug for the treatment of narcolepsy and belongs to a group of psychostimulant drugs. Both methylphenidates and Modafinil (Vigil) are increasingly propagated resp. misused as "brain doping" or "brain boosters" due to their wakefulness-keeping and concentration-promoting effect. Consumers also expect to improve their cognitive performance by taking them before exams or at work. The sales of the corresponding preparations have increased more than tenfold within a few years, so that there is a certain indication for a considerable amount of off-label use and, in the case of purchase via internet sources, also considerable use without any prescription.

11.2.1 Constitutional Framework

With regard to neuroenhancement, legal problems arise both at the level of the constitutional order and in various contexts of statutory law. To understand the constitutional implications, it is crucial to first become aware of the different dimensions of fundamental rights. While the fundamental rights guarantees (dignity, life, science, profession, property, etc.) laid down in the Basic Law are conceived in their "classical" function as defensive rights and thus aim to prohibit the state from violating individual positions, objective value decisions can also be inferred from fundamental rights. This objective effect of fundamental rights means that the state powers—legislative, judiciary and executive—must always act "in the light of fundamental rights" when exercising their activities. In this context, it may also be the case that the state's (for example, the court's) assessment of a matter concerns the dispute between private individuals and that this private legal relationship must also

[5] On this: Spranger (2009a, b), p. 1033 et seq.

[6] Kipper (2010), p. 189.

[7] See Sect. 11.2.2.

[8] So, for example: Kern (2019), § 6 recital 23.

be settled "in the light of fundamental rights"; this is then referred to as the "indirect horizontal effect" ("mittelbare Drittwirkung") of fundamental rights.

With regard to neuroenhancement, it is therefore true that the state, when dealing with distortions of competition arising from neuroenhancement, for example, must act against the backdrop of the principle of equal treatment laid down in Article 3 (1) of the German Basic Law (GG). However, it is difficult to draw the line between this and other forms of "enhancement" of cognitive abilities. To put it exaggeratedly: Why should the state regulate or even ban neuroenhancers if strong coffee, nicotine, guaraná products, energy drinks, meditation or methods with a placebo effect, or the so-called brain boosters remain completely unregulated before exams or in other contexts?[9] In the context of an examination situation, for example, the performance of the test as such is not affected by the use of the substances mentioned in all of the above-mentioned examples. Rather, it is exclusively a matter of compensating for upstream or accompanying deficits that are usually not included in the examination assessment. (Neuro-)enhancers therefore do not impart any higher insights and certainly no additional knowledge, but at best improve the recall or presentation of knowledge acquired elsewhere. Wisdom, reasoning and judgement thus remain unimpaired. The scope for state intervention is reduced not insignificantly by this circumstance.

However, certain special features arise in oral examinations: here, the form of verbal presentation as well as aspects of appearance, quick-wittedness and responsiveness as "key qualifications" are recognised as playing a not insignificant role and also significantly influencing the assessment, so that from a constitutional point of view, there would probably be a sufficient starting point for corresponding regulations. Nevertheless, the problem of differentiation from performance enhancement by other, culturally more accepted substances remains. The declaratory approach of describing energy drinks, etc. as more or less unproblematic "softhancers"[10] is in this respect merely of a semantic nature and cannot convince in this generality as an empirically unsupported demarcation.

Furthermore, the question of the detectability of a corresponding intake (with special consideration of examination candidates who have a medical indication for the corresponding medication) would have to be clarified. In this respect, there are not inconsiderable difficulties: Such tests would have to take place immediately before the corresponding examinations and would have to be measured not only against the standard of general personal right or—with regard to the generated data—against the right to informational self-determination, but—in the case of the necessity of a physical intervention (blood or hair sample)—also against the right to physical integrity. In addition, as in competitive sports, there would probably be evasion scenarios that would raise the question of the efficiency of an examination-related "doping test".

---

[9] On this point of view, which on closer examination is not so easy to invalidate, also: Schiess Rütimann (2016), p. 191.

[10] Like that: Scientific Services of the German Bundestag (2018).

A more viable approach for regulating neuroenhancement thus turns out to be another dimension of fundamental rights, which is addressed under the umbrella term of the "state's duty to protect". According to settled case law of the Federal Constitutional Court, the state must not only refrain from any interference with the rights of its citizens, but must rather "actively protect these rights". Such a duty to protect is recognised in particular with regard to the right to life and physical integrity under Article 2 (2) of the Basic Law. If—which seems to be disputed at present[11]—the non-indicated use of neuroenhancers entails health risks or has an increased addiction potential, a practical starting point for state regulation from the point of view of a state duty to protect would undoubtedly exist.[12] It seems worth mentioning here that the health risks with regard to potential third party dangers also include possible increases in aggression or forms of reduced ability to control (overestimation of self, development of manias or psychoses).

11.2.2 Considerations at the Sub-constitutional Level

At the level of statutory law, the legal findings are comparatively sparse: the instruments of narcotics and medicinal products law are only capable of normatively limiting psychopharmacological "improvement" to a minimum degree. The decisive factor here is the fact that the meaning and purpose of these regulations point in very specific directions: Methylphenidate is listed in Annex III to Sec. 1 (1) of the German Narcotics Act ("Betäubungsmittelgesetz", BtMG) and is therefore considered a marketable and prescribable narcotic. Modafinil (Vigil) is part of Annex 1 to Sec. 1 (1) of the Regulation on the Prescription of Medicinal Products ("Verordnung über die Verschreibungspflicht von Arzneimitteln", AMVV) and may therefore only be dispensed in the presence of a medical or dental prescription. Both the BtMG and the AMVV have an effect on the question of the marketability of a product, but do not say anything about how actions are to be evaluated that are carried out while taking preparations that were obtained in disregard of the corresponding restrictions. This clarification is particularly important because the institutionally hardly controllable procurement of corresponding active substances via the internet—also taking into account the distribution of placebos—apparently "works". The law on medicinal products and narcotics is therefore de lege lata hardly suited to absorb the broad effect of neuroenhancement that is of interest here. Legislative intervention is required here in the event that specific neuroenhancement substances are developed in which a clear performance-enhancing effect is accompanied by the absence of adverse effects for the healthy organism.[13]

However, study and examination regulations have not yet been designed to counteract the use of (possibly) performance-enhancing substances.[14]

---

[11] Scientific Services of the German Bundestag (2018).

[12] Lindner (2010), p. 467.

[13] Volkmer (2019), Preliminary remarks on the AntiDoping Act, recital 18.

[14] Schiess Rütimann (2016), p. 183; Bublitz (2010), p. 306 et seq.

Possible effects at the workplace—for example, in the form of indirect pressure on employees to use such drugs to improve their performance—are at first glance as difficult to grasp as in the context of examinations. The dimension that the problem of "brain doping" has reached in the meantime is made clear by the current relevant DAK health report from 2015, according to which it must be assumed, including the number of unreported cases, that 12% of employees in Germany take prescription drugs to increase their performance at work.[15] The permanent nature of many employment relationships—in contrast to the more selective or temporary exam situations—makes it seem necessary for the legislator to take action if there is a risk of health disadvantages for the person concerned. The circumstance of a possible distortion of competition in the workplace, on the other hand, is hardly suitable to justify a legislative duty to act, for the reasons already mentioned.

If, against this background, the legislature were to decide in favour of sectoral or more comprehensive regulation, it would have to act—as already explained—in the light of fundamental rights. In doing so, in addition to the aforementioned guarantees in Art. 2 (2) and Art. 3 (1) of the Basic Law, it would also have to include in its decision-making, for example, the freedom of occupation (Art. 12 (1) of the Basic Law) of the manufacturing companies in particular or the freedom of research (Art. 5 (3)(1) of the Basic Law) of the neuroscientists working in this field. In this context, any form of (self-)regulation should take into account three fundamental areas of conflicts in the neuroenhancement discussion: the risk of possible steering of the discussion by industry interests,[16] the problem of partly false and suggestive citation of scientific studies on consumption behaviour even within the scientific expert discussion,[17] and the danger of the self-fulfilling prophecy of increased consumption of corresponding preparations due to increased media reporting.[18]

11.3 Specifics of Neuroenhancement Using Brain Organoids

The specific challenges of research on and with brain organoids are scarcely legally established[19] and characterised mainly by bioethics.[20] Questions as diverse as a possible consciousness of brain organoids,[21] the creation of human–animal chimeras,[22] patients' rights,[23] the brain death criterion[24] or the possible reduction of animal

[15] DAK (2015).

[16] Lieb (2010).

[17] Schleim (2010), p. 182 et seq.

[18] DAK (2009), p. 86.

[19] Lavazza and Pizzetti (2020); initial legal considerations can also be found at Molnár-Gábor (2020), p. 237 et seq.; Taupitz (2020), p. 212 et seq.

[20] Besides, this corresponds to the classical role of bioethics; for that: Spranger (2010), p. 31 et seq.

[21] Kaulen (2018), Lavazza and Pizzetti (2020), and Koplin and Savulescu (2019), p. 760.

[22] Chen et al. (2019), p. 462 et seq.; Loike (2018).

[23] Farahany et al. (2018), p. 429 et seq.

[24] Id.

experiments[25] are discussed. If and insofar as the term enhancement is used in the respective publications, it is only in relation to the possible improvement of animal brains by human brain organoids.[26] Thus, the question of possible enhancement through the use of brain organoids has not yet been visibly addressed. On the one hand, this silence could be due to the fact that no enhancement potential whatsoever can be attributed to organoid technology. On the other hand, there is also the theoretical possibility that there are no specifics with regard to brain organoids that would give cause for discussion beyond the general neuroenhancement debate. In detail:

The transplantation of organoids or organoid-derived cells in cell replacement and regenerative therapy is already being discussed as a future clinical application.[27] Transplants from the patient's own (autologous) as well as from foreign (allogenic) material are conceivable. Through the additional use of genetic engineering methods, disease-causing mutations could also be corrected in this respect in order to differentiate healthy organoids for transplantation.[28] The mere use in cell replacement and regenerative therapy thus already raises the questions of differentiation yet described in the introduction with regard to medically indicated therapy on the one hand and non-indicated enhancement or therapeutic overcompensation on the other. The combination of organoid technology with genetic engineering methods further increases the options for applications that could be subsumed under the term enhancement. It can thus be considered certain that brain organoids and their fields of application must also be discussed with a view to possible enhancement.

However, it is more difficult to answer the question of whether the conceivable risks here differ from those that determine the general (neuro-)enhancement discussion. Unlike psychopharmacological neuroenhancement, enhancement by means of brain organoids would be characterised by the fact that no chemical substances act "from the outside", but that more or less integral components of the human body then perform "improving" functions. It seems that such enhancement would also surpass the quality of the areas of application of bionic components.[29] Comparable uses, however, would be those that would be possible in the field of human genetics, especially with the use of genome editing technologies.

As an interim result, it can be stated that enhancement by means of brain organoids is possible, but currently no unique feature of such a form of enhancement is recognisable. From a normative point of view, such a unique feature also does not result from the fact that the brain—as is sufficiently known from the brain death debate—is "not an organ like any other", not only in the perception of most people. From a legal policy point of view, it would nevertheless be conceivable for the

[25] Bredenoord et al. (2017); see also: Tierversuche verstehen – Eine Informationsinitiative der Wissenschaft (2021).

[26] Chen et al. (2019), p. 462 et seq.; see also Schicktanz (2020), p. 201.

[27] Bartfeld (2020), p. 16 et seq.

[28] Id.

[29] About this: Spranger (2009a), p. 206 et seq.

legislature to make use of the leeway available to it and regulate or even prohibit the "optimising" use of the corresponding technology.

The constitutional parameters already mentioned in the introduction would be of particular relevance in this respect. The risks associated with organoid transplantation would have to be considered above all in the light of the state's duty to protect under Article 2 (2)(1) of the Basic Law. Depending on the focus of the debate, however, the duty to protect human dignity (Article 1(1) of the Basic Law) should also be taken into account. State restrictions planned on this basis would then have to be harmonised with conflicting fundamental rights, since the desire for "self-optimisation" also enjoys constitutional protection in the first instance, as does the implementation of this desire by the corresponding professional groups. Relevant here would be, above all, the general right of personality (Art. 2 (1) in conjunction with Art. 1 (1) of the Basic Law),[30] freedom of occupation (Art. 12 (1) of the Basic Law), freedom of science (Art. 5 (3) (1) of the Basic Law), general freedom of action (Art. 2 (1) of the Basic Law) and the equal treatment clause (Art. 3 (1) of the Basic Law).

Should there be a need on the part of the state to intervene in a regulatory manner due to the identified enhancement potentials, the aforementioned constitutional parameters would have to be brought to an appropriate balance. The necessity of such a balance does not mean that a state ban would be out of the question. Rather, even a ban implemented under criminal law can be "softly formulated", for example by means of exceptions. Whether a criminal prohibition of neuroenhancement by using brain organoids to protect a "right to mental self-determination" could be implemented in the Criminal Code[31] seems questionable, not so much with regard to the objective of such a request,[32] but rather from the point of view of legislative technique: The technical background, the potential areas of application, the diversity of the rights and interests affected, but also the linkage with other sub-areas of law[33] would, in the case of legislative action, speak more in favour of drafting a specific "anti-enhancement law". Certain parallels can be drawn here with sports doping, which has been regulated since 2015 by the Act against Doping in Sport (German Anti-Doping Act, "Anti-Dopinggesetz"—AntiDopG).[34]

The Anti-Doping Act explicitly serves to tackle the use of doping substances and doping methods in sport in order to protect the health of athletes, to ensure fairness and equal opportunities in sporting competitions and thus contributes to maintaining

[30] Albers (2016), p. 195 et seq.

[31] In general about neuroenhancement already Merkel (2009), pp. 950–951. Critically on this Volkmer (2019), Preliminary remarks on the AntiDoping Act, recital 18. Further: Beck (2016), p. 117 et seq.

[32] The question of whether such a procedure would protect the right described or, conversely, violate it, does not need to be examined further in this context.

[33] In this context data protection law is particularly relevant.

[34] From December 10, 2015 (BGBl. I p. 2210), last amended by Article 1 of the Regulation of July 3, 2020 (BGBl. I p. 1547).

the integrity of sport.[35] It goes without saying that the rules of the Anti-Doping Act cannot be directly transferred to the constellation of neuroenhancement that is of interest in this case, so that fundamental objections against the Anti-Doping Act,[36] that have been raised in the literature, do not need to be discussed further in this case. What is decisive for drawing a parallel here is rather the fact that in the case of an "anti-enhancement law", the intention of the legislator is likely to be comparable. Here, as there, the health of those affected and aspects of fairness and equal opportunities are at the centre of interest. Whether it would also be necessary—analogous to the integrity of sport[37]—to address "mental integrity" would ultimately be left to the assessment of the legislator. In general, it may be pointed out that the use of established or legally tangible categories always benefits the practical implementation of laws, so that conversely, too vague constructs should be avoided.

Conversely, the inadequacies that experience has shown to be associated with such a ban should be viewed critically. Apart from the differentiation between therapeutic application on the one hand and applications for the purpose of enhancement on the other, which must be guaranteed, problems of proof would arise despite the invasiveness of the necessary interventions.[38] Above all, however, it must be warned against overestimating the behavioural control effect of legal regulations. Even in the case of a sanctioned ban of the respective techniques, a sufficiently large incentive among potential interested parties leads to the development of evasion or circumvention strategies. Of course, such phenomena are just as little opposed to the enactment of legal obligations as they are to legal policy making.

11.4 Summary

Every form of enhancement requires an interdisciplinary approach: in addition to dealing with the scientific-medical foundations (and possibilities), there is also a debate on ethical legitimacy (for example, with regard to aspects of equal opportunities, personal authenticity or the moral status of the human condition). To a considerable extent, though, the law is also called upon, which—with the exception of a few prominent areas such as sports doping—has failed to fulfil its function in this respect so far.

The so-called neuroenhancement raises not only medical and ethical but also numerous legal problems. While any distortions of competition—for example, at school, at universities or at the workplace—are difficult to grasp from a legal point of view, possible health risks for the user represent a more suitable starting point for state measures. So far, the legislator has not prepared any specific regulations in this area, also against the background of the uncertain empirical starting position. If the

[35] § 1 AntiDopG.

[36] About that: Heger (2018), p. 61 et seq.; Lutz (2016), p. 21 et seq.; Winkler (2019), p. 14 et seq.

[37] About that: Dittrich (2017), p. 189 et seq.; Jansen (2017), p. 600 et seq.

[38] In this way about the psychopharmacological enhancement: Schiess Rütimann (2016), p. 199 et seq.

legislature wants to take action here, the various fundamental rights affected (Art. 2 (1) GG in conjunction with Art. 1 (1) of the Basic Law, Art. 12 (1) of the Basic Law, Art. 5 (3)(1) of the Basic Law, Art. 3 (1) of the Basic Law, Art. 2 (1) of the Basic Law) must be brought to an appropriate balance. In principle, it would also be possible to prohibit neuroenhancement in general or specifically in the form of the use of brain organoids. In this respect, the Anti-Doping Act could serve as a regulatory model. In the case of such a legal regulation, not inconsiderable challenges for law enforcement would have be to be expected. However, even a law suffering from certain enforcement deficits would still have a signal effect in terms of legal policy and society.

References

Albers M (2016) Grundrechtsschutz und Innovationserfordernisse angesichts neuartiger Einblicke und Eingriffe in das Gehirn. In: Lindner JF (ed) Die neuronale Selbstbestimmung des Menschen. Grundlagen und Gefährdungen. Nomos, Baden-Baden, pp 63–97

Bartfeld S (2020) Kernaussagen und Handlungsempfehlungen zu Organoiden. In: Bartfeld S et al (eds) Organoide. Ihre Bedeutung für Forschung, Medizin und Gesellschaft. Nomos, Baden-Baden, pp 13–28

Beck S (2016) Die neuronale Selbstbestimmung als Thema des materiellen Strafrechts. In: Lindner JF (ed) Die neuronale Selbstbestimmung des Menschen. Grundlagen und Gefährdungen. Nomos, Baden-Baden, pp 117–156

Bredenoord AL, Clevers H, Knoblich JA (2017) Human tissues in a dish: the research and ethical implications of organoid technology. Science 355(6322):9414. https://doi.org/10.1126/science.aaf9414

Bublitz JC (2010) Doping-Kontrollen im Staatsexamen? Über leistungssteigernde Substanzen und Chancengleichheit in Prüfungen. Z Juristische Stud 3:306–317

Chen HI et al (2019) Transplantation of human brain organoids: revisiting the science and ethics of brain Chiemras. Cell Stem Cell 25(4):462–472

DAK (2009) Health report 2009. https://www.iges.com/e6/e1621/e10211/e6061/e6064/e6199/e9682/e9684/attr_objs9687/DAK_Gesundheitsreport_2009_ger.pdf. Accessed 21 Jan 2021

DAK (2015) Health report 2015. https://www.dak.de/dak/bundesthemen/gesundheitsreport-2015-2109048.html#/. Accessed 19 Jan 2021

Deister S (2016) Das Potential einer erforderlichen Behandlungsalternative als zentrale Voraussetzung der Anwendbarkeit von Methoden im Krankenhaus. Neue Z Soz 25: 328–337

Deister S (2017) Potential und Wirtschaftlichkeit bei der Krankenhausbehandlung. Kranken Pflegeversicherung 69:59–65

Dittrich J (2017) Die Integrität des sports – Schutzzweck oder Gesetzesziel? Z Wirtschaftsstrafrecht Haftung Unternehmen 7:189–196

Eichberger R (2019) Liposuktion – Ein Dauerbrenner in der Sozialgerichtsbarkeit. Die Soziagerichtsbarkeit 66:214–219

Farahany et al (2018) The ethics of experimenting with human brain tissue. Nature 556:429–432

Federal Constitutional Court (2005) Resolution of the first senate of 6 December 2005 - 1 BvR 347/9. NJW 59:891–894

Fuchs M, Lanzerath D et al (2002) DRZE-Sachstandsbericht 1: enhancement. Die ethische Diskussion über biomedizinische Verbesserungen des Menschen. Dt. Referenzzentrum f. Ethik in den Biowissenschaften, Bonn

Heger M (2018) Das Strafrecht als scharfes Schwert bei der Sanktionierung von doping? Vorgänge 223:61–73

Jansen S (2017) Der Schutz der "Integrität des Sports" durch das Strafrecht? Goltdammer's Arch Strafrecht 164:600–614

Kaulen H (2018) Braucht dieses Hirn einen Vormund? In: FAZ.NET. https://www.faz.net/aktuell/wissen/medizin-ernaehrung/ethik-in-der-forschung-braucht-dieses-hirn-einen-vormund-15600821.html. Accessed 21 Jan 2021

Kern BR (2019) § 6 Die Elemente der Rechtfertigung ärztlichen Handelns. In: Laufs A et al (eds) Handbuch des Arztrechts, 5th edn. CH Beck, München, p 23

Kipper J (2010) Hirnforschung. In: Fuchs M et al (eds) Forschungsethik. Eine Einführung. JB Metzler, Heidelberg, pp 176–195

Koplin JK, Savulescu J (2019) Moral limits of brain organoid research. J Law Med Ethics 47(4):760–767

Kunte A, Kostroman A (2014) Das Qualitäts- und Wissenschaftsgebot am Beispiel der Liposuktion bei Lip- und Lymphödem. Die Soziagerichtsbarkeit 61:607–612

Lavazza A, Pizzetti FG (2020) Human cerebral organoids as a new legal and ethical challenge. J Law Biosci 7:5. https://doi.org/10.1093/jlb/lsaa005

Lieb K (2010) Was ist dran am Hirndoping? FAZ 158:27

Lindner JF (2010) Neuro-Enhancement als Grundrechtsproblem. Z Gesamte Medizinrecht 28:463–471

Loike JD (2018) Opinion: Should human-animal chimeras be granted "personhood"? In: The Scientist. https://www.the-scientist.com/news-opinion/opinion-should-human-animal-chimeras-be-granted-personhood-36664. Accessed 21 Jan 2021

Lutz J (2016) Die Kriminalisierung des Sports - Anti-Doping-Maßnahmen des Strafrechts und der Sportverbände im Vergleich. Onlinezeitschrift Höchstrichterliche Rechtsprechung Strafrecht 17:21–28

Merkel R (2009) Neuartige Eingriffe ins Gehirn. Verbesserung der mentalen condicio humana und strafrechtliche Grenzen. Z Gesamte Strafrechtswissenschaft 121:919–953

Molnár-Gábor F (2020) Organoide: Ein Fall für den Datenschutz? In: Bartfeld S et al (eds) Organoide. Ihre Bedeutung für Forschung, Medizin und Gesellschaft. Nomos, Baden-Baden, pp 237–256

Runkel T, Heinemann T (2010) Enhancement. In: Fuchs M et al (eds) Forschungsethik. Eine Einführung. JB Metzler, Heidelberg, pp 211–224

Schicktanz S (2020) Sind menschliche zerebrale Organoide moralisch schützenswert? Ein kommentierter Überblick über die aktuelle internationale Ethikdiskussion. In: Bartfeld S et al (eds) Organoide. Ihre Bedeutung für Forschung, Medizin und Gesellschaft. Nomos, Baden-Baden, pp 190–211

Schiess Rütimann PM (2016) Neuroenhancement an Schulen und Universitäten. Bemerkungen zum Bildungsrecht in Deutschland und der Schweiz. In: Lindner JF (ed) Die neuronale Selbstbestimmung des Menschen. Grundlagen und Gefährdungen. Nomos, Baden-Baden, pp 169–213

Schleim S (2010) Cognitive enhancement – Sechs Gründe dagegen. In: Fink H, Rosenzweig R (eds) Künstliche Sinne, gedoptes Gehirn. Mentis, Paderborn, pp 179–207

Scientific Services of the German Bundestag (2018) Informationen zu neuroenhancement. WD 9 – 3000 – 060/18. https://www.bundestag.de/resource/blob/572364/c62698c6de2b0c74477a44c-c24a83c97/WD-9-060-18-pdf-data.pdf. Accessed 21 Jan 2021

Spranger TM (2009a) Neuroprothetik und bildgebende Hirnforschung – neue Impulse für die Praxis des Betreuungsrechts. Betreuungsmanagement 5:206–208

Spranger TM (2009b) Der Einsatz neurowissenschaftlicher Instrumente im Lichte der Grundrechtsordnung. Juristen Zeitung 64(21):1033–1040

Spranger TM (2010) Recht und Bioethik. Mohr Siebeck, Tübingen

Taupitz J (2020) Organoide: Die deutsche Rechtslage. In: Bartfeld S et al (eds) Organoide. Ihre Bedeutung für Forschung, Medizin und Gesellschaft. Nomos, Baden-Baden, pp 212–236

Tierversuche verstehen – Eine Informationsinitiative der Wissenschaft (2021) Organide können Tierversuche nicht komplett ersetzen. https://www.tierversuche-verstehen.de/organoide-koennen-tierversuche-nicht-komplett-ersetzen/. Accessed 19 Jan 2021

Tiryaki KT (ed) (2016) Inverse abdominoplasty. An illustrated guide. Springer, Heidelberg
Volkmer M (2019) Vorbemerkungen zum AntiDopG. In: Körner H, Patzak J et al (eds) Betäubungsmittelgesetz, 9th edn. CH Beck, München, p 18
Winkler N (2019) Doping-Strafrecht als "Allheilmittel" für einen sauberen sport? Hamburger Rechtsnotizen 9:14–20

Part VI

The Harmonization Problem

12 Global Harmonization of Legal Standards for Brain Organoid Research and Therapy?

Silja Voeneky

12.1 Introduction

Let me start on a personal note[1]. Back in 2001, I had my first experience with organoids and organoid research as a young legal scholar at the Max Planck Institute for Comparative Public Law and International Law in Heidelberg. At this time, a heated discussion about embryonic stem cell research and human cloning took place, but also much hope was part of the exchange for what could be achieved by cultivating stem cells and growing organoids. Biotech companies, startups at the time, presented their future products at a kind of open marketplace in Heidelberg. I remember one CEO who explained how to grow and produce organs out of adult stem cells. When asked how long it would take to produce a kidney or liver, he answered

[1] This article is based on my earlier proposals of how to govern and regulate research and (high-risk) emerging technologies in ethically disputed areas on the basis of human and constitutional rights, especially Vöneky (2018a); Voeneky (2018b), pp. 151–160, concerning the impact of human dignity, cf. Wolfrum and Vöneky (2004) and with regard to the need for a framework for ethic commissions, cf. Vöneky (2010). For an overview of the European legislation governing organoids, cf. Faltus (2021). For the specific questions of printed organs, cf. Mihalyi and Müller (2016). I am very grateful for the valuable comments and insights into this field by Dr. med. Philipp Kellmeyer, neurologist and neuroethicist, and FRIAS Research Fellow, and the philosopher Andrea Lavazza (Arezzo) and for the support by Silke Weller, Tobias Crone, and Laura Tribess as researchers of my team.

S. Voeneky (✉)
Albert-Ludwigs-Universität Freiburg, Institut für Öffentliches Recht Abt. 2 (Völkerrecht und Rechtsvergleichung), Freiburg, Germany
e-mail: silja.voeneky@frias.uni-freiburg.de

H.-G. Dederer, D. Hamburger (eds.), *Brain Organoids in Research and Therapy*, Advances in Neuroethics, https://doi.org/10.1007/978-3-030-97641-5_12

that he hoped to achieve this in up to 5 years. Today, we know that this was a much too optimistic view since science and development often move much slower than we expect or hope, and there might be a bias for those involved in the development of biomedical products, especially if they want to sell products or emerging technology, to oversell the prospects of future achievements.

Nevertheless, during the beginning of the 2000s, the debate about the regulation of this area of research did speed up. Concerning stem cell research, the German Stem Cell Act[2] was enacted in 2002. For many, this law is still an example of national overregulation and hardly justifiable limitations in this area of research; for others, it aims to protect the human dignity of human embryos and is therefore justified by the specific interpretation of the concept of human dignity[3] as part of the German Basic Law.[4]

During that period, there was also an initiative for the regulation of research involving human embryos on the international plane, which Germany promoted. The aim was to negotiate an international treaty against the reproductive cloning of human beings. This initiative's outcome was a soft and unclear resolution by the United Nations General Assembly (UNGA).[5] It calls for all UN Member States to adopt a ban on human cloning, which is "*incompatible with human dignity* and the protection of human life."[6] The drawback of this resolution is that it remains unclear which cases of cloning are prohibited because they are incompatible with human dignity. Until now, it has proven impossible to achieve a meaningful consensus and to bridge the gap between States fearing overregulation as a limit to therapeutic cloning and those concerned by the lack of a prohibition of reproductive cloning. Hence, the resolution is an example of a failed initiative of global harmonization.

One may ask why the last example is relevant concerning the problem of the future governance of human brain organoids. It is relevant as there is comparable uncertainty in the field of brain organoids with regard to the scientific and empirical basis of current and future research. This is the reason for major dissent about the ethical foundation and even the need for regulation: on the one hand, it is a field of basic and applied research linked to research aims that can be seen as highly

[2] Gesetz zur Sicherstellung des Embryonenschutzes im Zusammenhang mit Einfuhr und Verwendung menschlicher embryonaler Stammzellen (Stammzellgesetz) (28 June 2002) BGBl. 2002 I 42.

[3] For different views concerning the interpretation of Article 1 German Basic Law, cf. the articles by Starck Ch, Ipsen J, Vitzthum W and Zypries B in Vöneky and Wolfrum (2004), pp. 63–125; Herdegen (2020), paras 33 et seq. For similar and other legal, ethical, and religious approaches from an interdisciplinary perspective, including the Islamic, Buddhist, and Jewish perspectives, cf. Vöneky and Wolfrum (2004), pp. 3–55.

[4] Grundgesetz für die Bundesrepublik Deutschland (23 May 1949) BGBl. 1949 I 1, BGBl. 2020 I 2048 (German Basic Law).

[5] The UN Declaration on Human Cloning (2005) was adopted by a vote of 84 States in favor (including Austria, Germany, Italy, Poland, Switzerland, and the USA) to 34 against (including China, France, Japan, the UK, Spain, Sweden, the Netherlands, New Zealand, and India) with 37 abstentions (including Iran, Israel, and Turkey).

[6] Emphasis added.

justified, as organoid research serves eminent objectives such as the understanding of the development of diseases, as Parkinson's disease, or reducing the number of animal experiments.[7] On the other hand, some scholars argue that there is the possibility of creating sentient entities of human origin that could have a moral status and should be treated accordingly:

> [I]f there's even a possibility of the organoid being sentient, we would be crossing that line. We don't want people doing research where there is potential for something to suffer.[8]

Before I discuss whether international law is providing a framework that might guide this kind of research some light shall be shed on the "brain in a vat" discussion. "The case of brains in a vat" has been discussed for many years, among others by the philosopher *Hilary Putnam*.[9] However, the starting point and also the problems to be solved with regard to the brains in a vat thought experiment are completely different from the questions to be answered in the actual human brain organoid cases. The starting point of the brain in a vat thought experiment, as a "science fiction possibility," is that a "human being (…) has been subject to an operation by an evil scientist," as *Putnam* writes, and the brain of this person is removed from his or her body and then kept alive in a vat of nutrients in such a manner that the nerve endings "have been connected to a super-scientific computer."[10] This causes the person to believe that everything is completely fine and it appears to him or her as if there are people, objects, etc., but that all these sensations or appearances are only the results of the electronic impulses of the computer. The aim of the thought experiment is, on the one hand, as *Putnam* writes, to discuss the classic problem of skepticism ("How do you know you are not in this predicament?")[11] and, on the other hand, to discuss issues of the mind/world relationship, especially if we assume that we imagine—as Putnam suggests—that all human beings are brains in a vat and we are all subject to a "*collective* hallucination."[12] The question is: "Could we, if we were brains in a vat in this way, *say* or *think* that we were?"[13] and *Putnam's* answer is:

> No, we could not. (...) [T]he supposition that we are actually brains in a vat, although it violates no physical law, and is perfectly consistent with everything we have experienced, cannot possibly be true. *It cannot possibly be true*, because it is, in a certain way, self-refuting.[14]

---

[7] Cf., for instance, the Max Planck researcher Hans Schöler as quoted in Rösch (2018), pp. 59–61.

[8] Statement of Ohayon (2019), Annual meeting of the Society for Neuroscience in Chicago, cited according to Sample, Scientists 'may have crossed ethical line' in growing human brains, The Guardian, 21 Oct 2019, https://www.theguardian.com/science/2019/oct/21/scientists-may-have-crossed-ethical-line-ingrowing-human-brains. Accessed 3 June 2022. Cf. as well Lavazza and Pizzetti (2020), p. 13; Schicktanz (2020).

[9] Putnam (1981), p. 6.

[10] Putnam (1981), p. 6.

[11] Putnam (1981), p. 6.

[12] Putnam (1981), p. 6.

[13] Putnam (1981), p. 7.

[14] Putnam (1981), p. 7.

These questions and Putnam's answers show that the brain in a vat thought experiment is not related to the questions of a normative framework of human brain organoid research that shall be discussed below, starting with rules and principles of international law.

12.2 International Law Framework

The first question addressed with regard to the governance framework concerns the tools and cornerstones of a possible global harmonization of legal standards. Governance shall be understood in this article as multilayered standard-setting by various actors such as private entities, States, and International Organizations. However, in this article I will not analyze lawmaking by private actors and private entities, such as the World Medical Association and its influential Declaration of Helsinki,[15] as this would be beyond the scope of this paper. I will instead stress the sphere of public international law. The rules that form the bases of international standard-setting are the rules laid down as part of international law *treaties* as binding and international soft law declarations as non-binding instruments.[16] However, there is neither a specific international treaty regulating brain organoids nor any international soft law declaration governing this field.[17] Additionally, there is neither a global harmonization of stem cell research nor a global harmonization of data protection law as part of binding international law either although these fields of regulation are highly relevant for research with regard to brain organoid research. Therefore, it is necessary to rely on general international human rights law.

12.2.1 Human Rights Treaties

This justifies an inquiry into the human rights framework as part of international law.[18] It can be seen that international legal human rights make it possible to spell

---

[15] WMA Declaration of Helsinki—Ethical Principles for Medical Research Involving Human Subjects, adopted by the 18th WMA General Assembly, June 1964; amended several times, as by the 64th WMA General Assembly, October 2013, cf. https://www.wma.net/policies-post/wma-declaration-of-helsinki-ethical-principles-for-medical-research-involving-human-subjects/.

[16] For international soft law rules, see Sect. 12.3.

[17] In the area of biomedical research, regional treaty law exists: the Council of Europe Convention on Human Rights and Biomedicine (adopted 04 Apr 1997, entered into force 01 Dec 1999) 2137 UNTS 171 is the only international legally binding instrument in the biomedical field. It is supplemented, *inter alia*, by an Additional Protocol concerning biomedical research (adopted 25 Jan 2005, entered into force 1 Sep 2007) 2494 UNTS 135—however, the latter one is ratified by only 5 States and the Convention itself is ratified by 29 States (and not by Germany). Hence, neither the Convention on Human Rights and Biomedicine nor its Protocols can be seen as a successful approach to achieve global harmonization in the regulated fields.

[18] For the human rights as the basis of a legitimate governance framework in other areas of research and emerging technologies, esp. gene drive research and high-risk technologies cf. in more detail Voeneky (2018a), pp. 134–137; Voeneky (2018b), pp. 151–160.

out the decisive values that have to be taken into account for assessing organoid research and therapy. The general human rights treaties do not bind directly private actors, as companies,[19] but oblige State parties at the global and regional level to respect, protect, and fulfill[20] human rights.[21] Thus, they contain not only negative obligations to refrain from disproportionate interference with human rights, but also obligations to protect (positive obligations) which impose duties to act on the Contracting States. The relevant treaties are, first and foremost, at the universal level, the 1966 International Covenant on Civil and Political Rights[22] (ICCPR) and the 1966 International Covenant on Social, Economic, and Cultural Rights[23] (ICESCR). These treaties and the existing regional human rights treaties, as the European Convention on Human Rights[24] (ECHR), include rights protecting the freedom of science, as the right of freedom of thought and the freedom of expression,[25] as well as the right to life and bodily integrity, the right to health, and the right to privacy.

First of all, the freedom of science and research is not only a justified (i.e. moral or ethical) value. It is also a legally binding human right. Although the freedom of science and research is not expressly mentioned in most relevant international human right treaties, it is the shared view that this freedom is protected as part of the right of freedom of thought and the freedom of expression in these treaties, as Articles 18 and 19 ICCPR and Articles 9 and 10 ECHR.[26] This is confirmed, *inter alia*, by decisions of the Human Rights Committee, the ICCPR treaty body. It stated in 2016 that a

> state party should carry out all necessary legal amendments to ensure that research may be carried out without state authorisation and fully respect, protect and promote academic freedoms.[27]

---

[19] This is stressed rightly inter alia by Monnheimer (2021), cf. pp. 13–29.

[20] For a critical review of these obligations, cf. Monnheimer (2021), p. 48 et seq.

[21] See for further discussion of positive obligations under the ECHR von Arnauld (2019), paras 665 et seq.

[22] International Covenant on Civil and Political Rights (adopted 16 Dec 1966, entered into force 23 Mar 1976) 999 UNTS 171.

[23] International Covenant on Economic, Social and Cultural Rights (adopted 16 Dec 1966, entered into force 06 Jan 1976) 993 UNTS 3.

[24] Convention for the Protection of Human Rights and Fundamental Freedoms (adopted 04 Nov 1950, entered into force 03 Sep 1953) 213 UNTS 221.

[25] For the broader context of the right to science as a human right, cf. Shaver (2018), p. 34 et seq.

[26] See further von Arnauld (2019), paras 703 et seq., 712 et seq. A different approach is taken by the Charter of Fundamental Rights of the European Union (26 Oct 2012) OJ 2012 C 326/391, Article 13 (Freedom of the Arts and Sciences). There it is expressly laid down that "[t]he arts and scientific research shall be free of constraint. Academic freedom shall be respected." Similar norms are included in national constitutions, see e.g. German Basic Law, Article 5 para. 3 which states that "[a]rts and sciences, research and teaching shall be free. The freedom of teaching shall not release any person from allegiance to the constitution."

[27] Hum. Rts. Comm (2016), p. 8.

As part of the ICESCR, the right to freedom of scientific research is enshrined in Article 15 para. 3, according to which "the States Parties [...] undertake to respect the freedom indispensable for scientific research." The scope of this provision was further clarified by the Committee on Economic, Social and Cultural Rights in 2020:

> In order to flourish and develop, science requires the robust protection of freedom of research. The Covenant establishes a specific duty for States to "respect the freedom indispensable for scientific research" (art. 15 (3)). This freedom includes, at the least, the following dimensions: protection of researchers from undue influence on their independent judgment; the possibility for researchers to set up autonomous research institutions and to define the aims and objectives of the research and the methods to be adopted; the freedom of researchers to freely and openly question the ethical value of certain projects and the right to withdraw from those projects if their conscience so dictates; the freedom of researchers to cooperate with other researchers, both nationally and internationally; and the sharing of scientific data and analysis with policymakers, and with the public wherever possible.

To protect the freedom of science and research as a human right does not mean that this freedom is absolute. According to international human rights law, legitimate aims can justify limitations of the right of freedom of science and research, provided that they are necessary and proportionate.[28] These legitimate aims are defined broadly as part of the human rights treaties (Articles 18 para. 3 and 19 para. 2 ICCPR; Articles 9 para. 2 and 10 para. 2 ECHR).[29] They include the interests of national security, territorial integrity or public safety, the prevention of disorder or crime, the protection of health or morals, the protection of others' reputation or rights, preventing the disclosure of information received in confidence, etc.[30] Hence, as long as States limit the freedom of science for those legitimate purposes, in a proportional way, this human right is not violated.

12.2.2 Human Dignity

Moreover, the principle of human dignity is a central one for a deontological approach for regulating brain organoids, and it is important for those scholars who

---

[28] Cf. von Arnauld (2019), paras 659 et seq.

[29] These are defined in much broader manner compared to Article 5 German Basic Law. German constitutional law allows limitations of the freedom of science and research only if there is the aim to protect high-ranking goods that are protected by the Basic law, as no express possibility to limit this freedom is mentioned in the wording of Article 5 para. 3 Basic Law (note 26), cf. Gärditz (2020), para. 151.

[30] Cf. the wording of Article 10 para. 2 ECHR: "The exercise of these freedoms, since it carries with it duties and responsibilities, may be subject to such formalities, conditions, restrictions or penalties as are prescribed by law and are necessary in a democratic society, in the interests of national security, territorial integrity or public safety, for the prevention of disorder or crime, for the protection of health or morals, for the protection of the reputation or rights of others, for preventing the disclosure of information received in confidence, or for maintaining the authority and impartiality of the judiciary."

want to deduce from the ethical debate arguments for a future global harmonization. From an international law view, however, it has to be noted that the decisive human rights treaties mentioned above—the ICCPR, the ICESCR, and foremost the ECHR—do not include a human dignity clause similar to Article 1 German Basic Law.[31] Only the non-binding Preambles to the two human rights Covenants mention the inherent dignity of the human person.[32] I have argued that there is an *opinio iuris* of States (i.e. one of the two constituent elements necessary for determining whether a practice constitutes customary international law)[33] that human dignity is a valid principle in international law, even though there is only very limited consensus about the scope or content of this principle in the area of biomedicine,[34] and I want to stress this in this article again. This means that the specific aspects of human dignity that are relevant in the field of brain organoid research cannot be deduced from the human rights treaties that exist today and currently are understood to protect human beings after they have been born. Therefore, for the governance of research and development of brain organoids, as a fast-moving field of research, a clearer understanding of the scope and content of human dignity should be developed by States and laid down in international soft law declarations.[35] This means that under the condition that, at some point in the future, there will be a consensus of States parties that the aim to develop a human brain in vitro that could feel pain would be a violation of human dignity, this can be laid down in a soft law declaration with reference to the principle of human dignity.

Besides, there is a "red line" for any medical research or treatment (biomedical or other) that is already laid down in Article 7 ICCPR and shall be discussed below with regard to the regulation of human brain organoids. It is closely linked to the principle of human dignity. Article 7 ICCPR reads:

> [...] In particular, no one shall be subjected without his free consent to medical or scientific experimentation.[36]

---

[31] This is different in EU law, where the Charter of Fundamental Rights of the European Union states (Article 1) that "Human dignity is inviolable. It must be respected and protected."

[32] Cf. first and second Recital of the Preamble to the ICCPR: "The States Parties to the present Covenant, Considering that [...] recognition of the inherent dignity [...] is the foundation of freedom, justice and peace in the world, recognizing that these rights derive from the inherent dignity of the human person [...]."

[33] The other element is the "State practice," i.e. a repeated behaviour of a relevant number of States, cf. von Arnauld (2019), paras 251 et seq.

[34] Cf. for the following Wolfrum and Vöneky (2004), pp. 133–143. Certainly, there is consensus that acts of torture and slavery constitute violations of human dignity, and most argue that the same is true for the reproductive cloning of human beings.

[35] Similar in the area of human cloning von Achenbach and Clados (2008), para. 31, stressing the potential of "extending existent or formulating further soft-law instruments" in order to "contribute to the gradual formation of international legal binding standards." Correctly Petersen (2020), para. 41, is mentioning the "dangerous tendency in the legal discourse to broaden th[e] narrow concept of dignity [...], ultimately diminish[ing] the authority of the concept."

[36] Vöneky (2018a), pp. 131–151.

This prohibition is to be seen as *ius cogens*[37] and must be obeyed and implemented without exception by all States. According to the current understanding, the notion of "no one" in Article 7 ICCPR refers to living human beings, i.e. in the case at hand, to the patients and donors involved in brain organoid research. However, if there will be at some point sufficient consensus by State parties, that "no one" can be interpreted broadly, including entities that are called brain organoids today, as, for instance, brain organoids are functionally equivalent to a human brain, and the human brain is seen as the essence of a human being, there would be the need to get informed consent by the brain organoid or by a kind of guardian, a representative for this entity.[38] This kind of informed consent could be laid down in a soft law declaration, too.

Such a dynamic interpretation of a norm of international law—here Article 7 ICCPR—is not per se excluded, as the interpretation of a notion in an international treaty is not limited to the meaning proposed during the drafting process.[39] In international law, the historical circumstances are only a supplementary means of interpretation.[40] State parties' practice can change the interpretation if there is consensus by the State parties about the new interpretation, and the interpretation is not overstepping the boundaries of the ordinary meaning of a notion.[41]

This does not mean that State parties of the international human rights instruments are free to decide what affects and violates human dignity. The UN Charter,[42] for instance, speaks of the "reaffirmation of the faith in the dignity" of the human person.[43] This shows that human dignity has a pre-normative character in international law. By engaging in a substantial dialogue—which should include scientists, scholars, and civil society—States can "elucidate" the scope and content of human dignity regarding the challenges that lie ahead of us in the field of research and use of brain organoids. This dialogue has to consider the very elements of the scope and notion of human dignity, especially whether and to what extent it covers (certain types of) brain organoids. In this respect, the scientific findings, the various cultural, religious, and ethical approaches have to be taken into account. Nevertheless, it is to be stated again that as long as there is no consensus of States, the mere concept of human dignity cannot be the basis for a prohibition of research or for State obligations and restrictions in the field of brain organoids.

Nevertheless, the protection of human dignity, understood also as the protection of the dignity of human brain organoids could be seen as a legitimate aim to restrict the freedom of science as a human right. This view could be deduced from the broad variety of legitimate purposes—mentioned above—as express justifications for

[37] See Frowein (2013), para. 7.

[38] Cf. Schicktanz (2020), pp. 202–203; cf. as well Farahany et al. (2018).

[39] On the interpretation of international treaties, see Articles 31–33 Vienna Convention on the Law of Treaties (VCLT) (adopted 12 May 1968, entered into force 27 Jan 1980) 1155 UNTS 331.

[40] Article 32 VCLT.

[41] Article 31 paras 1, 3 VCLT.

[42] Charter of the United Nations (adopted 26 June 1945, entered into force 24 Oct 1945) 1 UNTS 16.

[43] See second Recital of the Preamble to the UN Charter.

restrictions of the freedom of expression and enshrined in the human rights treaties. Thus, the "rights of others" that can justify to limit the freedom of expression could refer not only to international human rights but also, as in EU law, to human and constitutional rights as part of national law[44]—such as human dignity under Article 1 of the German Basic Law. This interpretation of "rights of others," however, is disputed; contrary to this view that *constitutional* rights might be seen as reasons to limit *international human* rights, the Human Rights Committee stated, concerning the right to freedom of opinion and expression as part of the ICCPR:

> The first of the legitimate grounds for restriction listed in paragraph 3 is that of respect for the rights or reputations of others. The term 'rights' includes human rights as recognised in the Covenant and more generally in *international* human rights law.[45]

12.2.3 Treaties in Other Fields of International Law: Cartagena Protocol

Another example of an international law treaty governing brain organoids if they constitute "genetically modified organisms" is the Cartagena Protocol on Biosafety.[46] This treaty aims to ensure the safe use of living modified organisms resulting from modern biotechnology and entered into force in 2003. It has been ratified by 173 Parties (not the US, however)[47] and governs all forms of biotechnology, including up to date technologies such as gene drives.[48] The Cartagena Protocol is not applicable to the modification of human beings but governs questions of genetic modification of organisms such as brain organoids.

The treaty's aim is expressly in line with the precautionary principle—as a legal or soft law principle.[49] The precautionary principle states, according to the version of the 1992 Rio Declaration on Environment and Development, that where

> there are threats of serious or irreversible damage, lack of full scientific certainty shall not be used as a reason for postponing cost-effective measures to prevent environmental degradation.[50]

---

[44] Grote and Wenzel (2013), para. 91.

[45] Hum. Rts. Comm (2011), para. 28. Emphasis added.

[46] Cartagena Protocol on Biosafety to the Convention on Biological Diversity (adopted 29 Jan 2000, entered into force 11 Sep 2003) 2226 UNTS 208.

[47] See Parties to the Cartagena Protocol and its Supplementary Protocol on Liability and Redress, Status of Ratification and Entry into Force. https://bch.cbd.int/protocol/parties/. Accessed 17 Mar 2021.

[48] For gene drive research and the international law framework, cf. Vöneky (2018a), pp. 131 et seq.

[49] Cf. Article 1 Cartagena Protocol stating that "[i]n accordance with the precautionary approach contained in Principle 15 of the Rio Declaration on Environment and Development, the objective of this Protocol is to contribute to ensuring an adequate level of protection in the field of the safe transfer, handling and use of living modified organisms resulting from modern biotechnology [...]."

[50] United Nations Conference on Environmental Development (1992), Rio Declaration on Environment and Development, Principle 15; cf. Proelß (2017), paras 24–48 (and forthcoming 2nd ed. 2022).

This "better safe than sorry" principle was traditionally limited to the aim of protecting the environment. But it is more and more linked to the protection of other values such as human rights if a State has to decide about action or inaction in a field where there is scientific uncertainty. Articles 10 para. 6 and 11 para. 8 Cartagena Protocol expressly state: "*Lack of scientific certainty due to insufficient relevant scientific information and knowledge* regarding the extent of the potential adverse effects of a living modified organism on the conservation and sustainable use of biological diversity [...]."[51] Hence, as there is much scientific uncertainty in the field of brain organoids, the application of this principle is in line with its object and purpose, even if the Cartagena Protocol is not applicable if, e.g. the brain organoid is—in a specific case—not a genetically modified organism.[52]

12.2.4 Merging the Precautionary Principle and Human Rights Law

Even if brain organoids cannot be seen as equivalent to a human being, it might be argued that they might possess a special status in the future, as it is argued by authors of this volume[53] because they might feel pain in a far or not so far future. To take into account this special status and the scientific uncertainty when or at which point the threshold is reached[54] (or if the threshold will be reached at all), one might propose that the precautionary principle[55] shall govern these cases and that human rights law and the precautionary principle have to be merged and interpreted in a coherent way, not as fragmented bodies of international law.[56]

From a doctrinal view, more and more voices argue that human rights law should be interpreted by taking into account the precautionary principle.[57] For instance, the Committee on Economic, Social and Cultural Rights stated in 2020 that

---

[51] Emphasis added.

[52] In case a brain organoid is a genetically modified organism, it is directly applicable according to current international law.

[53] This is argued by Lavazza in this volume. See Chap. 4, Sect. 4.5 and 4.6

[54] With regard to the threshold of the definition of "human," cf. Knoppers and Greely (2019), p. 1457.

[55] This principle (or approach) is discussed and justified by philosophers as well, cf. especially Steel (2014).

[56] For this important merger in areas of high-risk research cf. as well Vöneky and Beck (2017), paras. 105–109 and forthcoming 2nd ed. 2021.

[57] The discussion usually focuses on the application of the precautionary principle to specific human rights, cf., for instance, in the context of the rights to life and health in relation to environmental protection note 56; furthermore, see Seminara (2016). National constitutional courts increasingly refer to this relationship in the context of States duties and climate change; see, e.g. First Senate of the Federal Constitutional Court of Germany (2021), para. 229: "If there is scientific uncertainty regarding causal relationships of environmental relevance, Art. 20a GG imposes a special duty of care on the legislator. This entails an obligation to even take account of mere indications pointing to the possibility of serious or irreversible impairments, as long as these indications are sufficiently reliable."

> some scientific research can carry health-related risks [...] [and] States parties should prevent or mitigate these risks through careful application of the precautionary principle [...].[58]

Furthermore, in its 2017 advisory opinion, the Inter-American Court of Human Rights argued that

> in the context of the protection of the rights to life and to personal integrity, [...] States must act in keeping with the precautionary principle.[59]

Although some criticized the decision itself,[60] one might not draw from this criticism that a merger of human rights law and the precautionary principle is per se excluded or not convincing. In reference to a proposal spelled out earlier with regard to gene drive research and field trials, I want to stress the need for a merger of human rights and the precautionary principle.[61] In the field of brain organoid research, this means that at least the following conditions have to be met:

Firstly, a scientifically sound case-by-case risk assessment has to take place. This assessment should lead to the conclusion that the health benefits of the research or application outweigh the health risks and other risks, including pain that might be (plausibly) caused to the organoid, especially if they might become partially conscious.[62]

Secondly, as the main feature of the precautionary principle is to broaden the empirical basis in the areas that are uncertain, there should be an ongoing duty to collect data before the experiments take place and during the experiments to shed light on the question of the research's benefits and its risks, including pain that might be (plausibly) caused to the organoid, as well as a duty to share the collected data with other scientists.[63]

Thirdly, instead of a general prior, free and informed consent given by the organoid itself as an entity with an unclear or disputed status, a representative entity such

---

[58] United Nations Committee on Economic, Social and Cultural Rights (2020), p. 15; cf. Massu (2020), pp. 92–94; even for the fight against the pandemic this is mentioned to protect human health by Wissenschaftlicher Dienst des Deutschen Bundestages (2020), p. 17 without further reference, however.

[59] Inter-American Court of Human Rights (IACHR) (2017), para. 180; cf. also Ekardt (2020a), pp. 197–198; Ekardt (2020b), p. 36; ECHR (2009).

[60] Kahl (2019), pp. 125–126.

[61] For an analysis concerning gene drive research and the humanization of environmental law Vöneky (2018a), pp. 137–142.

[62] Such a risk–benefit assessment is enshrined in the 2005 Universal Declaration on Bioethics and Human Rights in a similar way, cf. its Article 4: "In applying and advancing scientific knowledge, medical practice and associated technologies, direct and indirect benefits to patients, research participants and other affected individuals should be maximized and any possible harm to such individuals should be minimized."

[63] Data sharing is a well-established part of research for peaceful purposes, cf., for instance, Article X of the Convention on the Prohibition of the Development, Production and Stockpiling of Biological and Toxin Weapons and on Their Destruction (adopted 10 Apr 1972, entered into force 26 Mar 1975) 1015 UNTS 163.

as an ethics commission should assess the research before it takes place; to enhance the protection of the organoid, one could even argue that prior to the assessment of the ethics commission, a guardian should give his or her consent on behalf of the brain organoid.[64]

These preconditions will enhance procedural and substantive legitimacy before a brain organoid experiment is taking place. This seems necessary if the experiment affects the brain organoids as living entities derived from humans and if there are "plausible threats of serious or irreversible damage," as mentioned in the Rio Declaration quoted above, because there are reasons to believe that the brain organoids involved feel pain and/or possess consciousness, even if there is a lack of full scientific certainty whether this is the case.

12.3 International Soft Law Framework

12.3.1 Soft Law as Governance Tool

Apart from international treaties, soft law rules are relevant for standard-setting in the area of biomedicine. There are specific soft law norms, rules, and principles that cover areas of biomedicine, most importantly those that are laid down in the 2005 Universal Declaration on Bioethics and Human Rights.[65] If rules are part of international soft law, it means that they are not binding as law in the strict sense, but nevertheless have a normative force since States parties agreed on these principles and with this declared and promised that they will not violate these principles and rules.[66]

Soft law declarations are relevant in two ways if we discuss international standard-setting in the area of biomedicine today: from a procedural and from a substantive point of view. They are decisive from a procedural point of view since they can be seen as effective tools to bridge the bottom-up/top-down norm creation gap, i.e. the gap resulting from rule-creating by private entities, including experts (bottom-up), and by States (top-down). They are relevant from a substantive point of view, as they spell out guidelines that have a normative force that is relevant for national rules and laws, and they can bridge the gap between (legally binding) human right norms and (non-binding) bioethical principles.

The procedural advantages can be illustrated by the drafting process of the 2005 Universal Declaration on Bioethics and Human Rights. In 1993, UNESCO established the International Bioethics Committee (IBC), an expert body that consists of 36 members who are independent experts in bioethics. The IBC can give advice and issue recommendations. Five years later, in 1998, the Intergovernmental Bioethics Committee (IGBC) was established as a counterbalance for the IBC, as the IGBC members are State representatives. Nevertheless, it was the IBC—the expert

[64] Cf. note 38 above; this threshold is, however, disputed.

[65] Vöneky (2010), p. 368 et seq.; 383 et seq.

[66] Thürer (2009), para. 37.

body—that was decisive in drafting the 2005 Universal Declaration on Bioethics and Human Rights.[67] The 2005 Universal Declaration on Bioethics and Human Rights is still a model for future developments in international standard-setting in biomedicine because it combines State-based regulation and norm creation by experts. The drafting of this Declaration shows that an international document can be created that comprises an overlapping consensus of experts in the field and State representatives in a short period of time.

From a substantive point of view, the advantages can be seen with regard to the 2005 Universal Declaration on Bioethics and Human Rights as well. The 28 articles entail key elements of biomedical and bioethical standards.[68] In substance, they stress human dignity and human rights, the principle of maximizing benefits and minimizing harm; the principle of prior, free, and informed consent; the respect for human vulnerability; the principles of personal integrity, privacy, equality, justice and equity, non-discrimination, respect for cultural diversity, and the principles of solidarity and cooperation, social responsibility, sharing of benefits, and protection of the environment (Articles 1–17).

Looking at the drafting history of the declaration, one has to notice that—although it is sometimes stated—it would be incorrect to say that during the drafting process, a bioethical (and utilitarian) document was changed into a human rights document because of the influence of the State representatives. The key elements, which are human rights-based, were already part of the IBC draft version of the Declaration (human dignity in Article 3; autonomy in Article 5; informed consent in Article 6, integrity in Article 8; privacy in Article 9; non-discrimination in Articles 10 and 11 draft version of the Declaration).[69]

12.3.2 A New Brain Organoid Soft Law Declaration?

In conclusion, I want to answer the question posed above, whether it would be feasible to have a new soft law declaration governing brain organoids: in my opinion, a soft law declaration governing brain organoids research could have advantages. This is true with regard to the normative gray areas in the field of brain organoid research and therapy and as far as there is a meaningful overlapping consensus by States. With regard to the content of such a future declaration, five elements seem to be most relevant:

First of all, the notion of organoid and brain organoid could be defined in such a declaration. One could argue that an entity is an organoid if it can fulfill the task of an organ, like the kidney, liver, or pancreas, and hence could replace a human organ.

[67] For this and the forthcoming para. cf. in more detail Vöneky (2010), pp. 361–377.

[68] For a link between bioethics and human rights law, cf. Murphy (2018), p. 74 et seq.

[69] United Nations Educational, Scientific and Cultural Organization (UNESCO), International Bioethics Committee (IBC) (2008).

This might not be possible for a brain organoid as there is not such a clear-cut function of a human brain, apart from the integrative function for the human body.[70]

Secondly, the soft law declaration could include elements that spell out criteria that should be applied to measure consciousness or pain if a brain organoid is created and used for research: different models are promulgated, and there should be a scientific discussion which model would be convincing to measure levels of consciousness, pain, etc.[71]

Thirdly, a declaration could help clarify the relevance of the precautionary principle with regard to the question of how to assess probabilities of levels of consciousness and grades of pain.

Fourthly, a new declaration should contain and stress the requirement to consult an ethics commission with the task to assess the research before the process begins.[72]

Fifthly, a new declaration should lay down criteria for valid informed consent by the donor and/or patient;[73] and—as some argue—criteria whether, or in which cases, the consent by a guardian on behalf of the organoid is a necessary condition before the research takes place.

If meaningful overlapping consensus can be reached on these elements, the rules of a new soft law declaration will significantly minimize normative gray areas in the field of brain organoid research and therapy.

12.4 Summary and Outlook

There is no sector-specific international treaty on brain organoid research, development, and use, and problems linked to brain organoid research and therapy remain only partially governed by already existing norms in a fragmented way or by non-binding international soft law.

Nevertheless, currently, there are human rights treaties that provide a relevant and meaningful basis for legitimate standards in the field of brain organoid research and therapy. Freedom of science and research is an important human and constitutional right, but it is not absolute. From the principle of human dignity, as a part of international law, a ban of brain organoid research cannot be deduced as this principle's content is too unclear in ethically disputed areas. In the future, a dynamic interpretation of human right treaties might be convincing if there is consensus that

[70] Cf. German Ethics Council (2015), p. 73.

[71] For a discussion on the different models and theories of consciousness, see Lavazza (2020), pp. 111–114; Schicktanz (2020), pp. 199–203; cf. Faltus (2021), pp. 132–133.

[72] This is already enshrined in the 2005 Universal Declaration on Bioethics and Human Rights, hence there is no governance gap, but this recommendation should probably be stressed with regard to brain organoids.

[73] Cf. the 2005 Universal Declaration on Bioethics and Human Rights; hence, there is no general governance gap, but one could stress the need for informed consent with regard to the aim to do research with brain organoids and answer specific questions of express and informed consent (to use it for this kind of research, to collect it as part of biobanks, to use it for commercial purposes, etc.).

brain organoids are functionally equivalent to human brains and should be protected the same way as an individual, as they feel the same pain as a human person.

Besides, even today, the Cartagena Protocol as an international environmental treaty with regard to biotechnological aspects is applicable if organoids are "living modified organisms" within the meaning of the Protocol. From this, we can conclude that the precautionary principle governs research, at least for some types of brain organoids.

Finally, to reduce gray areas and achieve more legal certainty, it could be feasible to draft a new soft law declaration that would add more specific rules to the existing principles. This should include, inter alia:

- a definition of the notions of organoid and human brain organoid;
- criteria that should be applied to measure consciousness or pain of brain organoids;
- criteria to apply the precautionary principle; especially how to assess probabilities of levels of consciousness and grades of pain; and the need to collect data in order to broaden the factual and empirical basis;
- the requirement to involve an ethics commission to assess the research;
- criteria for valid informed consent by the donor and/or patient;
- and—as some argue—criteria whether, or in which cases, the consent by a guardian on behalf of the organoid is a necessary condition before the research takes place.

However, this proposal of a new soft law declaration is based on the assumption that a meaningful consensus about these criteria and rules can be achieved on the international plane. Such a consensus can be achieved only if there is an open discourse within and between civil society, experts (scientists, philosophers, legal scholars), research organizations, and representatives of States. For this result, there is a long way to go—in the meantime, meaningful discourse, and exchange should be fostered.

References

ECHR (2009) Judgment in Tătar v. Romania, Appl. No. 67021/01

Ekardt F (2020a) Environmental humanities: transformation, governance, ethics, law. Springer, Rostock

Ekardt F (2020b) Menschenrechte und Klimaschutz. RphZ 6:20–39

Faltus T (2021) Organoide, Assembloide, Gastruloide, SHEEFs & organ-on-chip: Rechtliche und medizinethische Aspekte künstlich erzeugter Organ- und Embryomodelle. MedR 39:125–134

Farahany NA, Greely HT, Hyman S et al (2018) The ethics of experimenting with human brain tissue. Nature 556:429–432

First Senate of the Federal Constitutional Court of Germany (2021) Order of 24 March 2021, - 1 BvR 2656/18 -, paras. 1-270. http://www.bverfg.de/e/rs20210324_1bvr265618.html. Accessed 30 Mar 2021

Frowein JA (2013) Ius Cogens. In: Peters A (ed) The Max Planck encyclopedia of public international law. https://www.mpepil.com. Accessed 23 Mar 2021

Gärditz K (2020) Artikel 5 Abs. 3 [Recht der freien Meinungsäußerung]. In: Herzog R, Scholz R, Herdegen M, Klein H (eds) Grundgesetz Kommentar, Band I. C.H. Beck, München

German Ethics Council (2015) Brain death and decisions regarding organ donation, opinion. Deutscher Ethikrat, Berlin
Grote R, Wenzel N (2013) Kapitel 18: Die Meinungsfreiheit. In: Dörr O, Grote R, Marauhn T (eds) EMRK/GG Konkordanzkommentar zum europäischen und deutschen Grundrechtsschutz, 2nd edn. Mohr Siebeck, Heidelberg
Herdegen M (2020) Artikel 1 Abs. 1 [Schutz der Menschenwürde]. In: Herzog R, Scholz R, Herdegen M, Klein H (eds) Grundgesetz Kommentar, Band I. C.H. Beck, München
Hum. Rts. Comm (2011) General comment No. 34, Article 19, Freedoms of opinion and expression. UN Doc. CCPR/C/GC/34 https://www2.ohchr.org/english/bodies/hrc/docs/gc34.pdf. Accessed 30 Mar 2021
Hum. Rts. Comm (2016) Concluding observations on the second report of Namibia. UN Doc. CCPR/C/NAM/CO/2. http://tbinternet.ohchr.org/_layouts/treatybodyexternal/Download.aspx?symbolno=CCPR%2fC%2fNAM%2fCO%2f2&Lang=en. Accessed 23 Mar 2021
Inter-American Court of Human Rights (IACHR) (2017) Advisory opinion OC-23/17 requested by the Republic of Colombia. https://www.corteidh.or.cr/docs/opiniones/seriea_23_ing.pdf. Accessed 07 April 2021
Kahl V (2019) Ökologische Revolution am Interamerikanischen Gerichtshof für Menschenrechte: Besprechung des Rechtsgutachtens Nr. 23 "Umwelt und Menschenrechte" (OC-23/17). EurUP 2:110–131
Knoppers BM, Greely HT (2019) Biotechnologies nibbling at the legal "human". Science 366:1455–1457
Lavazza A (2020) Human cerebral organoids and consciousness: a double-edged sword. Monash Bioeth Rev 38:105–128
Lavazza A, Pizzetti FG (2020) Human cerebral organoids as a new legal and ethical challenge. J Law Biosci 5:1–22. https://doi.org/10.1093/jlb/lsaa005
Massu C (2020) International investment law and the right to health. Rev Chilena Derecho 47:73–100
Mihalyi J, Müller A-K (2016) The custom-made body – Legal aspects of bioprinted tissue and organs. In: Mayr HC, Pinzger M (eds) Informatik 2016, lecture notes in informatics (LNI). Gesellschaft für Informatik, Bonn, pp 1637–1647
Monnheimer M (2021) Due diligence obligations in international human rights law. Cambridge University Press, Cambridge
Murphy T (2018) Judging bioethics and human rights. In: Land MK, Aronson JD (eds) New technologies for human rights law and practice. Cambridge University Press, Cambridge, pp 71–92
Petersen N (2020) Human dignity, international protection. In: Peters A (ed) The Max Planck encyclopedia of public international law. https://www.mpepil.com. Accessed 23 Mar 2021
Proelß A (2017) Prinzipien des internationalen Umweltrechts. In: Proelß A (ed) Internationales Umweltrecht, 1st edn. De Gruyter, Berlin, pp 69–103
Putnam H (1981) Reason, truth, and history. Cambridge University Press, Cambridge
Rösch H (2018) A grain of brain. Max Planck Res 4:54–61
Schicktanz L (2020) Sind menschliche zerebrale Organoide moralisch schützenswert? Ein kommentierter Überblick über die aktuelle internationale Ethikdiskussion. In: Bartfeld S, Schickl H, Alev C, Koo B-K, Pichl A, Osterheider A, Marx-Stölting L (eds) Organoide – Ihre Bedeutung für Forschung, Medizin und Gesellschaft. Nomos, Baden-Baden, pp 190–211
Seminara L (2016) Risk regulation and the European Convention on Human Rights. EJRR 4:733–749
Shaver L (2018) Safeguarding human rights from problematic technologies. In: Land MK, Aronson JD (eds) New technologies for human rights law and practice. Cambridge University Press, Cambridge, pp 25–45
Steel D (2014) Philosophy and the precautionary principle. Cambridge University Press, Cambridge
Thürer D (2009) Soft law. In: Peters A (ed) The Max Planck encyclopedia of public international law. https://www.mpepil.com. Accessed 23 Mar 2021

United Nations Committee on Economic, Social and Cultural Rights (2020) General Comment No. 25 (2020) on science and economic, social and cultural rights (article 15 (1) (b), (2), (3) and (4) of the International Covenant on Economic, Social and Cultural Rights). E/C.12/GC/25
United Nations Conference on Environment and Development (1992) Rio declaration on environment and development. UN Doc. A/CONF.151/26 Vol. I
United Nations Educational, Scientific and Cultural Organization (UNESCO) (2005) Universal declaration on bioethics and human rights. General Conference of UNESCO, 33rd session
United Nations Educational, Scientific and Cultural Organization (UNESCO), International Bioethics Committee (IBC) (2008) Preliminary draft declaration on universal norms on bioethics. Jahrbuch Wissenschaft Ethik 10:381–390
United Nations General Assembly (2005) United Nations declaration on human cloning. UN Doc. A/C.6/59/L.27/Add. 1
von Achenbach J, Clados M (2008) Cloning, international regulation. In: Peters A (ed) The Max Planck encyclopedia of public international law. https://www.mpepil.com. Accessed 23 Mar 2021
von Arnauld A (2019) Völkerrecht, 4th edn. C.F. Müller, Heidelberg
Vöneky S (2010) Recht, Moral und Ethik. Mohr Siebeck, Tübingen
Vöneky S (2018a) International standard setting in biomedicine – foundations and new challenges. GYIL 61:131–151
Voeneky S (2018b) Human rights and legitimate governance of existential and global risks. In: Voeneky S, Neuman GL (eds) Human rights, democracy, and legitimacy in a world of disorder. Cambridge University Press, Cambridge, pp 139–162
Vöneky S, Beck F (2017) Umweltschutz und Menschenrechte. In: Proelß A (ed) Internationales Umweltrecht, 1st edn. De Gruyter, Berlin, pp 133–181
Vöneky S, Wolfrum R (eds) (2004) Human dignity and human cloning. Martinus Nijhoff Publishers, Leiden
Wissenschaftlicher Dienst des Deutschen Bundestages (2020) Die Corona-Pandemie im Lichte des Völkerrechts (Teil 2) – Völkerrechtliche Pflichten der Staaten und die Rolle der Weltgesundheitsorganisation. WD 2 – 3000 – 038/20. https://www.bundestag.de/resource/blob/708058/baa77392dc14d0dbdb99678cd9aaf69a/WD-2-038-20-pdf-data.pdf. Accessed 23 Mar 2021
Wolfrum R, Vöneky S (2004) Who is protected by human rights conventions? Protection of the embryo vs. scientific freedom and public health. In: Vöneky S, Wolfrum R (eds) Human dignity and human cloning. Martinus Nijhoff Publishers, Leiden, pp 133–143

GPSR Compliance

The European Union's (EU) General Product Safety Regulation (GPSR) is a set of rules that requires consumer products to be safe and our obligations to ensure this.

If you have any concerns about our products, you can contact us on ProductSafety@springernature.com

In case Publisher is established outside the EU, the EU authorized representative is:

Springer Nature Customer Service Center GmbH
Europaplatz 3
69115 Heidelberg, Germany

Batch number: 10726060

Printed by Printforce, the Netherlands